HIGH-VACUUM TECHNOLOGY

MECHANICAL ENGINEERING

A Series of Textbooks and Reference Books

Editor

L. L. Faulkner

*Columbus Division, Battelle Memorial Institute
and Department of Mechanical Engineering
The Ohio State University
Columbus, Ohio*

1. *Spring Designer's Handbook*, Harold Carlson
2. *Computer-Aided Graphics and Design*, Daniel L. Ryan
3. *Lubrication Fundamentals*, J. George Wills
4. *Solar Engineering for Domestic Buildings*, William A. Himmelman
5. *Applied Engineering Mechanics: Statics and Dynamics*, G. Boothroyd and C. Poli
6. *Centrifugal Pump Clinic*, Igor J. Karassik
7. *Computer-Aided Kinetics for Machine Design*, Daniel L. Ryan
8. *Plastics Products Design Handbook, Part A: Materials and Components; Part B: Processes and Design for Processes*, edited by Edward Miller
9. *Turbomachinery: Basic Theory and Applications*, Earl Logan, Jr.
10. *Vibrations of Shells and Plates*, Werner Soedel
11. *Flat and Corrugated Diaphragm Design Handbook*, Mario Di Giovanni
12. *Practical Stress Analysis in Engineering Design*, Alexander Blake
13. *An Introduction to the Design and Behavior of Bolted Joints*, John H. Bickford
14. *Optimal Engineering Design: Principles and Applications*, James N. Siddall
15. *Spring Manufacturing Handbook*, Harold Carlson
16. *Industrial Noise Control: Fundamentals and Applications*, edited by Lewis H. Bell
17. *Gears and Their Vibration: A Basic Approach to Understanding Gear Noise*, J. Derek Smith
18. *Chains for Power Transmission and Material Handling: Design and Applications Handbook*, American Chain Association
19. *Corrosion and Corrosion Protection Handbook*, edited by Philip A. Schweitzer
20. *Gear Drive Systems: Design and Application*, Peter Lynwander
21. *Controlling In-Plant Airborne Contaminants: Systems Design and Calculations*, John D. Constance
22. *CAD/CAM Systems Planning and Implementation*, Charles S. Knox
23. *Probabilistic Engineering Design: Principles and Applications*, James N. Siddall
24. *Traction Drives: Selection and Application*, Frederick W. Heilich III and Eugene E. Shube

25. *Finite Element Methods: An Introduction*, Ronald L. Huston and Chris E. Passerello
26. *Mechanical Fastening of Plastics: An Engineering Handbook*, Brayton Lincoln, Kenneth J. Gomes, and James F. Braden
27. *Lubrication in Practice: Second Edition*, edited by W. S. Robertson
28. *Principles of Automated Drafting*, Daniel L. Ryan
29. *Practical Seal Design*, edited by Leonard J. Martini
30. *Engineering Documentation for CAD/CAM Applications*, Charles S. Knox
31. *Design Dimensioning with Computer Graphics Applications*, Jerome C. Lange
32. *Mechanism Analysis: Simplified Graphical and Analytical Techniques*, Lyndon O. Barton
33. *CAD/CAM Systems: Justification, Implementation, Productivity Measurement*, Edward J. Preston, George W. Crawford, and Mark E. Coticchia
34. *Steam Plant Calculations Manual*, V. Ganapathy
35. *Design Assurance for Engineers and Managers*, John A. Burgess
36. *Heat Transfer Fluids and Systems for Process and Energy Applications*, Jasbir Singh
37. *Potential Flows: Computer Graphic Solutions*, Robert H. Kirchhoff
38. *Computer-Aided Graphics and Design: Second Edition*, Daniel L. Ryan
39. *Electronically Controlled Proportional Valves: Selection and Application*, Michael J. Tonyan, edited by Tobi Goldoftas
40. *Pressure Gauge Handbook*, AMETEK, U.S. Gauge Division, edited by Philip W. Harland
41. *Fabric Filtration for Combustion Sources: Fundamentals and Basic Technology*, R. P. Donovan
42. *Design of Mechanical Joints*, Alexander Blake
43. *CAD/CAM Dictionary*, Edward J. Preston, George W. Crawford, and Mark E. Coticchia
44. *Machinery Adhesives for Locking, Retaining, and Sealing*, Girard S. Haviland
45. *Couplings and Joints: Design, Selection, and Application*, Jon R. Mancuso
46. *Shaft Alignment Handbook*, John Piotrowski
47. *BASIC Programs for Steam Plant Engineers: Boilers, Combustion, Fluid Flow, and Heat Transfer*, V. Ganapathy
48. *Solving Mechanical Design Problems with Computer Graphics*, Jerome C. Lange
49. *Plastics Gearing: Selection and Application*, Clifford E. Adams
50. *Clutches and Brakes: Design and Selection*, William C. Orthwein
51. *Transducers in Mechanical and Electronic Design*, Harry L. Trietley
52. *Metallurgical Applications of Shock-Wave and High-Strain-Rate Phenomena*, edited by Lawrence E. Murr, Karl P. Staudhammer, and Marc A. Meyers
53. *Magnesium Products Design*, Robert S. Busk
54. *How to Integrate CAD/CAM Systems: Management and Technology*, William D. Engelke
55. *Cam Design and Manufacture: Second Edition*; with cam design software for the IBM PC and compatibles, disk included, Preben W. Jensen

56. *Solid-State AC Motor Controls: Selection and Application*, Sylvester Campbell
57. *Fundamentals of Robotics*, David D. Ardayfio
58. *Belt Selection and Application for Engineers*, edited by Wallace D. Erickson
59. *Developing Three-Dimensional CAD Software with the IBM PC*, C. Stan Wei
60. *Organizing Data for CIM Applications*, Charles S. Knox, with contributions by Thomas C. Boos, Ross S. Culverhouse, and Paul F. Muchnicki
61. *Computer-Aided Simulation in Railway Dynamics*, by Rao V. Dukkipati and Joseph R. Amyot
62. *Fiber-Reinforced Composites: Materials, Manufacturing, and Design*, P. K. Mallick
63. *Photoelectric Sensors and Controls Selection and Application*, Scott M. Juds
64. *Finite Element Analysis with Personal Computers*, Edward R. Champion, Jr., and J. Michael Ensminger
65. *Ultrasonics: Fundamentals, Technology, Applications: Second Edition, Revised and Expanded*, Dale Ensminger
66. *Applied Finite Element Modeling: Practical Problem Solving for Engineers*, Jeffrey M. Steele
67. *Measurement and Instrumentation in Engineering: Principles and Basic Laboratory Experiments*, Francis S. Tse and Ivan E. Morse
68. *Centrifugal Pump Clinic: Second Edition, Revised and Expanded*, Igor J. Karassik
69. *Practical Stress Analysis in Engineering Design: Second Edition, Revised and Expanded*, Alexander Blake
70. *An Introduction to the Design and Behavior of Bolted Joints: Second Edition, Revised and Expanded*, John H. Bickford
71. *High Vacuum Technology: A Practical Guide*, Marsbed H. Hablanian
72. *Pressure Sensors: Selection and Application*, Duane Tandeske
73. *Zinc Handbook: Properties, Processing, and Use in Design*, Frank Porter
74. *Thermal Fatigue of Metals*, Andrzej Weronski and Tadeusz Hejwowski
75. *Classical and Modern Mechanisms for Engineers and Inventors*, Preben W. Jensen
76. *Handbook of Electronic Package Design*, edited by Michael Pecht
77. *Shock-Wave and High-Strain-Rate Phenomena in Materials*, edited by Marc A. Meyers, Lawrence E. Murr, and Karl P. Staudhammer
78. *Industrial Refrigeration: Principles, Design and Applications*, P. C. Koelet
79. *Applied Combustion*, Eugene L. Keating
80. *Engine Oils and Automotive Lubrication*, edited by Wilfried J. Bartz
81. *Mechanism Analysis: Simplified and Graphical Techniques, Second Edition, Revised and Expanded*, Lyndon O. Barton
82. *Fundamental Fluid Mechanics for the Practicing Engineer*, James W. Murdock
83. *Fiber-Reinforced Composites: Materials, Manufacturing, and Design, Second Edition, Revised and Expanded*, P. K. Mallick

84. *Numerical Methods for Engineering Applications*, Edward R. Champion, Jr.
85. *Turbomachinery: Basic Theory and Applications, Second Edition, Revised and Expanded*, Earl Logan, Jr.
86. *Vibrations of Shells and Plates: Second Edition, Revised and Expanded*, Werner Soedel
87. *Steam Plant Calculations Manual: Second Edition, Revised and Expanded*, V. Ganapathy
88. *Industrial Noise Control: Fundamentals and Applications, Second Edition, Revised and Expanded*, Lewis H. Bell and Douglas H. Bell
89. *Finite Elements: Their Design and Performance*, Richard H. MacNeal
90. *Mechanical Properties of Polymers and Composites: Second Edition, Revised and Expanded*, Lawrence E. Nielsen and Robert F. Landel
91. *Mechanical Wear Prediction and Prevention*, Raymond G. Bayer
92. *Mechanical Power Transmission Components*, edited by David W. South and Jon R. Mancuso
93. *Handbook of Turbomachinery*, edited by Earl Logan, Jr.
94. *Engineering Documentation Control Practices and Procedures*, Ray E. Monahan
95. *Refractory Linings Thermomechanical Design and Applications*, Charles A. Schacht
96. *Geometric Dimensioning and Tolerancing: Applications and Techniques for Use in Design, Manufacturing, and Inspection*, James D. Meadows
97. *An Introduction to the Design and Behavior of Bolted Joints: Third Edition, Revised and Expanded*, John H. Bickford
98. *Shaft Alignment Handbook: Second Edition, Revised and Expanded*, John Piotrowski
99. *Computer-Aided Design of Polymer-Matrix Composite Structures*, edited by Suong Van Hoa
100. *Friction Science and Technology*, Peter J. Blau
101. *Introduction to Plastics and Composites: Mechanical Properties and Engineering Applications*, Edward Miller
102. *Practical Fracture Mechanics in Design*, Alexander Blake
103. *Pump Characteristics and Applications*, Michael W. Volk
104. *Optical Principles and Technology for Engineers*, James E. Stewart
105. *Optimizing the Shape of Mechanical Elements and Structures*, A. A. Seireg and Jorge Rodriguez
106. *Kinematics and Dynamics of Machinery*, Vladimír Stejskal and Michael Valášek
107. *Shaft Seals for Dynamic Applications*, Les Horve
108. *Reliability-Based Mechanical Design*, edited by Thomas A. Cruse
109. *Mechanical Fastening, Joining, and Assembly*, James A. Speck
110. *Turbomachinery Fluid Dynamics and Heat Transfer*, edited by Chunill Hah
111. *High-Vacuum Technology: A Practical Guide, Second Edition, Revised and Expanded*, Marsbed H. Hablanian
112. *Geometric Dimensioning and Tolerancing: Workbook and Answerbook*, James D. Meadows

Additional Volumes in Preparation

Heat Exchanger Design Handbook, T. Kuppan

Handbook of Thermoplastic Piping System Design, Thomas Sixsmith and R. Hanselka

Handbook of Materials Selection for Engineering Applications, edited by George T. Murray

Mechanical Engineering Software

Spring Design with an IBM PC, Al Dietrich

Mechanical Design Failure Analysis: With Failure Analysis System Software for the IBM PC, David G. Ullman

HIGH-VACUUM TECHNOLOGY
A PRACTICAL GUIDE

SECOND EDITION, REVISED AND EXPANDED

MARSBED H. HABLANIAN
*Varian Associates, Inc.
Lexington, Massachusetts*

MARCEL DEKKER, INC.   NEW YORK • BASEL

Library of Congress Cataloging-in-Publication Data

Hablanian, M. H.
   High-vacuum technology : a practical guide / Marsbed H. Hablanian.
—2nd ed., rev. and expanded.
     p. cm. — (Mechanical engineering ; 111)
   Includes index.
   ISBN 0-8247-9834-1 (alk. paper)
   1. Vacuum technology. I. Title. II. Series: Mechanical
engineering (Marcel Dekker, Inc.) ; 111.
TJ940.H33 1997
621.5'5—dc21
                                                                                                    97-4017
                                                                                                      CIP

The author and his former employer, Varian Associates, disclaim any and all liability and any warranty whatsoever relating to the practice, techniques, safety and results of procedures or their applications decribed in this book.

The publisher offers discounts on this book when ordered in bulk quantities. For more information, write to Special Sales/Professional Marketing at the address below.

This book is printed on acid-free paper.

Copyright © 1997 by Marcel Dekker, Inc. All Rights Reserved.

Neither this book nor any part may be reproduced or transmitted in any form or by any means, electronic or mechanical, including photocopying, microfilming, and recording, or by any information storage and retrieval system, without permission in writing from the publisher.

Marcel Dekker, Inc.
270 Madison Avenue, New York, New York 10016

Current printing (last digit):
10 9 8 7 6 5 4

PRINTED IN THE UNITED STATES OF AMERICA

Preface to the Second Edition

The first edition of this book was prepared nearly ten years ago, and this new edition therefore contains important changes. Approximately half of the text reflects developments in technology. For example, Chapter 4, on vacuum systems, has a new section for estimating the evacuation time of vacuum chambers through the entire pressure region. The section on oil-free rough vacuum pumps has been expanded to include information on new equipment. Chapter 7, on turbomolecular pumps, has been more than doubled to reflect the emergence of these pumps as the main method for creating high vacuum and the use of hybrid or compound turbine-types pumps and new bearing technologies.

The term *diffusion pump* has been replaced by *vapor jet pump* throughout the book, because this unfortunate misnomer has created an enormous amount of confusion about the mechanism of pumping, causing serious system design errors and costly industrial misapplications. Partly in association with that issue, a new chapter, Chapter 10, has been added that is devoted entirely to the subject of overloading of high vacuum pumps. Most textbooks on high vacuum devote much attention to the mechanisms of molecular capture by various pumps but completely neglect energetic considerations required for the prevention of overloading.

The combined chapter on ion-gettering pumps and ultrahigh vacuum in the first edition has been separated into two independent chapters to avoid associating ion-getter pumps only with ultrahigh vacuum. Additions have also been made to the chapters on cyrogenic pumps and leak detection, as well as to most bibliographic listings and the appendix.

The general scope and style of the book remain the same. The book provides a practical introduction and an overview of engineering aspects of high-vacuum technology with an emphasis on conceptual understanding of issues involved in designing and operating high-vacuum systems.

I would like to acknowledge the use of various internal company publications of the Vacuum Products Division of Varian Associates, particularly in Chapters 8, 11, 12, and 13. In many cases, it was difficult or impossible to trace the original writers and editors, but, at the risk of having some unintentional omissions, the following should be mentioned: P. Forant, S. Burnett, P. Fruzzetti, W. Worthington, F. Turner, D. Harra, M. Audi, J. Kirkpatrick, J. DeRijke, W. Briggs, H. Steinherz, J. Maliakal, K. Welch, and Y. Strausser.

Marsbed H. Hablanian

Preface to the First Edition

This book deals with achieving and maintaining an environment of extremely low gas densities, the instrumentation for measuring extremely low pressures, the design of high vacuum systems, and the associated requirement for highly sensitive leak detection techniques.

It is intended to be an essentially nonanalytical introduction to high vacuum technology for engineers and technicians who would like to have a general orientation before proceeding to a more technical and scientifically rigorous text.

The book should also be useful for technical and nontechnical industrial and laboratory personnel who may not have a direct interest in high vacuum science and technology, but would like to develop some familiarity with the subject.

The book develops a conceptual understanding, scope, limitations and appreciation of basic relationships between pumps, instrumentation, and system performance. It is not intended to be a design handbook or a compilation of all relevant details. The many industrial and laboratory applications of high vacuum equipment are not treated here because each one would require a separate chapter and that would double the size of the book.

Concerning the organization of the text, I have tried to make each chapter as self-sufficient as possible, while even accepting some repetition in the process. It is virtually impossible to arrange the sequence of the subjects with unassailable logic. For example, pressure gauges are placed at the end although pressure is discussed throughout the text. The reader may prefer to review the sections on pumps (Chapters 5 to 9) before returning to Chapter 4, "Vacuum Systems."

The chapter on diffusion pumps (vapor jet pumps) is longer than the chapters on other pumps. Partly, this is due to my particular familiarity with the subject. However, much of the discussion in the diffusion pumps chapter is applicable to other pumps and, having been placed in the beginning, according to history, diffusion pumps are discussed in greater detail.

I would like to acknowledge the use of various internal company publications (Varian Associates, Vacuum Products Division), particularly in Chapters 8, 9, 10, and 11. In many cases, it was difficult or impossible to trace the original writers and editors, but, at the risk of having some unintentional omissions, the following should be mentioned: P. Forant, S. Burnett, P. Fruzzetti, W. Worthington, F. Turner, D. Harra, M. Audi, J. Kirkpatrick, and J. DeRijke.

Marsbed H. Hablanian

Contents

Preface to the Second Edition   *iii*
Preface to the First Edition   *v*

Chapter 1   **Introduction**   **1**
      1.1   Nature of Vacuum   1
      1.2   Scope of Vacuum Science and Technique   2
      1.3   Uses and Applications of Vacuum Technology   3
      1.4   Production of Vacuum   4
      1.5   Brief History of Development   5
      1.6   History of Vacuum Pumps   6
            References   12

Chapter 2   **Properties of Gases**   **15**
      2.1   Pressure and Density   15
      2.2   Atomic Numbers and Distances   16
      2.3   Basic Gas Law   17
      2.4   Velocities and Temperature of Gases   19
      2.5   Vapors and Vapor Pressure   22
      2.6   Evaporation   24
      2.7   Adsorption and Desorption   25
      2.8   Gas Content of Materials   30
      2.9   Outgassing   31

	2.10	Water Vapor	32
		References	39

Chapter 3 Fluid Flow and Pumping Concepts — 41
- 3.1 Pressure and Flow — 41
- 3.2 Mass Flow and Volume Flow — 43
- 3.3 Flow Regimes — 45
- 3.4 Flow Through Nozzles and Diffusers — 47
- 3.5 Molecular Flow — 50
- 3.6 Conductance — 55
- 3.7 Pumping Speed — 58
- 3.8 Pumps and Compressors — 62
- 3.9 Fluid Flow in Small Passages — 66
- References — 75

Chapter 4 Vacuum Systems — 77
- 4.1 Vacuum Chamber Design — 77
- 4.2 Evacuation and Process Gas Pumping — 81
- 4.3 Evacuation Time — 83
- 4.4 Ultimate Pressure — 95
- 4.5 Conductance Calculations — 96
- 4.6 Outgassing Effects — 115
- 4.7 Pumping System Design — 117
- 4.8 Operation of High-Vacuum Systems — 121
- 4.9 Estimation of Pressure-Time Progress During Evacuation — 129
- References — 134

Chapter 5 Coarse Vacuum Pumps — 137
- 5.1 Introduction — 137
- 5.2 Rotary Vane Pumps — 142
- 5.3 Performance Characteristics — 144
- 5.4 Ultimate Pressure — 148
- 5.5 Condensable Vapors and Gas Ballast — 153
- 5.6 Oils and Backstreaming — 154
- 5.7 Operation and Maintenance — 157
- 5.8 Other Coarse Vacuum Pumps — 162
- 5.9 Oil-Free Vacuum Pumps — 170
- References — 203

Chapter 6 Vapor Jet (Diffusion) Pumps — 207
- 6.1 Introduction — 207
- 6.2 Pumping Mechanism — 208
- 6.3 Basic Design — 213
- 6.4 Basic Performance and Operation — 216
- 6.5 Pumping Fluids — 218
- 6.6 Performance Characteristics — 221
- 6.7 Other Performance Aspects — 243

CONTENTS

6.8	Baffles and Traps	250
6.9	Maintenance	262
	References	267

Chapter 7 Turbomolecular Pumps — 269
- 7.1 Introduction — 269
- 7.2 Molecular Drag Pumps — 271
- 7.3 Turbomolecular Pumps — 286
- 7.4 Summary of Properties — 314
- References — 317

Chapter 8 Cryogenic Pumps — 319
- 8.1 Basic Principles of Operation — 319
- 8.2 Cryosorption Pumping — 321
- 8.3 Gaseous Helium Cryopumps — 322
- 8.4 Large Cryopumps — 345
- 8.5 Water Vapor Pumps — 345
- References — 351

Chapter 9 Gettering and Ion Pumping — 353
- 9.1 Gettering Pumps — 353
- 9.2 Sputter-ion Pumps — 360
- 9.3 Basic Performance of Sputter-ion Pumps — 363
- 9.4 Pump Types and Performance with Different Gases — 365
- 9.5 Operation of Ion Pumps — 369
- 9.6 Nonevaporated Getter (NEG) Pumps — 372
- 9.7 Combination Pumps — 374
- 9.8 Clean Roughing Systems — 376
- References — 381

Chapter 10 Overloading of Vacuum Pumps — 383
- 10.1 Introduction — 383
- 10.2 Volume Flow and Mass Flow — 385
- 10.3 Cross-over Pressure — 388
- 10.4 Throughput Limits of Vapor Jet Pumps — 394
- 10.5 Mass Flow Limits of Turbine-type Pumps — 397
- 10.6 Similarities Between Pumps — 401
- 10.7 Cryopumps and Ion-getter Pumps — 402
- References — 404

Chapter 11 Ultrahigh Vacuum — 405
- 11.1 Introduction — 405
- 11.2 Degassing by Baking — 406
- 11.3 All-Metal Systems and Components — 407
- 11.4 Throughput-type Pumps and Ultrahigh Vacuum — 411
- 11.5 Capture Pumps — 427
- References — 429

Chapter 12 Vacuum Gauges and Gas Analyzers   **431**
    12.1   Introduction   431
    12.2   Force-measuring Gauges   432
    12.3   Heat Transfer Gauges   438
    12.4   Spinning Rotor Gauges   442
    12.5   Ionization Gauges   444
    12.6   New Trends of Gauges and Sensors   460
    12.7   Mass Spectrometers or Partial Pressure Gauges   462
            References   469

Chapter 13 Leak Detection   **471**
    13.1   Introduction   471
    13.2   Sizes of Leaks and Units of Measurement   474
    13.3   Methods of Leak Detection   479
    13.4   Helium Mass Spectrometer Leak Detectors   493
            References   513

Appendix   *515*

Index   *547*

1
Introduction

1.1 NATURE OF VACUUM

The word "vacuum" is used to describe a very wide range of conditions. At one extreme, it refers to nearly complete emptiness, a space without matter, or more specifically, space in which air and other gases are absent.

At the other extreme, vacuum is any air or gas pressure less than a prevailing pressure in an environment or, specifically, any pressure lower than the atmospheric pressure. An example of conditions approaching the first meaning is intergalactic space. Examples of the second meaning are pressures existing at the inlet of an ordinary vacuum cleaner and in a straw used for drinking.

The basic property involved here is the density of the gas. The degree of vacuum can easily be described in terms of gas particle density instead of pressure. Scientific and engineering interest in vacuum spans an enormous range of gas densities—15 orders of magnitude—in other words, it involves density changes of a million billion times. Corresponding to this wide variation in density, there are industrial and laboratory processes in which certain ranges in this wide variety of conditions are used. Production of a high degree of vacuum and an understanding of high-vacuum technology were closely associated with the development of physics as a science beginning approximately 350 years ago.

At atmospheric pressure, there are about 2×10^{19} molecules in a cubic centimeter of air. At the altitude of earth-orbiting satellites, the corresponding number is 10^9. Resistance to motion is reduced accordingly, letting satellites stay in orbit for many years. At atmospheric conditions, molecules can move a distance of only a few millionths of a centimeter without colliding with one another, while at the altitude of orbiting satellites, this distance between collisions can be several miles. In a high-vacuum chamber, collisions between molecules are much less frequent than with walls of the chamber.

The existence of great travel distances without collision is used, for example, in a television tube, where a focused beam of electrons must travel from the electron gun on one end to the picture screen in the front without dispersion or scattering. Inside the television picture tube, the degree of vacuum is about one billionth of an atmosphere.

1.2 SCOPE OF VACUUM SCIENCE AND TECHNIQUE

The commonest example of partial vacuum involves respiration. When we breathe we expand our lungs and increase their volume. This produces a pressure inside which is slightly lower than in the outside air and permits additional air to enter the lung. Similarly, the suction in an ordinary drinking straw is created by inhaling the air from the straw, thereby reducing the pressure on the liquid surface inside the straw relative to the pressure above the liquid in the drinking cup. The excess pressure pushes the liquid into the straw. If anyone doubts this process, let the person try to drink through a straw after taking a very deep breath.

Partial vacuum or rarefied gas conditions are used in many common devices. For example, vacuum exists in automobile engine pistons when they are withdrawn to let in atmospheric air. A much higher degree of vacuum is used in a thermos bottle but for a very different purpose. The air is removed between the double walls of the bottle to reduce the amount of heat conduction from the hot to the cold wall. In an electric light bulb, vacuum is used during a stage of its manufacture to remove atmospheric oxygen and prevent a chemical reaction with very hot tungsten filaments.

A television picture tube is a vacuum tube. In addition to the increased travel distance of a coherent stream of electrons, vacuum environment also prevents rapid oxidation of the electron emitters (cathodes). The same two basic requirements are used for thin-film coating applications. For example, metallic films deposited on various substrates (ranging from plastic toys to microelectronic circuits) are produced in vacuum chambers instead of electroplating. This assures that the evaporated metal does not become oxidized and that it can reach the substrate without colliding with a large number of air molecules.

INTRODUCTION

1.3 USES AND APPLICATIONS OF VACUUM TECHNOLOGY

These few examples just described illustrate a variety of possible applications and the importance of high-vacuum technology. Many commonly observed and utilized properties of gases are changed substantially when the gas is highly rarefied. Life, with the possible exception of some very hardy viruses, cannot be supported.

Animals do not survive exposure to high vacuum longer than a few seconds. Volatile liquids, including water, would evaporate at a very high rate without an atmosphere surrounding our planet. Lakes and rivers would evaporate within a few days if the resulting "steam" were to be removed and condensed elsewhere. Ordinarily, water and other fluids evaporate slowly because the evaporating molecules can return into the fluid after colliding with air molecules present above the liquid surface.

The boiling temperature of a liquid is reduced if the liquid is placed in a vacuum. Water will boil at room temperature if it is subjected to a pressure of about 1/40 of an atmosphere. An astronaut working outside of a space station cannot hear the "noise" of the hammer or wrench. Sound cannot be transmitted through high vacuum because sound is created by pressure perturbations in the air.

The transmission of electromagnetic radiation, however, light, radio waves, and so on, is not affected and is even improved by the absence of air. This is the reason, for example, for placing telescopes and other instruments in space.

These examples indicate the basic uses of vacuum: removal of gas to impede heat transfer, removal of gas to permit travel of particles through a required distance, evaporation or drying, particularly at a lower-than-usual temperature, and removal of chemically active gases. In vacuum metallurgy, removal of oxygen and reduction of the amount of dissolved gases in the melt are of primary importance. Many metals, such as tungsten, molybedenum, tantalum, and titanium, cannot be molten in atmospheric air. They react rapidly and sometimes violently when heated to high temperatures in the presence of oxygen.

Large accelerators built for investigations in the field of high-energy physics are essentially gigantic vacuum tubes. Even though vacuum tubes have almost disappeared from radio sets and television receivers, they are employed in devices used for the generation and transmission of radio, radar, and television signals.

With the advent of space exploration, large vacuum chambers have been built to simulate conditions existing in interplanetary space. Such chambers may be as large as 10 m in diameter and 30 m high. Entire smaller rockets, subassemblies of large rockets and satellites, can be placed in such chambers for long periods of time. The performance of mechanisms and instruments can be tested without the presence of air or the presence of absorbed layers of water vapor on sliding surfaces. Even the presence of unshielded radiation of the sun can be simulated.

In the microelectronic industry, high-vacuum chambers are used for precisely controlled deposition of thin films of desired materials on various substrates. The rapid development of sophisticated electronic computers in recent years would not be possible without the existence of refined techniques in high-vacuum technology.

1.4 PRODUCTION OF VACUUM

Vacuum can be produced by a few basic methods. It can be produced by mechanical displacement of gases from an enclosed space by means of pistons, rotating vanes, lobes, and so on. Any compressor, in principle, can be converted into a vacuum pump simply by connecting the chamber to the inlet of the device instead of the discharge. This includes air and steam ejectors, vapor jet pumps, and turbocompressors. To be effective in high-vacuum work, such compressors are usually redesigned to enhance certain required performance characteristics. A vacuum can also be produced simply by condensing and freezing the gas in the enclosed space. Normally, this requires very low temperatures (cryogenic pumping).

A vacuum can also be created by chemical reactions with the gas, which produce solid residues, thereby removing the gas from space (chemical pumping or gettering). Physical absorption can also be used as a means of removing gases from space. Highly porous substances are used for this purpose (e.g., activated charcoal, activated alumina, or minerals called zeolites), often in conjunction with cryogenic temperatures. Such substances have extremely large internal surface areas which can act as gas "sponges."

The final method of producing a vacuum, developed approximately 30 years ago (primarily at Varian Associates), involves ionization of the gas, accelerating the ions in a high-voltage field, and driving the gas directly into a wall or target electrode.

To produce a high vacuum in a chamber, the evacuation process must start at atmospheric pressure. Due to the enormous range of gas densities involved, a single pumping device cannot be used efficiently throughout the evacuation process. Usually, a sequence of at least two different devices is used. A coarse vacuum is produced either by mechanical means or by cryosorption, the higher degree of vacuum is developed by vapor jet pumps (called diffusion pumps), turbomolecular pumps, cryosorption pumps, and finally, by ion-gettering pumps.

The instruments used to measure the degree of vacuum must also span a very wide range of gas densities (or pressures). At least two different vacuum gauges

INTRODUCTION 5

are used in most vacuum systems. At the higher pressures, gauges based on force measurement can be used, such as liquid-filled manometers and various diaphragm gauges.

At a high degree of vacuum, the pressure force and force difference between degrees of vacuum become so small that direct force measurements become impractical and indirect ways must be used. For example, one of the most common methods involves ionizing the gas and measuring the intensity of the ion current, which depends on the amount of gas present. Such gauges, called ionization gauges, are used at gas densities as low as a million billion times less than 1 atm.

1.5 BRIEF HISTORY OF DEVELOPMENT

Vacuum technology is a relatively new technology although speculations about the possibility of creating a vacuum go back perhaps 2000 years. Ancient Greek philosophers speculated about the nature of vacuum. For a variety of reasons, the existence of vacuum was doubted and even rejected. Aristotle (384–322 B.C.) thought, for example, that vacuum was not possible because he assumed that the concept of empty space would invite the concept of motion without resistance (i.e., motion at infinite velocity). Such opinions persisted for many years and were held by many notable writers, including Roger Bacon (1214–1299) and René Descartes (1596–1650).

Galileo (1564–1642) was among the first to conduct experiments attempting to measure forces required to produce vacuum with a piston in a cylinder. Many of these early experiments were associated with the pumps used to remove water from mines. Torricelli (1608–1647), an associate of Galileo's, was first to use mercury instead of water for such experiments, thereby reducing to convenient dimensions the size of the apparatus required. He produced a vacuum by filling a glass tube (closed at one end) with mercury and submerging the open end in a pool of mercury. He demonstrated that the mercury column "suspended" in the tube was always 76 cm above the level in the pool, regardless of the size, length, shape, or degree of tilt of the tube. He correctly explained that the pressure of atmospheric air acting on the surface of the mercury pool supports the mercury column in the tube. In honor or Torricelli, the unit of measure of degree of vacuum, the unit "millimeters of mercury" (mmHg) was called torr. Recently, however, the International Standards Organization established a new unit, called the pascal in honor of Blaise Pascal (1623–1662), who contributed to understanding of vacuum physics, carried out many early experiments, demonstrated his findings to large audiences, and was among the first to devise a barometer.

Otto von Guericke (1602–1686) was first to modify water pumps for direct pumping of air. He produced sufficient vacuum inside two gasketed copper hemispheres (essentially large suction cups, about 50-cm-diameter) to conduct his famous demonstration in Magdeburg (1654), where horses hitched to each side of the sphere could not pull the hemispheres apart until air was readmitted. (In retrospect, it may be observed that this experiment was not conducted correctly because the horses did not apply a maximum force simultaneously [Ref. 33].) Many such early experiments were also made by Robert Boyle (1627-1691), who made an improved vacuum pump.

The understanding of vacuum phenomena stimulated many branches of science and engineering, such as hydraulics, hydrostatics, pneumatics, and physics of rarefied gases. Vacuum technique was associated with many important discoveries in physics: for example, electrical discharges in gases, x-rays (1895), and the existence of electrons (1897).

In the twentieth century, the development of vacuum science and technology has been associated with radio tubes, high-energy particle physics, atomic energy, isotope separation, optical and microelectronic coating, processing of heat-sensitive fluids such as photographic emulsions, and vacuum metallurgy. Important recent advances have been associated with the requirements in processing microwave tubes that led to the development of ion-getter pumps and ultrahigh-vacuum techniques.

Many modern instruments and tools would be unthinkable without high vacuum. Examples are electron and ion microscopes, mass spectrometers, ion implanters, and extremely sensitive leak detectors. Even the landing of human beings on the moon could not have been accomplished without the existence of space simulation technology and an understanding of vacuum science.

1.6 HISTORY OF VACUUM PUMPS

Most writers on the history of vacuum pumps attribute the first invention to Otto von Guericke, in about 1650, following Torricelli's experiments with vacuum in mercury-filled tubes in 1644 (or 1643). However, others mention Galileo and his attempts to measure the force of vacuum formation using cylinders, pistons, and weights (1640).

The initial attempts to create vacuum were made by pumping water out of filled containers using a variety of well-known water pumps. Water pumps go back at least to ancient Alexandria and were used in ancient Roman mines. Many sixteenth-century mining engineers probably produced 0.3-atm vacuum without knowing it when they tried to pump water from above. They staged their pumps at about 20-ft levels. Galileo reported in his book (Leyden, 1638) that he first heard about the 33- or 34-ft height limit from a mine worker.

This discovery must have been before 1630 because in that year G. B. Baliani reported to Galileo the failure of a water pipe siphon at Genoa where water

INTRODUCTION

would run down the hill on both sides of the summit even though no leaks were discovered. Galileo replied that he had long before discovered the impossibility of operating the siphon higher than 34 ft. He thought this was due to the break in the "rope of water," associating the phenomenon with his experiments of breaking wires and ropes in tension. Baliani appears to have disagreed with that and suggested the possibility of vacuum above the water column.

Galileo attempted to measure the force needed to create vacuum by using a cylinder open at the bottom and suspending weights or sand in the bucket attached to the piston. This apparatus was made of wood and was difficult to seal despite the use of water. The idea of pumping air directly did not occur until Guericke's third pump version. However, interestingly enough, air pumps for pipe organs existed in ancient Greece.

The thought that the force of atmospheric air had something to do with the action of water pumps had been recorded as early as 1615 in the writings of Isaac Beekman (b. 1588 in Middlebourg, d. 1637 in Dodrecht). According to de Waard, Beekman was a man of genius who did not publish, and is, therefore, not well known. In his thesis (University of Caen, 1618) he did not hesitate to state: "Aqua suctu sublata non attrahitur vi vacui sed ab aere incumbente in locum vacuum impellitur," which can be translated as "water raised by suction is not pulled by the force of vacuum but is driven into the empty space by the still air." It is not clear whether Beekman's observation was completely original. In those days, they were not always careful to give sources.

The ancient "fear of vacuum" was beginning to be questioned perhaps as early as 1600. Kepler (1603) realized the limited extent of the atmosphere (estimated 4 km from astronomical studies of bending of light). The rejection of the possibility of vacuum by ancient philosophers was based mainly on the premise that if vacuum were possible, the absence of resistance would permit infinite speed. In retrospect, it is difficult today to appreciate the confusion. Notable in the ancient writings is the absence of discussion on the degree of vacuum. They appear to have thought of air as a given substance and did not seem to realize that gases can have variations in density (the kinetic theory of gases was developed in the eighteenth century and for a long time was, and still is, called "theory").

Guericke was apparently the first to realize that just like water, air can be pumped directly by appropriately designed pumps, and his third-generation pump was built specifically for that purpose. His pump is essentially similar to the water pump and has manual valves. It is more carefully built, to make a tighter seal around the cylinder and tighter valves. In principle, the only difference between such pumps used for creating vacuum rather than compressing air (or lifting water) is that the working stroke is pulling instead of pushing, with corresponding valve sequences. The inlet is still at low pressure and the discharge at the higher pressure. The basic difference is that a sealed container is attached to the inlet instead of the discharge.

1.6.1 Mechanical Pumps

Due to difficulties associated with ancient concepts about vacuum and lack of understanding of properties of rarefied gases, the development of mechanical vacuum pumps often deviated from the main course, which, perhaps, should have been simply the adaptation of known pumping mechanisms for rarefied gas pumping.

In general, the history of the development can be traced as follows: first, modification of existing water pumps with pistons and valves, which lasted to the end of the nineteenth century, then a return to a more mechanically primitive concept of liquid mercury "piston" pumps, later the establishment of rotary mechanical pumps, followed by adaptations of vapor jet pumps, turbomachinery, and finally, pumps based on ionization, chemical combination, and cryogenic adsorption.

For a long time, vacuum pumps were not even called vacuum pumps. Guericke called them syringes and Boyle called them pneumatic machines; later the term "air pumps" was established. The *Engineering Index*, for example, used the "air pump" designation until the 1920s, then switched to "vacuum pump." The use of the word "pump" for the device, instead of "rarefied gas compressor" (which it is), is due to its association with water pumps.

The publication of Schott's books (1657 and 1664) and, later, Guericke's own book (1672) spread the new knowledge in Europe, and better pumps were built by others. Notable was Boyle, who was in Italy in 1640–1641, heard about Guericke's experiments in 1657, and published his own designs in 1660. The pump was, in principle, the same as Guericke's but had an added rack-and-pinion drive. Engineering improvements by Hooke and Hauksbee, among other's, followed. Pumps of that general type were in use until almost the twentieth century, when an Englishman, H. A. Fleuss, introduced a more-or-less industrial version of a piston pump which remained in use until the development of the incandescent lamp industry sparked the early advances in vacuum technology at the turn of the century. Before the twentieth century, vacuum pumps were used for laboratory work and were often made by local instrument makers.

A historically important development was Sprengel's pump (1865), which was based on the continuous application of Torricelli's method of obtaining vacuum. Sprengel's pump, with improvements and modifications, was used by Edison in 1879 for evacuation of the first incandescent lamps. The evacuation was started with a Geissler pump and finished by Sprengel pump. Edison and his associates designed a unique new Sprengel–Geissler pump which was used until 1896. A person continuously lifted and lowered mercury bottles, connected to the pump by flexible tubes, to trap and eject volumes of air. Good vacuum was obtained when the recompressed air volumes turned into small bubbles. A solid mercury column when the bubbles became invisible was an indication of high vacuum.

The first pump that was adaptable for use with the emerging electric motors was introduced by W. Kaufman in 1905. It consisted of a "Torricelli" tube twisted into a helix and a return line for mercury, so that the device resembled a tubular Archimedes screw in which captured volumes of air betweens slugs of mercury would transfer from the inlet to the discharge side. A much more practical design was developed by Gaede in the same year.

Gaede demonstrated his rotary mercury vacuum pump in 1905. This pump is representative of attempts to reach low pressure by mechanical pumps alone. It was used in series with water jet pumps, and produced 7×10^{-5} torr at 20 rpm. It was 10 times faster in pumping a 6-L volume than were the Sprengel-type pumps used then. In principle, it was essentially a rotary liquid piston device. Such pumps enjoyed a brief period of popularity until a line of oil-sealed rotary mechanical pumps was developed (before 1910).

Mechanical pumps existed in Edison's time, but interestingly enough, the vapor pressures of the oils used in those days were higher than that of mercury. However, by 1910 mechanical vacuum pumps, more or less similar to present designs, were fully developed. Of course, the basic designs are very old (at least 1588), and they were adapted for vacuum pumping using tighter mechanical tolerances and better oils.

For successful industrial or scientific laboratory device development, there must often be the confluence of a pressing need, available technology, an inventive idea, and some scientific understanding of underlying principles. Many early inventive ideas were impractical because of the lack of available technology. Thus Guericke struggled with the problem of fitting a piston tightly into a cylinder, just as before him Galileo had struggled to seal his cylinder when he tried to measure the force of formation of vacuum, just as 100 years later James Watt struggled to produce a tight fit between the piston and the cylinder of his steam engine. Similarly, Gaede was ahead of his time with the molecular pump.

An interesting pattern can also be traced regarding the attempts to reach a better degree of vacuum without any apparent regard to pumping speed. Until the twentieth century, the pumping speed of most vacuum pumps appears to have been less than 1 L/s. Writing in 1922, Dunoyer gives pumping speeds for many early pumps. Geissler and Toepler pumps, which Edison used for his incandescent lamps (1880), were essentially arrangements for continuous repetition of Torricelli's upside-down mercury-filled tube, with a pumping speed of an astonishing 4 cm^3/s. An early rotary mechanical pump produced 2 L/s;

another mechanical pump, 120 cm^3/s; Gaede's rotary mercury pump, 0.15 L/s maximum; Gaede's first molecular pump, 1.4 L/s; Holweck's molecular pump, 2.5 L/s; the first Gaede diffusion pump 80 cm^3/s maximum; the Langmuir diffusion pump, less than 4 L/s; and Crawford's diffusion pump, 1.3 L/s.

Only in the 1930s did larger pumps appear producing pumping speeds of a few hundred liters per second. Characteristic of the time is a description of a "high-capacity" vacuum installation in 1929 with 12 glass and three steel Langmuir "condensation" pumps (used in parallel), producing a total pumping speed of 40 L/s at 5×10^{-3} torr.

1.6.2 Vapor Jet (Diffusion) Pumps

The advantage of the vapor jet pumps compared to the mechanical pump for the creation of high vacuum is that they can produce much higher pumping speed at low pressure for the same size, weight, and cost. This was not realized at first.

Gaede, thinking in terms of kinetic theory, correctly deduced that gases can be pumped by the molecular drag effect, only to be defeated by mechanical difficulties. In the case of vapor jet pumps, he assumed that the pumping action was due to the classical picture of the diffusion of air into a "cloud" of mercury vapor. Actually, if the mercury cloud were immobile, no pumping action would result. This confusion kept vapor jet pump speeds at very low levels, and the full realization of their capabilities was not achieved until 20 to 25 years later.

Langmuir followed with a better design (1916) but he overemphasized the condensation aspects of the pumping mechanism. In fact, condensation is basically not required for the pumping action, only for the separation of pumped gas and pumping fluid. The first correct design from the point of view of fluid mechanics was made by W. W. Crawford in 1917, but the maximum pumping speed was only about 2 L/s for a 7-cm-diameter pump (roughly 50 times lower than modern designs). Such pumps were improved continuously during the next 20 years.

Low-vapor-pressure oils were introduced by C. R. Burch in 1928. In 1937, C. M. Van Atta reported a design with 200 L/s for an 18-cm-diameter pump, about five times lower than that for present pumps. Large pumps were designed in the early 1940s for large gas separation plants and later for vacuum metallurgy, industrial coating, and space simulation chambers. Modern designs date from about 1960.

1.6.3 Molecular Pumps

The first mention of Gaede's molecular pump occurs in 1912 in *The Engineering Index Annual*. The pump was not successful because in 1912 the art of

INTRODUCTION

making high-speed rotating machinery had not yet been developed. It took 45 to 50 years before molecular pumps were made as an industrial device to adequate reliability. Other attempts, by Holweck and Siegbahn, were hardly more successful. They utilized, respectively, a cylinder and a disk as the molecular drag surface. One advantage of the bladed, multistaged, axial-flow compressor-type vacuum pump is that each rotor surface is exposed only to a particular vacuum level. In cylindrical designs the same surface passes from the high-pressure to the low-pressure side.

The idea of using the axial turbine as a vacuum pump seems to have occurred simultaneously to at least two persons (H. A. Steinherz and W. Becker), the former attempting simply to utilize an existing compressor device as a vacuum pump. The closed cell design by Becker was published in 1958. A. Shapiro (Massachusetts Institute of Technology) and his associates analyzed the open-blade design, which was found to be entirely practical and efficient. Later, the SNECMA Company in France developed industrial versions of such pumps. The open-blade designs are generally used now by most manufacturers.

1.6.4 Ion-Getter and Sputter-Ion Pumps

Many phenomena utilized today for high- and ultrahigh-vacuum pumping have been known, and basically understood, for a long time. These include outgassing, adsorption, gettering, sputtering, and cryogenic temperatures. They were not used earlier because the technology was not ready and the need was not pressing.

A vacuum of a few micrometers of mercury (10^{-3} torr) was required in Edison's original electric lamps. In 1881, it took five hours to achieve. To handle the outgassing of water vapor, bulbs were heated to about 300°C during evacuation. Gettering by phosphoric anhydride was introduced from the beginning. Sprengel pumps are associated with the initial successes of the lamp industry. Improvements in pumping speed reduced evacuation time to 30 minutes. This process was used until 1896, when Malignani (Udine, Italy) chemical exhaust processing was adopted. This consisted of placement inside the lamp of red phosphorus, which would vaporize when the filament was heated. The evacuation times was reduced to a minute. The effect of phosphorus is partly chemical and partly due to adsorption.

The cleanup of gases by an electrical discharge has been observed for over a century. Gettering for lamp making is at least that old. T. M. Penning mentions sputtering as a method for reducing gas pressure at least as early at 1939. The mention of ionic pumping appears at least as early as 1953, associated with the initial development of ultrahigh-vacuum techniques.

In the early 1950s, R. Herb (University of Wisconsin) became determined to eliminate oil vapors in vacuum systems. One of his associates suggested titanium for gettering. A container was pumped only by a forepump and a titanium sponge evaporator. "It was successful but had a low-pressure limit which we suspected was argon. Then an ionizer was built to ionize residual gas and drive ions into the wall receiving titanium. It worked and I believe we had the first practical getter ion pump." Such "Evapor-ion" pumps were rather difficult to operate and maintain. Sputter-ion pumps developed by L. D. Hall and associates at Varian Company were simpler and achieved universal acceptance a few years later.

The first ion pumps were single-cell appendage pumps developed for maintaining high vacuum in large, high-power microwave tubes. They resembled Penning gauges with the cathode made of titanium. Multicell designs and triode pumps for better argon pumping followed quickly. Later, orbitron pumps, also introduced by Herb and his associates, enjoyed a brief popularity but were withdrawn due mainly to problems with reliability.

Most modern ultrahigh-vacuum instruments would not be possible without the development of sputter-ion pumps. A group of engineers and scientists at Varian Associates pioneered industrial utilization of sputter-ion pumps and other ultrahigh-vacuum pumping devices, such as sorption pumps, as well as valves and metal gasket flanges.

1.6.5 Cryopumping

Cryopumping effects with and without high-surface-sorption materials were understood at least 100 years ago. Early in the twentieth century, one of the methods for obtaining vacuum was to use activated charcoal cooled with liquid air. With the invention of vapor jet pumps, liquid nitrogen traps were generally used. Although the intended use of the traps was for suppression of the mercury vapor, they did pump water vapor remaining in the chamber after initial evacuation.

During the late 1950s and 1960s, cryogenic shrouds were often used in the construction of space simulation chambers. Occasionally, liquid helium was used for enhanced cryopumping in early, extreme-high-vacuum chambers. The industrial art of cryopumping had to wait until the development of reliable closed-loop gaseous helium cryorefrigerators. These refrigerators were developed for other than high-vacuum uses, but have recently found wide acceptance in the vacuum industry.

REFERENCES

1. E. N. Da C. Andrade, in *Advances in Vacuum Science and Technology*, Proc. 1st Int. Cong. Vac. Tech., Namur, Belgium, E. Thomas, ed., Pergamon Press, Elmford, N.Y., 1960.

2. J. D. Bernal, *The Extension of Man*, The MIT Press, Cambridge, Mass., 1972.
3. J. K. Finch, *The Story of Engineering*, Doubleday & Company, Inc., New York, 1960.
4. S. Drake, *Galileo at Work*, University of Chicago, Chicago, 1978.
5. G. Sarton, *Isis*, 23(1), 212 (French). Book review of *L'expérience barométrique, étude historique*, by C. de Waard, Thouars, Deux Sèvres, France, 1936.
6. Isaac Beekman, *Journals*, C. de Waard, ed., Martinus Nijhoft, The Huage, 1939.
7. G. Sarton, *A History of Science*, Harvard University Press, Cambridge, Mass., 1952.
8. A. Brachner, *Kultur Tech.*, 3, 2 (1979).
9. G. Schott, *Technica Curiosa*, Nuremberg, 1664.
10. R. T. Gunther, *Early Science in Oxford*, Vol. 1, pp. 16, 250, Oxford University Press, Oxford, 1923.
11. J. M. Lafferty, *Physics Today*, p. 211, November 1981.
12. *Encyclopedia Brittanica*, 1947 Edition.
13. L. Dunoyer, *Vacuum Practice*, G. Bell & Sons, Ltd., London, 1926.
14. S. Dushman, *Scientific Foundations of Vacuum Technique*, John Wiley & Sons, Inc., New York, 1949.
15. R. Jaeckel, *Kleinste Drucke*, Springer-Verlag, Berlin, 1950.
16. M. Dunkel, *Vac.-Tech.*, 24(5), 133 (1975).
17. R. Conot, *A Streak of Luck*, Seaview Books, New York, 1979.
18. W. Gaede, *Trans. Ger. Phys. Soc.*, 3(21), 287 (1905).
19. F. Klemm, *A History of Western Technology*, The MIT Press, Cambridge, Mass., 1984.
20. W. Gaede and W. H. Keeson, *Z. Inst. Kunde*, 49(6), 298 (1929).
21. W. W. Crawford, *Phys. Rev.*, 10, 557 (1917).
22. W. Becker in *Advances in Vacuum Science and Technology*, Proc. 1st Int. Cong. Vac. Tech., Namur, Belgium, 1958, E. Thomas, ed., Pergamon Press, Elmsford, N.Y., 1960.
23. M. H. Hablanian in *Advances in Vacuum Science and Technology*, Proc. 1st Int. Cong. Vac. Tech., Namur, Belgium, 1958, E. Thomas, ed., Pergamon Press, Elmsford, N.Y., 1960.
24. C. H. Kruger and A. H. Shapiro, in *Seventh National Symposium on Vacuum Technology Transactions*, C. R. Meissner, ed., Pergamon Press, Elmsford, N.Y., 1961.
25. J. W. Howell and H. Schroeder, *History of the Incandescent Lamp*, The Magna Company, Schenectady, N.Y., 1927, p. 123.
26. T. M. Penning, U.S. Patent 2,146,025, 1936.
27. R. H. David and A. S. Davatia, *Rev. Sci. Instrum.*, 25(12), 1193 (1954).
28. L. D. Hall, *Rev. Sci. Instrum.*, 29(5), 367 (1958).
29. W. F. Westendorp, U.S. Patent 2,755,014, filed April 1953.
30. L. D. Hall, U.S. Patent 2,993,638, filed July 1957.
31. G. F. Weston, *Vacuum*, 28(5), 209 (1978).
32. W. E. Gifford and H. O. McMahon, *Adv. Cryog. Eng.*, 5, 354, 1959.

33. T. E. Madey, *History of Vaccum Science and Technology,* American Vacuum Society, New York, 1984.
34. P. A. Redhead, ed., *Vacuum Science and Technology, Pioneers of the 20th Century,* AIP Press (1994).

2
Properties of Gases

2.1 PRESSURE AND DENSITY

It is customary to specify level of vacuum in terms of pressure. In technical literature on high vacuum technology the pressure is always considered to be an absolute pressure. This means that the absence of gas is taken as zero pressure without any direct reference to the atmosphere. In the rough vacuum field the pressure is sometimes used in the same absolute sense and sometimes it is referred to the atmospheric pressure. Thus, there are designations such as psia and psig, where a stands for "absolute" and g for "gauge." In old Anglo-American units, the atmospheric pressure is 14.7 pounds per square inch at sea level. Then, if a chamber contains 100-psig gas, it means that the absolute pressure is 114.7 psi(a). There is another even more unfortunate system of units using inches of mercury (or inches of water) which are referenced to atmospheric pressure. Here the atmospheric pressure is taken to be zero vacuum and the best achievable vacuum becomes about 29.92 inches of mercury.

To avoid all the confusion and to avoid any reference to atmospheric pressure, a new pressure unit has been established by the International Standards Organization (ISO) the pascal (Pa). It is defined as 1 newton per square meter, pressure being force per unit area. In this system the atmospheric pressure is 101,000 Pa. There has been some reluctance to use the pascal in the vacuum engineering field, both in the United States and in Europe.

The ISO units standard permits, parenthetically, the use of bar or millibar for pressure because of their wide use in the field of physics, 1 bar being the atmospheric pressure. In current European usage, use of the millibar is very common. In the United States the unit torr, which is a direct substitute for the old millimeters of mercury (mmHg), is widely used and is used almost exclusively in this book. Conversion to millibars is simple: 1000 mbar is equivalent to 760 torr. For conceptual and qualitative understanding, mbar and torr can be used almost interchangeably. Other conversions that should be noted are: 1 torr = 133 Pa and 1 Pa = 7 mtorr (see the Appendix for conversion tables).

In high-vacuum technology, pressure is not the most interesting property of the rarefied gas environment. Near the atmospheric end, the knowledge of forces on the walls of the vacuum chamber is important for design. At 10^{-5} torr, the internal forces are so small as usually to be of no interest. The more significant parameters are the molecular density and perhaps the amount of absorbed gases on the surfaces. Therefore, beginning somewhere near 10 mtorr, a mental adjustment should be made to think in terms of density rather that pressure as such. In other words, the designation 10^{-7} torr should convey a degree of vacuum, a degree of emptiness, or molecular density rather than a force. Several suggestions have been made in recent years to abandon the pressure units and use molecular density or to use such units as the decibar, in analogy to decibels in sound measurements, or even to specify, instead, the condition of the surfaces. These concepts have some advantages, but it seems prudent to maintain a single continuous system going from high to low pressure (following the ISO system) and make the mental adjustment to think in terms of density ratios when going, for example, from 10^{-7} torr to 10^{-9} torr, rather than have any association with pressure forces.

2.2 ATOMIC NUMBERS AND DISTANCES

Simplified, approximate estimates of molecular number densities associated with a certain high vacuum level can be made by remembering only two items. First, it should be recalled that most atoms and simple molecules of common gases have diameters near 3 Å (i.e., 3×10^{-8} cm). Despite rather large differences between atomic weights (e.g., from hydrogen to mercury), the extent of the atom or molecule is nearly the same. Thus a row of molecules 1 cm long will contain approximately 3.3×10^7 molecules. Therefore, a tightly arranged molecular structure in a solidified gas contains 3×10^{22} molecules per cubic centimeter.

The second item to remember is that when solids are evaporated into gas *at atmospheric pressure*, their volume increases by about three orders of magnitude. For example, the ratios of density of solid (or liquid) to gas at atmospheric

PROPERTIES OF GASES

pressure are: for helium, 700; for hydrogen, 790; for neon, 1340; for nitrogen, 650; for air, 715; for oxygen, 800; and for argon, 800. Somewhat higher numbers are obtained for water (1600) and for mercury (3000), but the order of increase is still near 1000 times. Therefore, the number of molecules in a cubic centimeter of air is near 3×10^{19}. For rough estimates, atmospheric pressure can be taken to be 1000 instead of 760 torr. Then it can be quickly estimated that at the best ultrahigh-vacuum level normally achieved, 10^{-12} torr, there are 30,000 molecules per cubic centimeter. Starting from a solid with 3×10^{22}, we reduce three powers of 10 for transformation from solid to gas at atmospheric pressure, additional three powers of 10 from atmosphere to 1 torr, and finally, 12 powers of 10 to reach 10^{-12} torr (see Table 2.1).

It is interesting to note that even at the highest vacuum level normally achieved in a laboratory, a surface is exposed to molecular bombardment of 400 million times per second, at room temperature and cm^2. Note also that the mean free path between collisions of a molecule with other molecules at 10^{-12} torr surpasses the diameter of the earth. It is clear that at pressures below 10^{-5} torr, most of the collisions in a vacuum chamber are with the walls and other surfaces present in the chamber rather than between molecules.

2.3 BASIC GAS LAW

In its simplest form, the general gas law can be stated as follows: For a given amount of gas (i.e., given mass of gas), the pressure, volume, and temperature have a definite relationship:

$$\frac{P_1 V_1}{T_1} = \frac{P_2 V_2}{T_2} = \text{constant} \tag{2.1}$$

The subscripts refer to two different conditions and the temperature is the absolute temperature, such as kelvin. From this it follows immediately that if the temperature is kept constant, the product of pressure and volume remains constant. If in a piston the volume is reduced to half of the previous volume, the pressure will double, and so on. This relationship is not exactly true through the entire temperature range and it does not exactly hold for gases that have a complex molecular structure. For this reason, the law is called the *perfect* gas law. The law is, however, entirely sufficient for the occasional simplified calculations used in this book.

It also follows from Eq. (2.1) that if pressure or volume is kept constant, the remaining variable will be changed in a definite proportion. Thus if a gas in an isolated chamber (i.e., constant volume) is heated to a higher temperature, the pressure will increase in the same proportion. In courses on vacuum

Table 2.1 Some Molecular Relationships (at Room Temperature)

	Pressure (torr)				
	760	10^{-3}	10^{-6}	10^{-9}	10^{-12}
Number of molecular collisions on cm² of surface	3×10^{23}	4×10^{17}	4×10^{16}	4×10^{11}	4×10^{8}
Average distance between molecular collisions (cm)	6.5×10^{-6}	5	5×10^{3}	5×10^{6}	5×10^{9}
Number of molecules per cubic centimeter	2.3×10^{19}	3×10^{13}	3×10^{10}	3×10^{7}	3×10^{4}

PROPERTIES OF GASES

technology, it is often assumed that the temperature remains constant. In most cases, this assumption is reasonable, but caution is in order when dealing, for example, with furnaces or cryogenic temperatures.

Equation (2.1) pertains to a given amount of gas. If the law is stated in terms of mass, the gas law can be written as

$$PV = RWT \tag{2.2}$$

where W is the mass and R the constant of proportionality, which depends on the units chosen for all quantities.

It is often convenient to use the molecular weight instead of a mass. Then

$$PV = R_0 \frac{WT}{M} \tag{2.3}$$

It should be recalled that molecular weight is the ratio of the mass of a given molecule to the mass of hydrogen. R_0 is the universal gas constant per mol of gas. The mol is the mass of gas related to its molecular weight. That is, a gram mol of oxygen has a mass of 32 g (O_2), a gram mol of hydrogen (H_2) is 2g, and so on. In other words, in eq. (2.3), $W/M = 1$ if we deal with 1 mol of gas. Then we can write

$$PV = R_0 T \text{ (for 1 mol)} \tag{2.4}$$

If we choose standard atmospheric conditions, $P = 760$ torr, $T = 273$ K (i.e., 0°C), and also note that under standard (atmospheric) conditions, 1 g mol of any gas occupies 22.4 L of volume. We obtain from Eq. (2.4),

$$R_0 = \frac{PV}{T} = 62.4 \text{ torr} \cdot \text{L/K} \cdot \text{g mol} \tag{2.5}$$

2.4 VELOCITIES AND TEMPERATURE OF GASES

Gas molecules are in constant motion. The velocity of this motion depends on the molecular weight and temperature. At a given temperature, the higher the molecular weight, the lower the velocity. From the kinetic theory of gases, the average velocity of a gas molecule can be obtained as

$$v = 146\sqrt{\frac{T}{M}} \text{ m/s} \tag{2.6}$$

Near room temperature, this will give for hydrogen, about 1800 m/s; for water, 600 m/s; for nitrogen, 480 m/s; and for oxygen, 450 m/s. Note that these velocities are independent of pressure and that they are greater than the speed of sound in air (330 m/s).

The concept of temperature appears to be simple in our daily experience, but it cannot easily be associated with a single molecule except in terms of the kinetic energy and velocity of motion. Human beings (and most animals) cannot exist unless they are surrounded by air, which has molecular velocities roughly between 400 and 500 m/s. It is either too cold or too hot outside that range.

The constant motion of gas molecules produces collisions between molecules and at any surface exposed to the gas (see Table 2.1). The average distance traveled between collisions is called the mean free path. It depends on the size of the molecule and the number of molecules present or, in other words, density or pressure. The average mean free path, λ, is according to kinetic theory:

$$\lambda = \frac{1}{\sqrt{2}\ \pi\ d^2\ n} \qquad (2.7)$$

where d is the molecular diameter, cm, and n is the number of molecules per cubic centimeter. For air at room temperature, the mean free path (MFP) can be found from

$$\text{MFP} = \frac{5 \times 10^{-3}}{P} \qquad (2.8)$$

if P is in torr and MFP is in centimeters. The range of variation in the mean free path for common gases is about a factor of 10. Instead of the 5 used in the expression for the mean free path for air, for helium it should be 14, for hydrogen 9, for water vapor 3.2, and the values are proportional to the absolute temperature.

It should be apparent that when the mean free path of the molecules becomes greater than the dimensions of the vessel containing them, the collisions between molecules become less frequent and the molecules begin to collide mostly with the walls of the vessel. For most high-vacuum work it is often assumed that the collisions between molecules have negligible effect on gas behavior. Usually this is a safe assumption, but there are cases where some secondary effects may depend on the infrequent collisions between molecules.

There is some association between the mean free path and the conductivity and viscosity of gases. In regard to thermal conductivity, a distinction must be made between the high- and low-pressure ranges encountered in high-vacuum technology. Above approximately 10 torr, the heat transfer through a gas inside a relatively small chamber (a few centimeters in size) is dominated by convection. The heat transfer by convection is due to bulk motion of gas in a gravity field because warm air expands (at constant pressure) and rises relative to surrounding colder gas. At the higher pressures, the thermal conductivity of the gas (divorced from convection) does not change significantly with changes in pressure. Gas molecules colliding with a hot surface acquire a higher velocity and deliver this additional energy to the cold surface. If the pressure between the two surfaces decreases, the number of energy carriers (i.e., the number of molecules) decreases, but they travel proportionately longer distances and transfer their energy to the cold surface with fewer collisions.

When the mean free path becomes greater than the dimensions of the container, the heat transfer depends only on the number of carriers and there develops a linear relationship between conductivity and pressure. The complete effect is illustrated in Figure 2.1, which shows the variation of heat transfer from a heated wire with changing pressure of the surrounding gas. At pressures below 10^{-4} torr, radiation and conduction through the wire supports become significant and mask the actual conduction by the gas.

The variation of conductivity with pressure can be utilized for indirect measurement of pressure by monitoring the heat transfer rate from a heated filament to the surrounding walls kept at ambient temperatures. Changes of viscosity with pressure can be utilized in a similar manner; for example, by

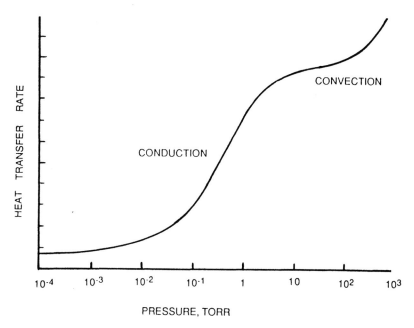

Figure 2.1 Relative heat transfer from a heated wire depending on pressure.

monitoring the deceleration rate of a rotating body due to collisions with the surrounding gas within a stationary container.

Heat transfer at high-vacuum conditions is very low compared to the gas at atmospheric pressure. It decreases by a factor of 100 between about a few torr and a few millitorr. There is very little heat flow between two metal plates placed together in a high-vacuum chamber because the real area of contact between two rigid bodies is usually only near 0.1% of the apparent area. Heat is transferred only through a few touching high points between the plates. To increase the flow, the actual area of contact must be increased. One (or both) of the plates should be soft and high clamping forces must be used. Even then, the heat transfer by conduction will be limited to the small areas of contact created near the clamping bolts.

2.5 VAPORS AND VAPOR PRESSURE

The common term "vapor" does not have a precise meaning. If the temperature is low enough, any gas can be termed "vapor," and if the temperature is high enough, any vapor can be described as a gas.

The term "vapor pressure" appears very often in a discussion of high-vacuum technology and its meaning should be clearly understood. If a liquid is placed in a closed container (Figure 2.2) kept at a constant temperature, sooner or later it develops a certain pressure of its own vapor, which is called saturated vapor pressure or equilibrium vapor pressure. The words "saturated" and "equilibrium" mean that the number of molecules leaving the liquid surface and the number returning to the liquid are the same. Liquids evaporate at any temperature. This process has little to do with commonly observed boil-

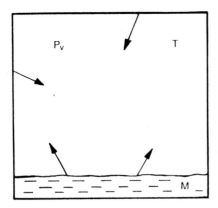

Figure 2.2 Concept of vapor pressure.

ing. For simplicity, we refer to "vapor pressure," omitting the word "saturated." This vapor pressure, obtained in the system in Figure 2.2, depends only on (absolute) temperature and the particular species of the liquid. It does not depend on the presence of other gases or vapors in the container. Note that the condition is a constant temperature, not the absence of heat transfer. For any substance, a certain temperature has a certain corresponding vapor pressure. If the temperature is increased, the vapor pressure will increase (i.e., more liquid will evaporate until a new equilibrium is reached). The opposite will occur if the temperature is decreased.

The general form of the dependence is

$$\log P = A - \frac{B}{T} \qquad (2.9)$$

where P is the vapor pressure, T the temperature, and A and B are constants associated with each species. Therefore, a plot of the logarithm of the vapor pressure as a function of the reciprocal of the absolute temperature gives a straight line. Vapor pressures of many substances can be found in handbooks of chemistry and physics. The vapor pressures of elements are given in the Appendix.

Vapor pressure has a steep dependence on temperature. It also depends strongly on the molecular weight of the substance, but the function is not smooth. For example, water has a lower vapor pressure than alcohol, even though its molecular weight is lower. Generally, however, the higher the molecular weight, the lower the vapor pressure. For example, the vapor pressure of water at room temperature is about 20 torr, and the vapor pressure of some pure oils used in vacuum pumps is 10^{-10} torr (at room temperature, molecular weight near 500). Vapor pressures of most metals at their (atmospheric) melting points are in the millitorr range, but there are important exceptions. For example, aluminum, gallium, and indium have unusually low vapor pressures at their melting temperatures.

If we supply heat to the system shown in Figure 2.2 and provide a hole in the wall of the container, we can maintain a certain value of the pressure of the vapor by controlling the amount of heat and the size of the hole. The rate of vapor escaping through the hole will depend on the latent heat of evaporation of the liquid and the characteristics of the hole, which may be called a nozzle. The external pressure may also play a role. Note that the temperature will be that associated with the saturated vapor pressure resulting in the system. If we double the amount of heat energy, the temperature will increase somewhat but only because the pressure of the vapor will increase. Unless the container is completely open (as in boiling in the atmosphere), the rate of vapor escape will to some degree be influenced by the flow characteristics of the nozzle. The rise in temperature will be small because vapor pressure is a very steep function of temperature.

The rate of evaporation depends on the rate of removal of the vapor from the vicinity of the liquid surface. In still air, diffusion rates will be involved; otherwise, natural or forced convection. A hot-air blower or an electric hair dryer utilize the latter. The higher temperature increases the rate of evaporation (or desorption) and the "wind" carries the vapor away. It may be well to note, as a rule of thumb, that for water near room temperature, a $10°C$ change affects the vapor pressure by a factor of 2. At partial pressures in the high-vacuum region, however, $10°C$ will produce a tenfold change in vapor pressure.

If sufficient heat is supplied, the liquid may boil. It is interesting to observe that the mechanical action of the boiling process is not the primary phenomenon. Boiling occurs when the heat transfer rate to the liquid is high and the surface area is too low for the vapor, which is created under the surface, to escape. If we heat a fluid from above using radiative heat (to which the liquid is opaque), there will be no boiling in the normal sense. At a given pressure, once the corresponding temperature is reached (the "boiling point"), the rate of evaporation will depend on the amount of heat supplied in excess of the heat required just to maintain the surface at the temperature. It is important not to confuse this evaporation rate with the normal evaporation rate associated with the saturated vapor pressure. The vapor pressure does not change when the fluid is slightly below or at the boiling point. Vapor pressure versus temperature curves do not change their slope at the boiling point or the freezing point. Note also that at high vacuum the boiling temperature is independent of pressure.

2.6 EVAPORATION

In the preceding section it was noted that the saturation vapor pressure does not depend on the presence of other gases over the liquid surface. However, the time to reach the new equilibrium when the temperature is increased does depend on the presence of other gases. The molecule leaving the liquid surface is likely to collide with gas molecules and even return to the liquid. The rate of evaporation depends on the diffusion characteristics of the vapor in the surrounding gas.

It should be apparent, then, that the rate of evaporation is the greatest in high vacuum. If the evaporation conditions are such that the evaporating molecules are entirely removed from circulation, the maximum rate of evaporation will be obtained. For example, a cube of ice at $-10°C$ could be placed in a high-vacuum chamber and surrounded by surfaces kept at liquid-nitrogen temperature. The condensation coefficient of water molecules striking the cold surface is essentially 100% and the evaporation rate is at the maximum corresponding to $-10°C$ (assuming that the temperature is kept constant).

The maximum evaporation rate is

$$Q_{max} = 0.058 P_v \sqrt{\frac{M}{T}} \qquad (2.10)$$

where Q is in g/cm²·s, P_v is the vapor pressure in torr, and T is the temperature in kelvin (M is the molecular weight). Note that the dependence on temperature is not due primarily to the square root in the expression but due to the variation of vapor pressure P_v with temperature (which is nearly logarithmic). The actual rate of evaporation in atmospheric conditions may be 10^5 or 10^6 times lower.

2.7 ADSORPTION AND DESORPTION

The terms "evaporation" and "condensation" refer to the phenomena associated with bulk substance. When the amount of liquid remaining on a solid surface is reduced to one molecular layer or less, the corresponding terms used to describe the arrival and departure of gas are "adsorption" and "desorption." The behavior changes because the attractive forces between molecules of a liquid or gas and the solid surface are different (usually greater) than the forces between the molecules of the gas itself. Adsorbed gases can exist on the surface of a solid in a form that may be thought of as a monomolecular sheet of liquid. The number of atoms per square centimeter of surface is approximately 10^{15} (see Section 2.2), and if each atomic site is occupied by an adsorbed gas molecule, we obtain a "film" with a density of a liquid.

The relationship between the number of molecules adsorbed on the surface and the pressure of the vapor above that surface is usually called the adsorption isotherm. This is perhaps not a very descriptive term because it emphasizes only the fact that the relationship is considered at constant temperature. Another way to think about the data shown in adsorption isotherms, for example in Figure 2.3, is perhaps to consider the graph as adsorption vapor pressure in analogy with condensation vapor pressure. It is reasonable to expect that in the situation shown in Figure 2.2, the vapor pressure would be reduced if the amount of liquid were to be depleted to the point where the surfaces would not be fully covered. In Figure 2.3 the saturation occurs at a coverage somewhat higher than 10^{15} atoms because the surface area is larger than 1 cm². Such data can be plotted versus percentage surface coverage, or in some cases, versus the number of molecular layers on the surface. The actual number depends on the material. For light gases, saturation may occur after a few monolayers have

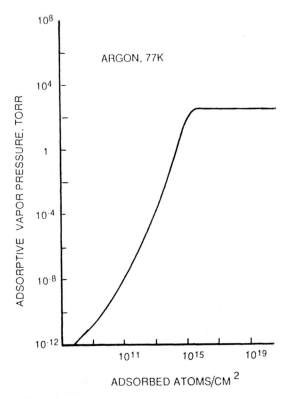

Figure 2.3 Adsorption isotherm relating the pressure of vapor to its adsorbed amount.

been adsorbed. For larger molecules such as oils of low vapor pressure, saturation may begin after the first monolayer.

It should be appreciated that the surfaces used in construction of vacuum apparatus are not smooth. The real surface can be greater than the apparent (projected) area. The magnification factor can range from about 5 for a smooth, shiny aluminum foil to nearly 1000 for an anodized aluminum surface. Values near 50 can be expected for structural materials.

The amount of surface gas is an important quantity in high- and ultrahigh-vacuum systems. Very small quantities of adsorbed gas can significantly affect the surface properties of materials. For example, bonding, wettability, and catalytic action are affected by layers ranging from a fraction of a monolayer to a few monolayers. The same is true for work functions associated with electron emission, lubrication by 1 to 10 monolayers; electrical contacts, 2 to 10 monolayers. Visual effects, such as color, involve thicker layers of surface films—200 to 2000 atomic or molecular layers—but optical adsorption is influenced by 1 to 200 layers. Oxides normally occurring in nature may be 5 to 100 layers thick; corrosion and surface treatment processes involve similar numbers.

In discussion of sorption and desorption, only physisorption is considered here. Chemical bonding (or chemisorption) is neglected. When an atom or a molecule approaches a surface, it experiences at first an attractive force, but after a certain distance is reached, further approach produces a strong repulsive force. The position maintained, as an interatomic distance, is a position where the net force is zero. The potential energy, relative to the surface associated with this force, will be at a minimum when the atom is in equilibrium. In addition to the potential energy associated with the position, there is also the kinetic energy associated with oscillations related to temperature. Due to these oscillations, atoms can leave the surface or approach the surface more closely. In the latter case, there is a possibility of producing a chemical bond if there is an appropriate chemical affinity. Chemical bonds usually have higher "adsorption" energies. By convention, the attractive forces are taken to be negative and repulsion positive. The chemisorbed atom is said to be in a deeper potential well.

The adsorption energy (or heat of adsorption, in analogy to heat of evaporation) depends on the molecular weight and temperature. For example, for hydrogen this may be 100 cal per gram mole at room temperature, for argon (on glass), 4 kcal/mol; for water, 8 to 10 kcal/mol; and for oils with molecular weights of 400 to 500, as much as 25 kcal/mol.

Another way to illustrate the strength of the adsorption bond is to consider the average time the atom or molecule remains in the adsorbed state. This period is called sojourn time or residence time. The residence time can be expressed (Frenkel, 1924) by

$$t = t_0 \exp\left(\frac{Q}{RT}\right) \qquad (2.11)$$

where t_0 is the time of oscillation in the adsorbed state, Q the heat of adsorption, R the gas constant, and T the absolute temperature. The magnitude of t_0 is 10^{-12} to 10^{-13} s and t is proportional to M (molecular weight) and the atomic distance. For example, for hydrogen, at 100 cal/mol, the residence time is 10^{-13} s, but if the heat of adsorption is 30 kcal/mol, the residence time becomes 4×10^{19} s, almost a century. This is why a water spot on a concrete floor will dry out in a few hours but an oil spot will remain for many years.

In addition to the possibility of leaving the surface (if sufficient energy is available), the molecule can also move on the surface. It takes less energy to hop from one site to another (from one potential well to another) than to leave the surface. Values quoted in technical literature for energy associated with surface mobility range from 0.25 to 0.75 of the desorption energy. From this, the corresponding residence time can be estimated.

To illustrate the great variation of residence time, depending on the substance and the temperature, examples of values from Eq. (2.11) are plotted in Figure 2.4. The three temperatures represent liquid nitrogen cooling, room temperature, and temperature used in baking vacuum systems to enhance desorption and to reduce the gas content on the surfaces exposed to high vacuum. It should be noted that light gases such as hydrogen and helium do not remain on the surface even at liquid nitrogen temperature. For practical considerations, we may think of them as rebounding from the surface after a collision. However, even after such a short time of residence, some energy interaction occurs, for example, in heat transfer.

When a molecule desorbs, the direction of its path is not related to the direction from which it came. The reflection may be assumed to follow the cosine law, illustrated in Figure 2.5a. The more likely direction is perpendicu-

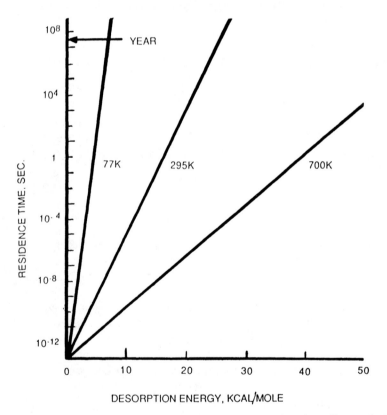

Figure 2.4 Residence time of gas molecules on a surface depending on the heat of adsorption and temperature.

PROPERTIES OF GASES

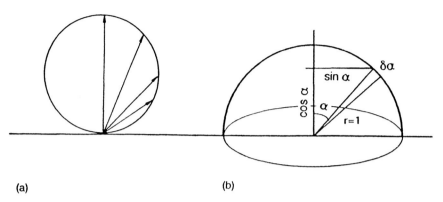

Figure 2.5 Cosine law of reflection. (a) The lengths of the arrows are proportional to the probability of reflection in that direction. (b) The solid differential angle as defined by Rozanov.

lar to the surface (proportional to the diameter in the figure); it is only half as probable to leave at a 30° angle to the surface; and so on. This applies even to hydrogen because usually surfaces are very rough compared to the size of the molecule. Mirrorlike reflection (accompanied by some scatter) may occur when a molecular beam of nonadsorbing gas is directed toward a very smooth crystal surface.

Cosine law is often mentioned but rarely fully explained in textbooks on high-vacuum technology, especially because the "angle of reflection" is often not defined. In optics, the angle of incidence and the angle of reflection usually refer to the surface, but in the cosine law of molecular reflection the angle refers to the normal to the surface. L. Rozanov defines it as "the solid differential angle"

$$d\omega = 2\pi \sin\alpha \, d\alpha$$

where the probability P of a molecule reflecting from a surface in a particular direction, as demonstrated by experimental evidence, is proportional to the cosine of the angle between the normal to the surface and the direction of flight (see Figure 2.5b).

$$dP = (d\omega/\pi) \cos\alpha \tag{2.12}$$

Integration of Eq. 2.12 with limits 0 to α gives the number of molecules contained inside the angle α

$$\xi = \int \sin 2\alpha \, d\alpha = \sin^2\alpha \tag{2.13}$$

From Eq. 2.13, the angle corresponding to a given portion of the molecular flow ξ can be expressed by

$$\alpha = \text{arc sin } \sqrt{\xi} \qquad (2.14)$$

This formula is often used for mathematical modeling of molecular flows in vacuum systems.

Associated with adsorption is the concept of monolayer formation time. When a molecule collides with a surface, the probability of retention depends on the two substances (molecular structure, molecular weight, heat of adsorption), the temperature of the incoming gas molecules, and especially the temperature of the surface. For example, for helium on glass the adsorption probability is near 0.24 at 0°C and 0.13 at 100°C. Corresponding numbers for hydrogen are 0.64 and 0.5, for nitrogen 0.81 and 0.7, and for argon 0.9 and 0.81.

The monolayer formation time can be associated with the vapor pressure of the adsorbing gas. For qualitative appreciation, a typical value for common gases at room temperature can be said to be a few seconds for a vapor pressure of 10^{-6} torr. The same ratio will produce a few hours of 10^{-10} torr, and so on. This immediately suggests that to perform a process involving a surface relatively free of adsorbed gases, it is necessary to obtain pressures lower than about 10^{-8} torr.

2.8 GAS CONTENT OF MATERIALS

As noted in Section 2.2, the ratio of atomic density of liquids and solids to that of gas at atmospheric pressure is about 1000. This suggests that if one out of 1000 atoms in a solid is an occluded gas atom, the gas content of the solid is equivalent to an atmospheric pressure inside it. In Figure 2.6 this idea is illustrated by means of the soap bubble analogy of a crystal structure. There are approximately 450 "atoms" of a crystalline solid in the picture and perhaps a dozen "foreign" smaller bubbles concentrated at the boundary between two adjacent grains. The solubility of gases in solids depends on species, temperature, history of exposure, crystal structure, and chemical reactivity. Regarding diffusion rates, only hydrogen may have significant diffusion rates through metals at normally encountered temperatures. In some metals, hydrogen can be present at much higher quantities compared to the amount equivalent to the atmospheric pressure. Inert gases are essentially insoluble and do not diffuse in metals.

Elastomeric materials (plastics, rubbers, etc.) may contain as much as 1% by weight of various gases. Water vapor is usually the most important gas contained in such materials. Plastics, in general, have a more porous structure than metals and can contain more gas than that equivalent to an atmospheric density because of internal adsorption.

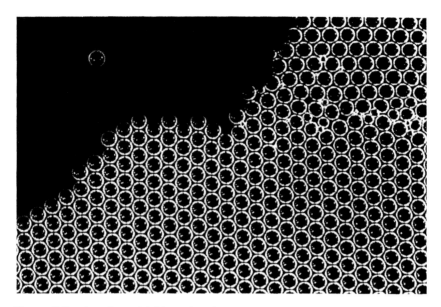

Figure 2.6 Atomic model illustrating the gas content of solid materials.

Glass also has a certain amount of porosity at least for helium. Most glasses can be expected to contain the same amount of helium as exists in air (about 5×10^{-3} torr). In general, the image of solids and liquids containing gas with densities comparable to that of atmospheric air is close to reality. It may be observed, then, that there is usually a sufficient amount of gas present near the surface exposed to vacuum. Even if the gases adsorbed on the surface are removed by elevating the temperature or other means of supplying energy to the adsorbed molecules, there is still a sufficient amount present in grain boundaries, porous domains, oxide layers, and so on, so that the amount of gas entering the space in a high-vacuum chamber is never exhausted in practical situations. Therefore, when testing the performance of high-vacuum pumps, it is essentially impossible to achieve a condition of zero flow. The ultimate, lowest-achievable pressure of a high-vacuum pump is usually a combination of the limitation of the pump itself and the result of the "parasitic" gas emanation from the surfaces of the testing system.

2.9 OUTGASSING

Adsorbed gases normally found in atmospheric conditions begin to evolve from the surfaces when exposed to high vacuum. This gas evolution, called outgassing, depends on the characteristics of the particular gas (or vapor), the temperature of the surface, and time of exposure to vacuum. The rate of evolution varies depending on the nature of the surface and the desorbing substance. In case of elastomeric materials, it is difficult to separate the outgassing from the surface and the evolution from the bulk of the material.

The various mechanisms of outgassing may produce the time dependence according to $t^{-0.5}$, t^{-1}, t^{-2}. For outgassing from metal surfaces, t^{-1} dependence may be used for measurement extrapolation. For time periods of several hours, the time dependence can be expressed as

$$q = q_0 \exp(-at) \tag{2.15}$$

where q_0 is the initial outgassing rate, a is a constant associated with the slope of the outgassing decay, and t is time. In practice, none of the mathematical expressions can be relied upon to give precise results. Generally, the behavior of gases on the surface can be complex. It depends on short and long history of the surface, presence of contamination, roughness, and so on. Knowledge of the outgassing rate is, of course, very important for design of vacuum systems because the amount of evolved gas determines the size of pumps necessary to maintain a given pressure.

Unfortunately, the outgassing rates published in reports and textbooks often vary by as much as a factor of 10. One reason for this situation is that the history of the surface prior to test is rather difficult to establish. Another reason is that the precise beginning of outgassing is not clearly coordinated (i.e., the evacuation of the apparatus is not standardized). Also, the surface conditions of samples and the method of cleaning are not always the same. Additional errors can occur due to the measurement technique; imprecise knowledge of pumping rate in the apparatus and vacuum gauge errors. In addition, most outgassing data are based on total pressure measurement, disregarding consideration of the different gases that may be evolved. In Figure 2.7, some outgassing rates are shown to illustrate some general trends. Additional examples, (compiled by Y. E. Strausser), are given in Figures 2.8 to 2.11.

2.10 WATER VAPOR

As noted before, a fluid in equilibrium with its vapor has a characteristic pressure associated with a given absolute temperature and the properties of the fluid (mainly its molecular weight). Water is a highly unusual substance. It is much less volatile than would be expected from its low molecular weight (e.g., water, 18; ethyl alcohol, 46).

The saturated vapor pressure does not depend on the presence of other gases. Conceptually, this is easy to see at higher vacuum, because all molecules act independently under molecular flow conditions. At high pressure, however, it is well known that water will boil at 100°C regardless of the humidity in the

PROPERTIES OF GASES

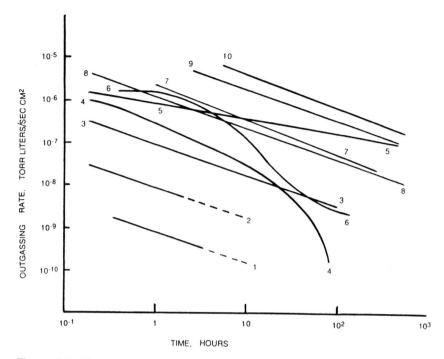

Figure 2.7 Examples of outgassing rates: 1 and 2 stainless steel; 3, cold-rolled 1020 steel; 4, aluminum; 5, 6, and 7, epoxy resins; 8, Teflon; 9, neoprene; and 10, Hycar.

room. At room temperature, the saturated vapor pressure of water is 18 to 20 torr. We should expect, then, that the vaporization activity should be near boiling conditions if the partial pressure of water vapor in the room is less than the saturated vapor pressure. The reason the presence of the air above the liquid seems to make a difference is that the "humidity" is much higher in the immediate vicinity of the liquid surface. The evaporating molecules are subject to a very high frequency of collisions and they remain near the surface of the liquid.

The bulk liquid behavior may not be the same as the invisible thin films of water vapor clinging to all the surfaces in a high-vacuum system. Water on the surface may have desorption energy near 8 to 10 kcal/mol. However, this energy can be much higher inside crystals (such as zeolites). Water adsorption can easily be in the realm of energies usually associated with chemisorption. There is, of course, no clear barrier between physisorption and chemisorption. For example, some vapor jet pump fluids have desorption (or evaporation) energies near 25 kcal/mol, which are usually associated with chemisorption.

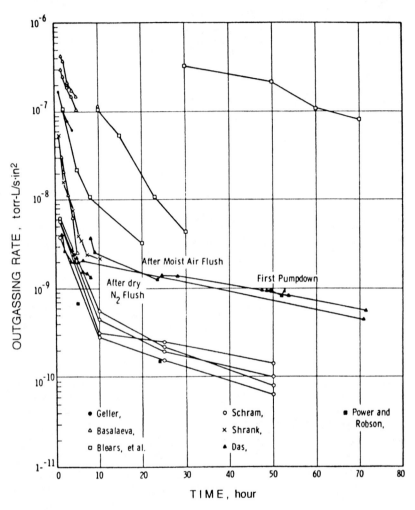

Figure 2.8 Compilation of aluminum outgassing rates.

It is well known that in some materials water will not remain on the surface but will penetrate the structure of the material itself. Plastics, composite materials, and some ceramics are good examples. "Permanent" gases, too, penetrate such materials (as well as metals) but they are usually less "sticky" than water. This is why it is necessary to bake ultrahigh-vacuum systems at temperatures as high as 400°C. The elevated temperature provides the desorption energy required to remove adsorbed layers, and it also enhances the diffusion rates required to bring the molecules to the surface exposed to vacuum. Consider,

PROPERTIES OF GASES

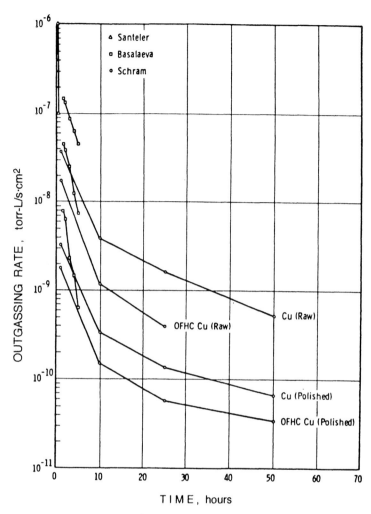

Figure 2.9 Compilation of copper outgassing rates.

for example, the amount of water vapor that may be held in the grain boundaries of a metal, where the atomic structure is disoriented enough to permit water penetration and also to provide tighter holding conditions for water molecules. Assuming an average grain size of 0.025 mm and a monolayer of water molecules penetrating into the boundaries near the surface, the additional amount of water will be equivalent to a monolayer covering the entire apparent surface area. In addition, oxide layers on metal surfaces can contain water vapor amounts equivalent to several monolayers.

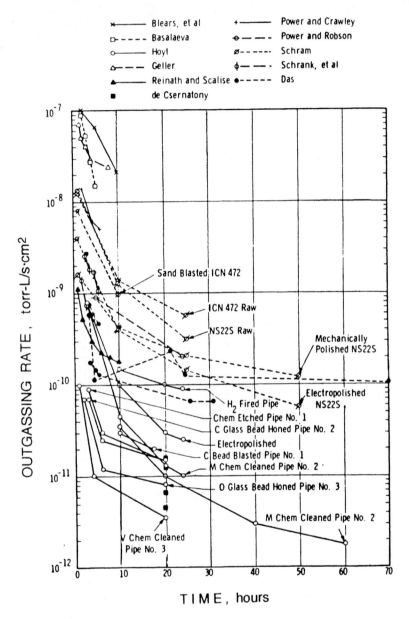

Figure 2.10 Compilation of stainless steel outgassing rates.

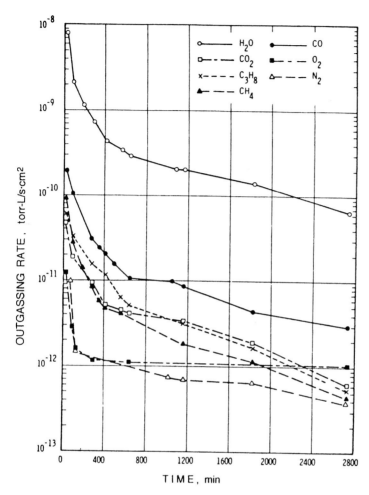

Figure 2.11 Outgassing rates for a stainless steel sample cleaned with Diversey DS-9.

The outgassing associated with surfaces in vacuum is particularly significant for relatively porous materials. Thus the amount of elastomer gaskets and their temperature can significantly affect the ultimate pressure of the system. A typical water absorption quantity associated with rubber or plastics is 0.1%, but this will vary by at least a factor of 10. This represents sufficient water content in the material to produce the commonly observed outgassing rate. The outgassing rate from elastomeric materials per unit area exposed to vacuum can be 1000 times higher than from metals. A typical 15-cm-diameter O-ring will have a water content equivalent to a few monolayers of water vapor covering all other surfaces in the vacuum system. The need to minimize the amount of water-absorbing materials becomes obvious. This is particularly important in and near vacuum gauges or other instruments, such as residual gas analyzers (mass spectrometers), if they are expected to indicate a measurement representative of the conditions in the system. Permeability of various polymeric materials to water can range from 10,000 to 1 (from natural rubber to Kel-F). There are materials with microporosity in their structure that will produce a higher rate of leakage for water than for helium. For example, silicone rubber has 100-fold higher permeability for water than for helium. Other solvents can have similar characteristics. This is why high-vacuum seals such as O-rings should not be cleaned by solvents and leak testing with acetone should be avoided. Neither should rubber O-rings be cleaned in ultrasonic baths because of enhanced absorption. The mold release powder sometimes found in O-rings can be wiped with a damp cloth, and if the system can tolerate it, the O-rings can be wiped using the smallest amount of grease suitable for high-vacuum use, just to make the surface appear shiny.

Water is a part of the atmosphere. The composition of the atmosphere is normally listed in textbooks for dry air (Table 2.2). This gives a somewhat distorted picture. At moderate temperatures, the saturated vapor pressure of water is about 20 torr. This constitutes about 2.5% of the atmosphere. Thus, even without considering rain and snow, there can be two or three times more water vapor in the air than argon and the oxygen and nitrogen content should be reduced by about 1 and 2%.

Because of the adsorption properties of water vapor, "liquid" layers of water are usually present on all surfaces exposed to atmospheric air. After the ini-

Table 2.2 Composition of Dry Air at 760 torr (Sea Level, 0°C)

Gas	Percent by volume	Partial pressure (torr)
Nitrogen	78.08	593.40
Oxygen	20.95	159.20
Argon	0.93	7.10
Carbon dioxide	0.03	0.25
Neon	0.0018	1.38×10^{-2}
Helium	0.0005	4.00×10^{-3}
Krypton	0.0001	8.66×10^{-4}
Hydrogen	0.00005	3.80×10^{-4}
Xenon	0.0000087	6.60×10^{-5}

tial evacuation of a vacuum chamber, the residual gas remaining in the vessel contains 90 to 95% water vapor, which is slowly evolved from the surfaces.

REFERENCES

1. H. A. Steinherz and P. A. Redhead, *Sci. Am.,* (March 1962).
2. J. H. deBoer, *The Dynamical Character of Adsorption*, Clarendon Press, Oxford, 1953.
3. L. B. Loeb, *The Kinetic Theory of Gases*, Dover Publications Inc., New York, 1961.
4. F. Pagano, and D. J. Santeler, *Vacuum Engineering*, D. H. Holkeboer, D. W. Jones, Boston Technical Publishers, Cambridge, Mass., 1967.
5. J. P. Hobson, *J. Chem. Phys.*, 73, 2720 (1969).
6. G. Lewin, *Fundamentals of Vacuum Science and Technology*, McGraw-Hill Book Company, New York, 1965.
7. B. B. Dayton, in *Sixth National Symposium on Vacuum Technology Transactions*, C. R. Meissner, ed., Pergamon Press, Elmsford, N.Y., 1960, p. 101.
8. B. B. Dayton, in *Transactions of the Ninth National Vacuum Symposium of the American Vacuum Society*, G. K. Bancroft, ed., MacMillan Publishing, 1962, p. 293.
9. B. B. Dayton, in *Transactions of the Second International Congress on Vacuum Science and Technology*, L. Preuss, ed., Pergamon Press, Elmsford, N.Y., 1962, p. 42.
10. S. Dushman and J. M. Lafferty, *Scientific Foundations of Vacuum Technique*, John Wiley & Sons, Inc., New York, 1962.
11. R. S. Elsey, *Vacuum*, 25(7), 299 (1975).
12. R. J. Elsey, *Vacuum*, 25(8), 347 (1975).
13. H. F. Dylla et al., *J. Vac. Sci. Technol.*, A11, 2623, (1993).
14. Y. E. Strausser, *Varian Vacuum Products, VR-51,* (1969).
15. P. A. Redhead, *J. Vac. Sci. Technol., A13*(2), 467 (1995).
16. B. B. Dayton, *J. Vac. Sci. Technol., A13*(2), 451 (1995).
17. L. N. Rozanov, *Vacuum Technique* (in Russian), Vysshaya Shkola, Moscow, 1990.

3

Fluid Flow and Pumping Concepts

3.1 PRESSURE AND FLOW

The basic characteristic of fluids is that they do not retain their shape. They flow easily under the influence of relatively low forces. If we avoid cases of extremely high viscosity (such as for glass at elevated temperatures), we can say that fluids flow readily from one place to another according to pressure differences.

In simple cases, typical velocities of liquids are near a few meters per second when subjected to small pressure differences. Gases, which have much lower viscosities than liquids, will flow typically with velocities of tens to hundreds of meters per second at relatively small pressure differences. When dealing with simple cases of gas flow, it is generally permissible for engineering calculations to disregard body forces (such as gravity or centrifugal force fields). Then the flow of gases can be likened to the flow of heat or electricity in a conductor. In analogy, the driving force, which is the pressure difference, can be associated with a voltage difference and the flow rate with an electrical current. The concept of resistance will be analogous in both cases. The flow rate is directly proportional to the potential difference and inversely proportional to resistance. In the case of electricity,

$$I = \frac{V}{R} \quad \text{or, more correctly,} \quad I = \frac{\Delta V}{R} \tag{3.1}$$

In the case of fluid flow,

$$Q = \frac{P_1 - P_2}{R} \quad \text{or} \quad Q = \frac{\Delta P}{R} \tag{3.2}$$

Introducing the reciprocal of resistance, which may be called conductance, we have

$$I = C(V_1 - V_2) \text{ and } Q = C(P_1 - P_2) \tag{3.3}$$

A well-known similar relationship is in heat transfer:

$$H = Ak\frac{\Delta t}{L} \tag{3.4}$$

where H is the thermal energy, k the conductivity, t the temperature, L the length, and A, the cross-sectional area of the conductor. This equation can be rewritten as

$$H = k\frac{A}{L}(t_1 - t_2) \tag{3.5}$$

and $k(A/L)$ can be called conductance.

In some gas flow relationships the flow appears to be proportional to the difference of the squares of pressures, but this is due to the expansion of the gas as it enters the region of low pressure. In other words, when the conductance depends on pressure, the average pressure $(P_1 + P_2)/2$ multiplied by $(P_1 - P_2)$ produces the difference of squares.

There is energy associated with flow: the potential energy of pressure and the kinetic energy of motion. The principle of preservation of energy can be applied to fluid flows. This is known as the Bernoulli principle, which may be expressed as follows: In a fluid of negligible viscosity the sum of potential energy and kinetic energy, associated with pressure and velocity along a streamline, remains constant. In other words, acceleration is accompanied by a fall in pressure along a streamline, and conversely, there must be a rise in pressure when the fluid decelerates.

Consider the simple case of flow in an elbow (Figure 3.1). How do particles 1 through 6 negotiate the elbow? Do they stay in line or does 1 lag behind? How do velocities V_1 and V_6 differ? A simple reasoning would suggest that the flow deflects in response to the higher pressure from the curved wall. Therefore, P_1 must be higher than P_6 and V_1 must be then lower than V_6. Whenever the streamlines of flow are curved there is a pressure gradient associated with

FLUID FLOW AND PUMPING CONCEPTS 43

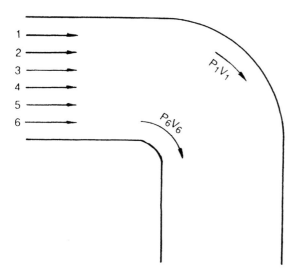

Figure 3.1 Flow of a fluid through an elbow.

the curvature. For example, Figure 3.2 shows two pressure gauges attached to a duct through which a fluid is flowing. In regard to the pressure in the bulk of the fluid, the gauge on the left will indicate a higher pressure and the gauge on the right a lower pressure than they should. Note: gauges should not be placed near elbows. [This effect will not occur in molecular flow (Section 3.5).]

Accelerations and decelerations of gases in various flow conditions are important in understanding the pumping mechanisms of jet pumps (vapor stream pumps or ejectors), flow through valves, and the general flow patterns of process gases in vacuum chambers. For rarefied gas flows the mass flows and pressure differences are rather low, but volumetric flows, pressure ratios, and velocities are rather high.

3.2 MASS FLOW AND VOLUME FLOW

The rate of flow can be described in volumetric units or mass units. Volumetric flow can be called perhaps a "kinematic flow." It describes the bulk velocity in a duct only if its cross-sectional area is known. Units used to describe volume flow (cubic feet per minute, liters per second, etc.) do not indicate any information about the actual amount of substance flowing through. It is impossible to know what the mass flow is unless the density (or pressure and temperature) is also known. The volume per time flow description is easy to visualize if we consider a unit volume in Figure 3.3 moving at a certain bulk velocity. When the imaginary plane A moving together with the fluid arrives

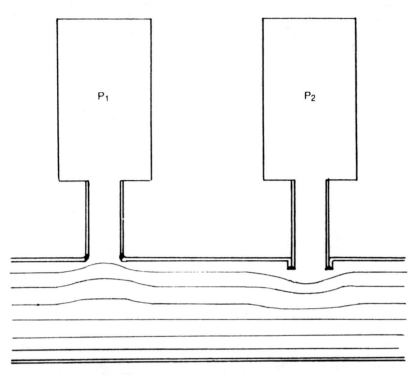

Figure 3.2 Effect of streamline curvature on accuracy of pressure measurements (viscous flow).

at the imaginary plane B, we can say that the designated volume has moved in certain time through the plane B. The volume flow rate can be written as

$$S = vA \qquad (3.6)$$

where A is the cross-sectional area of the pipe and v is the average or bulk velocity. Also, of course,

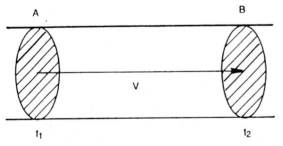

Figure 3.3 Concept of volumetric flow.

$$S = \frac{V}{t} \tag{3.7}$$

where V is volume and t is time. We can obtain mass flow from this simply by multiplying the volume flow and density,

$$G = \rho v A \quad \text{or} \quad G = \frac{\rho V}{t} \tag{3.8}$$

In high-vacuum technology, the mass flows are so low compared to the mass of the walls of a pipe or vessel that the gas temperature usually quickly accommodates to the wall temperature. If the assumption is made that in high-vacuum systems the remaining rarefied gas is always at constant temperature, the mass flow can be measured in units of "throughput," such as torr · L/s. Throughput is obtained by multiplying the volume flow rate by pressure (instead of density),

$$Q = pS \tag{3.9}$$

where Q is the throughput, p the pressure, and S the volumetric flow rate. Once the temperature and the gas species are known, throughput units can be converted to real mass flow units, such as grams per second.

The mass flow rate has an energy associated with it. When we "elevate" an amount of substance to a higher-pressure region in a given time, we expend energy. Throughput is equivalent to power because it represents the rate of expenditure of energy. The correspondence can be seen by a simple substitution of units:

$$\frac{\text{torr} \cdot \text{L}}{\text{s}} \rightarrow \frac{(\text{g/cm}^2)\text{cm}^3}{\text{s}} \rightarrow \frac{\text{g} \cdot \text{cm}}{\text{s}} \rightarrow \frac{\text{J}}{\text{s}} \rightarrow \text{W}$$

The substitution of actual values leads to

1 W = 7.5 torr · L/s

From this association it should be clear that pumping devices that have a certain limited power have also a limited mass flow.

3.3 FLOW REGIMES

Fluid flow can assume several different modes, which have significantly different relationships between various physical parameters. The basic flow types are turbulent, choked, laminar, molecular flow, and permeation.

Laminar flow, with its characteristic streamlines, was mentioned in Section 3.1. Laminar flow occurs when for a given diameter, the ratio of mass flow to viscosity of the fluid is low. The characteristic ratio is called the Reynolds number

$$\rho VD/\mu$$

where ρ is the density, V the velocity, and D a characteristic lateral dimension (such as a diameter for a circular tube). The Reynolds number has no dimensions. It is a ratio of body forces to the viscous forces in the fluid. Laminar flow occurs when the Reynolds number is below approximately 2000. The flow-pressure relationship is

$$Q \sim P_1^2 - P_2^2 \qquad (3.10)$$

and for common gases the flow will be inversely proportional to viscosity.

When the Reynolds number is greater than approximately 3000, the flow becomes turbulent (i.e., the streamlines break up into small eddylike patterns with local curvatures, accelerations, and decelerations). For the turbulent flow, the flow-pressure relationship becomes

$$Q \sim (P_1^2 - P_2^2)^{0.5} \qquad (3.11)$$

When the value of the Reynolds number is between 2000 and 3000 the flow can be either turbulent or laminar, depending on the smoothness of the duct and abrupt changes in size and direction.

Choked flow occurs when sonic velocity is reached in a flow restriction between two pressure regions. For example, if an orifice or a nozzle is installed between two chambers, flow will begin when the pressure in the downstream chamber is lowered. When the ratio of the pressures is approximately 2:1 (for common gases) sonic velocity is established at the orifice or at the throat of the nozzle. Then the volumetric flow through the restriction will remain constant. If the pressure in the upstream chamber is kept constant, the mass flow through the orifice will be at a maximum (for that pressure). The flow-pressure relationship is then simply $Q \sim P_1$.

When the gas becomes more and more rarefied the average distance between molecular collisions grows (see Table 2.1). When this mean free path is greater than the dimensions of the duct, the molecules collide mostly with the walls of the vessel, and their flow will be governed by the interaction with the walls and their own thermal velocity. This condition is called molecular flow (see Section 3.4). The pressure-flow relationship for molecular flow is

$$Q \sim P_1 - P_2 \qquad (3.12)$$

There is, as may be expected, a transitional region between molecular and laminar flows.

FLUID FLOW AND PUMPING CONCEPTS

There are three more types of gas transfer that do not belong to the category of pneumatic flows: Surface diffusion (see Section 3.9), permeation, and diffusion of one gas through another. Surface diffusion refers to molecules moving from one adsorption site to another without leaving the surface. Permeation is mass transfer of a substance through a solid. It is well known, for example, that helium permeates through glass. The permeation rates depend on the particular gas and the material through which it penetrates. The flow-pressure relationship is based on partial pressure difference or, in other words, on the concentration gradient of the gas (see Section 3.9),

$$Q \sim P_1^1 - P_2^1 \tag{3.13}$$

The diffusion of one gas through the other also occurs according to a concentration gradient. It is a slow process compared to pneumatic flows but not as slow as permeation. It is related to what is often referred to as Brownian motion. If a droplet of perfume is introduced at the end of a tube containing still air, the odor will eventually diffuse evenly through the entire tube, although it might take a few minutes to travel a few meters. The diffusion coefficients depend on the molecular size, pressure, molecular weight, and temperature (see Section 3.9). In Figure 3.4 an attempt is made to illustrate the various types of flows. It is not intended to render a realistic picture, only to provide quick graphic recognition.

3.4 FLOW THROUGH NOZZLES AND DIFFUSERS

The pressure and flow relationships noted in Section 3.1 occur in liquid jet pumps, air ejectors, or vapor jet pumps. Figure 3.5 shows a basic jet pump. A nozzle is a conduit leading a fluid from a high-pressure to a low-pressure region. A fluid accelerates at the nozzle exchanging the potential energy of high pressure to a kinetic energy associated with high velocity. This is analogous to a roller coaster where height and velocity are similarly exchanged. Encountering resistance of downstream fluid the stream will decelerate and its pressure, accordingly, will increase. If there were no friction and no external fluid load, the final pressure would be as high as the original (just as the final height of a frictionless roller coaster). The high-velocity jet will mix at its periphery with the surrounding fluid, imparting some acceleration to it, which results in a pumping action. If the external loading is high, the final pressure will not be as high as without loading.

The behavior is illustrated in Figure 3.6 using an inclined plane with spheres of different mass. The large sphere represents the pumping fluid. The top of the inclined plane on the left is the nozzle. The small spheres represent the pumped gas. They can be assumed to be moving in any direction, including

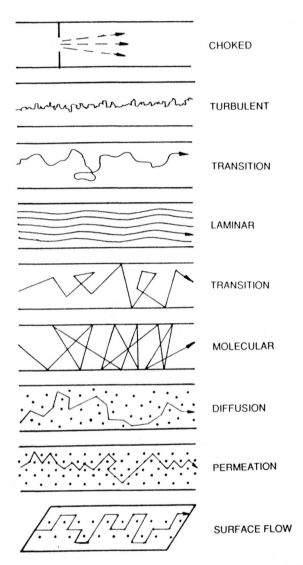

Figure 3.4 Molecular trajectories at various types of flow.

coming down from the right in the opposite direction to the "jet" of large spheres. If the large spheres encounter more and more small ones, they will reach successively lower height after deceleration. If they encounter too many spheres from the opposite direction they can either stop entirely or even go backwards, which indicates the overload condition.

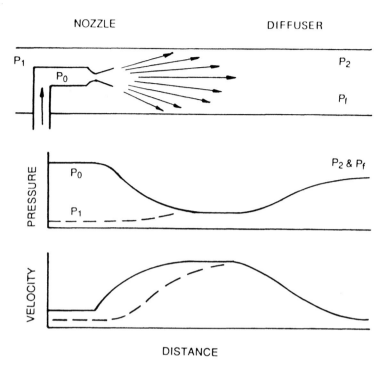

Figure 3.5 Pumping effect of a fluid jet. Solid lines are for pumping fluid, dashed line for pumped gas. P_0, P_1, initial; P_f, final pressure of the motive fluid; P_2, final pressure of pumped gas.

The mixing between the two streams can occur by turbulence in the mixing layers, but the exact nature of the mixing is not important for defining and measuring the basic performance characteristics of ejectors. To help visualizing how the slow or stagnant fluid is accelerated by the high-velocity jet, consider the flow velocity profile inside a pipe or plane-parallel conduit (Figure 3.7a). The velocity of the fluid layer immediately adjacent to the stationary walls is nearly zero. This is because compared to the dimensions of a molecule, the surface roughness of the wall is extremely great. Also, the molecules of the fluid can adhere to the solid wall more strongly than they adhere to each other. The profile shown in Figure 3.7a represents a laminar flow condition and it is established in this form, depending on the viscosity of the fluid. Now imagine one of the walls in Figure 3.7a begin to move at a velocity equal to or higher than the maximum velocity of the fluid. Then imagine this wall to be a high-velocity stream of another fluid. The flow profile is likely to change as shown in Figure 3.7b. This change illustrates the basic pumping mechanism of fluid jet pumps.

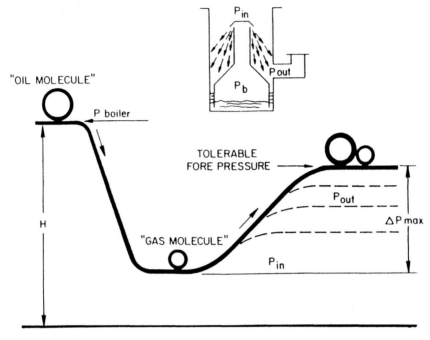

Figure 3.6 Inclined plane analogy for a jet pump.

Similar events occur in high-vacuum pumps known as diffusion pumps (Chapter 6). The simple analogies discussed here should enhance the understanding of performance aspects of pumps and the significance of terms such as maximum throughput and tolerable discharge pressure, used in reference to high-vacuum pumps.

3.5 MOLECULAR FLOW

Under molecular flow conditions gas molecules behave almost completely independent of each other. In the absence of collisions between molecules there cannot be a true pressure gradient in the gas stream even though pressure variation exists at the walls. However, even at the lowest pressures obtained in the high-vacuum technology there are a very great number of molecules present in space. Thus there is a definite density distribution that can be used in visualizing and computing flow phenomena.

Consider a small chamber containing gas attached through a valve to a larger chamber which has been evacuated, Figure 3.8. In viscous flow conditions (laminar or turbulent), when the valve is opened, all the molecules in the cham-

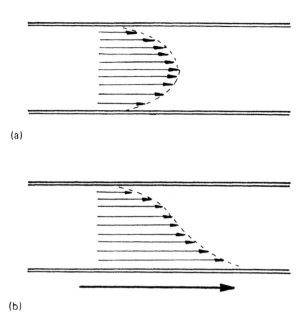

Figure 3.7 (a) Velocity distribution in laminar flow; (b) velocity distribution between a stationary and a moving surface.

ber will "feel" the sudden reduction of pressure. The speed of propagation of this pressure "information" will be essentially the velocity of sound. Streamlines will form leading to the valve until there is so little gas left in the small chamber that the mean free path of the gas becomes larger than the dimensions of the chamber. At this point the molecules lose contact with their neighbors, the streamlines disappear, and the "knowledge" of the existence of an exit is lost. The molecules then begin to bounce from one wall to another according to their velocity (associated with the temperature). The reflections from the walls are completely unrelated to the presence of an exit. This is, then, molecular flow.

Compared to the experience near atmospheric pressure, the molecular flow of gases is distinguished by very low mass flow, rather high-volume flow, high-pressure (or high-density) gradients, and a dependence on adsorption phenomena. The only mechanism by which a molecule can then leave the chamber is due to the statistical chance that its trajectory may happen to be directed through the exit.

In a sense, in molecular flow, the concepts of upstream and downstream do not apply in the common meanings of these words. A molecule reflecting from

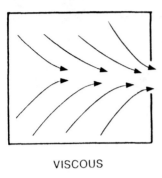

VISCOUS

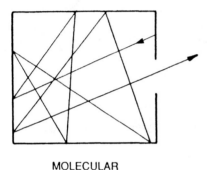

MOLECULAR

Figure 3.8 Viscous and molecular flow patterns.

the wall "has no knowledge" of where the pump is. A molecule can move upstream or downstream with equal probability. A mental adjustment has to be made: although we speak of pressure distribution, we should, perhaps, think of a density distribution. The use of pressure is simply a convenient way to provide a continuity of calculations in various flow regimes.

As noted earlier, in molecular flow environment, gas molecules rarely collide and, therefore, move independently of each other. Thus there is no true pressure gradient that would "guide" the molecules toward the region of lower pressure. An outgassing or evaporating molecule can move as easily away from the pump as into it. There exists a density (number of molecules per unit volume) gradient toward the pump, but for condensable species, even this cannot be assured, as temperature distribution may influence the flow more than the effect of a pump.

It is often assumed that residual collisions between gas (or vapor) molecules are completely negligible in high vacuum. This is not always true. Sometimes the remaining collisions may be the predominant method by which certain molecules pass through barriers such as cryogenic baffles and traps. Also, in some cases introduction of one gas into the system may affect flow conditions of another. For example, in high-vacuum pumps based on momentum transfer, an improvement of compression ratio for lighter gases can be obtained by introducing a certain amount of heavier gas into the pump.

3.5.1 Thermal Transpiration

An example of the difference between pressure and density considerations is the phenomenon known as thermal transpiration. Consider two vessels connected together with a tube. At higher pressures (viscous flow region) the pressure in both vessels will be the same no matter what their temperatures may be. At steady state, without flow, the pressure forces in the tube must be equal. However, under conditions of molecular flow the steady state demands that the number of molecules entering the tube from both chambers be equal. This leads to a pressure difference in the two chambers whenever the temperatures are different. From kinetic theory of gases it is known that the average velocity of gas molecules is

$$v_{ave} = \left(\frac{8kT}{\pi M}\right)^{0.5} \tag{3.14}$$

where k is Boltzmann's constant and M is the molecular weight (see Section 2.4). Also, the number of molecules striking a 1 cm² of surface per second, n, is

$$n = \frac{NV}{4} \tag{3.15}$$

where N is the number of molecules per cubic centimeter and V is the average velocity. The molecular flux on the surface or through an imaginary orifice in the space (of a chamber) depends on the average velocity, which, in turn varies as the square root of absolute temperature. Combining Eqs. (3.14) and (3.15), we obtain for equal flux between connected chambers, which are kept at different temperatures:

$$\frac{P_1}{P_2} = \left(\frac{T_1}{T_2}\right)^{0.5} \tag{3.16}$$

This condition is significant for the design of vacuum furnaces and for using pressure gauges that measure actual pressure forces (rather than the density) whenever the gauge is at a different temperature than the gas.

3.5.2 Flow Through an Orifice

It is useful to convert the flux of molecules striking 1 cm² of surface into a more common measure of flow. It is traditional in high-vacuum technology to translate the flux into volumetric units instead of a number of molecules. The usual geometry assumed for such calculations is shown in Figure 3.9. Integrating the number of molecules from all sites in the chamber which strike any unit area (cm²) gives, from Eq. (3.15),

$$n = \frac{3.5 \times 10^{22} P}{(MT)^{0.5}} \quad \text{per cm}^2 \cdot \text{s}$$

where P is pressure in torr (see Table 2.1). To convert this into volumetric units—in other words, to determine the volume that strikes the unit area—we must divide the number flux by the number density:

$$S = \frac{n}{N} = \frac{V}{4} = 3640 \left(\frac{T}{M}\right)^{0.5} \text{cm}^3/\text{s} \cdot \text{cm}^2 \tag{3.17}$$

For air at room temperature (22°C) this gives

$$3640 \left(\frac{295}{28.7}\right)^{0.5} = 11.6 \ L/s \cdot cm^2 \tag{3.18}$$

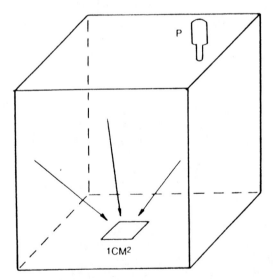

Figure 3.9 Unit surface exposed to molecular "bombardment."

It is important to appreciate that this conversion of number of molecules to volume flow units is based on total density (or pressure) rather than being associated only with the molecules that are moving toward the surface (as opposed to those which are moving away from the surface). The volumetric value of 11.6 L/s · cm² (for air at 22°C) is entrenched in the technical literature and it is important to note how it is derived. Consider the flow through an orifice shown in Figure 3.10. Assume that the pressure in the upper chamber is kept constant and the pressure in the lower chamber is zero or negligibly low. The flow through the orifice will be

$$A \times 11.6 \text{ L/s}$$

where A is the area in square centimeters. Note that in the vicinity of the orifice the rebounding molecules are absent. The flow is associated with the isotropic pressure, which exists far away from the orifice in a chamber that is much larger than the orifice. This is different from the methods used in the subject of continuum (viscous flow) fluid mechanics, where all gas properties (density, pressure, temperature, velocity) are measured at the same place.

Although molecules in the system shown in Figure 3.10 come toward the orifice from all angles, it is interesting to associate the volume flow rate with a (somewhat artificial) bulk velocity:

$$\frac{11.6 \text{ L}}{\text{cm}^2 \cdot \text{s}} = \frac{11{,}600 \text{ cm}^3}{\text{cm}^2 \cdot \text{s}} = 116 \text{ m/s} \tag{3.19}$$

which is nearly one-third the velocity of sound.

In practice, to obtain the value of 11.6 L/s · cm², the orifice should be at least 20 times smaller than the chamber, and its thickness should be at least 20 times smaller than its diameter (or other smaller characteristic dimension). On the other hand, the orifice should be much larger than the size of the molecule.

3.6 CONDUCTANCE

The concept of conductance has been discussed in Section 3.1. Conductance and pumping speed are associated with description of gas flow in volumetric terms. The units used for conductance (L/s) or resistance (s/L) may not be immediately apparent. They are derived from the definition

$$C = \frac{Q}{P_1 - P_2} \tag{3.20}$$

which can be written as

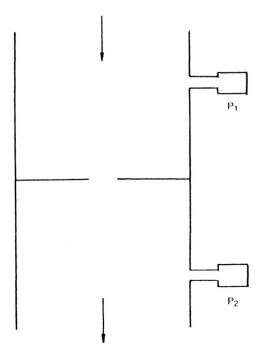

Figure 3.10 Orifice geometry.

$$C = \frac{pvA}{\Delta p} \qquad (3.21)$$

where p is the pressure, v the velocity, A the cross-sectional area, and Δp the pressure difference. Note that p is associated with density rather than with pressure as such, although the units of pressure can be canceled algebraically. The units of conductance should be torr · L per second of (mass) flow per torr of pressure difference. In other words, conductance determines a pressure drop in a duct at a given mass flow. In molecular flow conductance is independent of pressure. It is given entirely by the size and shape of the duct.

Conductance values for various ducts of common shapes are discussed in Chapter 4. When using conductance values for given flow elements it must be remembered that it is conventional for high-vacuum technology to associate the value of a conductance element with the existence of large chambers on both ends of the element, as shown in Figure 3.11. The conductance values [associated with Eq. (3.20)] are based on pressures P_1 and P_2, *not* on P_3 and P_4. When large chambers are absent, corrections must be made to account for pressure difference between P_1 and P_3 and P_4 and P_2.

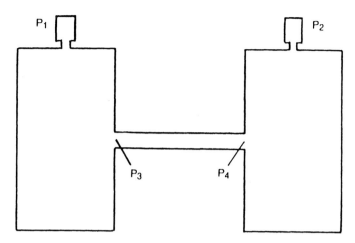

Figure 3.11 Conductance element between two large chambers.

The actual amount of gas flowing through a duct or into a pump expressed in volume units cannot be determined unless its density is also known. The more important quantity often is the mass flow, expressed in units such as grams per second or as a number of molecules per second. Usually, the conversion from volume flow to mass flow can easily be made using the well-known gas laws. However, these laws may lose their usual meaning when the gas approaches molecular beam conditions, which may exist on occasions at the entrance to a high-vacuum pump. It is not a simple matter to convert flow in g/s into L/s for a gas that does not have isotropic molecular velocity distribution (non-Maxwellian gas). For this reason, great precision and accuracy are hardly important for considerations of the design of most high-vacuum systems.

Because mass flows are very low, it is customary to use throughput instead of mass flow. The assumption is made that heat transfer effects are negligible and that the gas quickly accommodates to the temperature of the vessel or pumping ducts. When temperature is constant, the "mass flow" can be measured in units of throughput, torr · L/s. It is well to remember, however, that a given throughput value does not represent the same true mass flow for all gases. Throughput is simply the product of pressure and the volume flow of a gas ($Q = Sp$) for any given location (cross-sectional plane) in the flow passages.

Most analytical discussion of molecular flow found in textbooks on vacuum technology is associated with noncondensable gases and steady-state flow conditions. At ordinary temperatures, this may apply only to helium and neon. Surface effects that govern the transient flow of condensable gases are often neglected. Usually, vapors of higher molecular weight are troublesome, but water vapor, and even such "volatile" substances as acetone or alcohol are difficult to work with at room temperature. For example, if we attempt to measure the pumping speed of a high-vacuum pump for water vapor, measurements of the flow rate have to be done by careful consideration of the available surface area for evaporation from the liquid, heat transfer necessary for evaporation, the temperature of the tubes leading the vapor into the pump, and so on. Even then it is difficult to separate the amount of water vapor actually pumped by the pump and that which is condensed inside the cold inlet areas. Pumping fluid of a vapor jet pump in a baked system may take a week to be noticed by an ionization gauge with a 4-cm-long, 2-cm-diameter entrance tube; a mass spectrometer can retain a strong "memory" of acetone for 2 to 3 weeks if its ion source is 30 cm away from the pumps (2-cm-diameter tube). In such cases, baking (heating) is indispensable.

3.7 PUMPING SPEED

It is customary in high-vacuum technology to express gas flow in volumetric units (liters per second). This is convenient in basic computation if the temperature is assumed to be constant. There exist two distinct geometries in which the flow parameters are associated ($Q = pS$, where Q is the throughput, p the pressure, and S the volume flow). One can be called orifice geometry (Knudsen geometry), the other pipe flow geometry. The two are not identical. In the case of pipe flow, the pressure and speed are associated with the same cross-sectional plane; in the case of the orifice, the pressure is measured far away from the orifice (Figure 3.10). The definition of orifice conductance, C, is derived from

$$Q = C(P_1 - P_2) \tag{3.22}$$

where P_1 and P_2 are upstream and downstream pressures. If P_2 is much smaller than P_1, we may speak of the orifice speed

$$S = C = \frac{Q}{P_1} \qquad \text{when } P_2 \ll P_1 \tag{3.23}$$

As noted before, P_1 is not associated with the vicinity of the orifice, but with an upstream location where the gas conditions may be considered to be isotropic.

The two geometries mentioned above occur in practice. Often the geometry represents a mixture of the two. For pipe flow and for a sequence of matched flow elements connected in series, the association of pumping speed at a cross section may be more convenient. For a chamber with a small pumping port, the orifice association is more helpful. This duality of approaches is responsible for the difference between the pumping speed measurement standards [American Vacuum Society (AVS) 4.1 and International Standards Organization 1607].

Pumping speed and throughput can be measured by arranging a flow meter at the inlet to the high-vacuum pump and measuring the amount of gas pumped away in a given period of time at steady-state conditions. The ISO standard tends toward the orifice treatment and gives 10 to 15% lower values for pumping speed, compared to the older AVS standard. The two can be related depending on the ratio of pumping port to chamber diameter.

The net pumping speed of the pumping system or the pumping stack can be determined by the use of the same standard measuring system. To relate this measurement to the vacuum chamber in order to obtain an effective speed for the chamber, we should take into account a pressure difference existing between the center of the chamber and at the inlet into the pumping port. In practice, such considerations are not of great importance and simple approximate calculations can be made by adding conductances in series as long as no abrupt changes in geometry are present (see Section 4.5.2).

These simple concepts are sufficient for basic interpretation of the pumping speed performance curves. In reality, the gas conditions at the inlet to a high-vacuum pump are such that the gas properties are not the same in all directions, and the concepts of density and volume at given pressure need qualification.

In cases where pronounced unisotropy exists, pumping speed considerations may become meaningless. Consider an imaginary molecular beam with parallel molecular trajectories (without collisions) entering directly into a pump. A gauge mounted on the wall in the vicinity of this beam may not know of its existence. In such cases, the meaning of $Q = pS$ is not obvious. Situations of this nature may arise in practice.

The rate of pumping expressed in the volumetric units is the pumping speed. In other branches of mechanical engineering, this is often called volumetric capacity. For molecular flow conditions, it is more meaningful to consider a pumping speed of each gas species independently from each other. In practice, total independence is probably not obtained. However, unlike positive-displacement pumps, high-vacuum pumps exhibit different speeds for different gases. Thus the speed for each gas is obtained by dividing the mass flow (throughput) of that gas through the pump over the partial pressure of the same gas at the inlet of the pump.

The above represents one way of defining the speed at any cross-sectional plane in a system of ducts, pump orifices, or any other flow passages. This definition is related to an assumption of bulk velocity of gas through the passage, taking the product of that velocity and the cross-sectional area to be the volumetric flow rate.

Often, another rather different definition is proposed by using the evacuation time of a chamber. This may be useful in the forepump region but is nearly useless for use in association with high-vacuum pumps. Sometimes, estimates of pumping speed are made simply by measuring the time of evacuation to a certain pressure. This practice usually leads to gross errors due to neglect of outgassing in such measurements, unless helium is used as the pumped gas because its adsorption rates are usually negligible. Using a nonadsorbing gas, the net pumping speed of a vacuum system can be defined, *assuming constant pumping speed*, as

$$S = \frac{V \ln(P_1/P_2)}{t_1 - t_2} \tag{3.24}$$

where V is the volume of the chamber, P_1 and P_2 are two pressure points obtained during evacuation, and t_1 and t_2 are times at which these pressure points occurred (see Section 4.3).

Pumping speed is usually measured by introducing a known steady-state flow of gas into a measuring dome of specified geometry and measuring the resulting pressure established in the dome (Figure 3.12), which shows the experimental setup recommended by the American Vacuum Society. The older Tentative Standard, AVS 4.1, has the gauge location at a distance from the pump inlet equivalent to 0.25 of the pump inlet diameter.

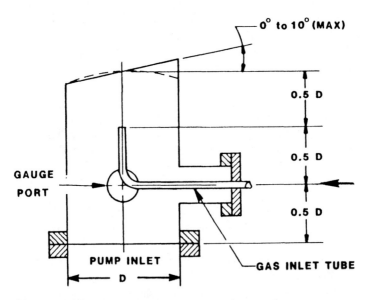

Figure 3.12 System for measuring pumping speed.

The pumping speed is determined by the AVS Recommended Practice as

$$S = \frac{Q}{P - P_0}$$

where Q is the flow rate (throughput) and P_0 is the "ultimate" pressure prior to the experiment.

The pumping speed measured by the AVS Recommended Practice refers to the inlet plane of the pump. Ducts connecting a pump to a chamber, baffles, and traps produce an impedance to flow, resulting in a pressure difference or pressure drop. Under molecular flow conditions, it is common for the conductance of a baffle or trap to be numerically equivalent to the speed of the pump. Thus the net speed at the pumping port in the chamber can easily be half or a third of the pump speed. The net speed is usually obtained with the approximation

$$\frac{1}{S_n} = \frac{1}{C} + \frac{1}{S_p} \tag{3.25}$$

In steady-state flow, throughput remains constant so that the maximum throughput capacity of the pump is not affected by baffles, orifices, or the like. The lower net speed at the chamber results, of course, in high pressure for any given gas load. Because of outgassing, the gas load in high-vacuum systems is never zero and the ultimate pressure in the vacuum chamber is always higher than the ultimate pressure of the pump itself.

3.7.1 Transient Effects

Most discussion of pumping speed in high-vacuum technology is based on the premise that pumping speed remains constant (or nearly constant) and on steady-state or quasi-steady-state conditions. There exist, however, cases where transient pumping speeds occur, which have little or nothing to do with the pumping speed of a pump. For example, when a valve placed between two chambers at different pressures is opened, the gas will rush into the chamber of lower pressure, with a rate depending on the geometry of the valve and the duct and the pressure ratio (or the pressure difference in case of molecular flow) existing across the valve. Typically, the time response in such cases is said to be

V/S

where V is the volume of the chamber at the higher pressure and S is the net speed of the pump. The geometry of flow and the ratio of the volumes (on both sides of the valve) are neglected. Precise calculations of time response in cases involving valves, intermediate chambers, manifolds, tubes, and preevacuated reservoirs may become so complex that it may be safer for engineering design to determine the behavior of the system experimentally.

The velocities of gas flowing across a suddenly opened valve into a pre-evacuated space can be very high. Supersonic velocity is likely to occur downstream of air admittance valves (i.e., valves used to readmit air into vacuum chambers). Similarly, when preevacuated load locks are used to introduce a process part into a vacuum chamber, a sudden opening of the internal load-lock valve can create high gas velocities. Typically, in load locks, the final lock pressure is near 50 mtorr, while the process chamber may be at high vacuum. High gas velocities are undesirable when the process is sensitive to deposition of particles on the surfaces of the parts, such as silicon wafers used for production of microcircuits. The air admittance valves are more likely to introduce particles because of the force of high-velocity jets entering the chamber at a density that is likely to be near 1/2 atm.

3.8 PUMPS AND COMPRESSORS

The gas flow behavior discussed in Chapter 2 is, of course, also relevant to the performance of pumps and used for the creation of rough and high vacuum inside a chamber. A pump is a device that uses power to transport a fluid from a region of low pressure to a region of high pressure. A compressor performs the same function except that it is associated with pumping a compressible medium. The word "pump" is normally associated with liquids except in high-vacuum technology. The reason is purely historical. Three hundred fifty years ago, to create vacuum, vessels were filled with water and water was pumped out with pumps used for fire extinguishing. Basically, there is no reason to think of vacuum pumps any differently than an ordinary compressor. The only basic difference is that the working chamber (the reservoir) is attached to the inlet instead of to the discharge.

Vacuum pumps actually have much higher compression ratios than ordinary compressors, although the pressure difference does not exceed 760 torr (14.7 psi). Vacuum pumps are therefore compressors designed for the pumping of rarefied gases.

High-vacuum pumps have two additional complications. One is due to the existence of free molecule flow, at least at the inlet of the pump; the other is due to the usual need for preevacuation to make the high-vacuum pump functional. Molecular flow exists when the mean free path is large enough that the molecules collide primarily with the walls at the inlet to the pump.

A high-vacuum pump is then a compressor designed for high-vacuum pumping conditions. It compresses a gas from any high-vacuum level to a maximum pressure typically of about 0.5 torr, although some pumps have been made with discharge pressures as high as 40 torr. In principle, they can be made to discharge even at atmospheric pressure, but this is impractical. High-vacuum pumps are usually used in series with a rough vacuum pump. Both pumps are capable of very high compression ratios, over 1 million:1, but the rough pump produces a much higher maximum pressure difference, nearly 760 torr, compared to only 0.5 torr for a typical high-vacuum pump, although some modern turbo-drag pumps, for example, can produce a pressure difference in excess of 40 torr.

High-vacuum technology deals with pumping conditions, including a very wide range of pressure and density changes. It also deals with pumping of gases that cannot be regarded with conventional continuum gas concepts. In the pressure range below about 10^{-3} torr (approximately one millionth of atmospheric pressure), depending on the size of the conduit or vessel, molecular flow conditions prevail. As far as compressor behavior is concerned, this means that in a mixture of gases each species may be pumped with different effects. It is possible to have different maximum compression ratios and different flow rates for gases having different molecular weights.

An additional complication arises from the adsorption and desorption effects. At high-vacuum conditions, there are generally more molecules adsorbed on the surfaces than are present in the volume space. Again, different molecules can have vastly different adorption properties, which also depend highly on temperature.

Due to the high ratio of surface-to-volume gas content and adsorption effects, several varieties of pumps have been developed which do not exhaust the pumped gas but hold it inside the pump until the capture surfaces are replaced or regenerated. For a thorough understanding of how some high-vacuum pumps work, their limitations, overload conditions, and pumping performance, it may be instructive for the reader to review the general characteristics of compressors and flow of fluids in nozzles.

All compressors, regardless of the type of construction, size, and the pumped medium can be characterized by two basic parameters, one of mass flow and one of pressure (Figure 4.3). The basic performance can be shown without precisely specifying the shape of the curve. Note that this is analogous to an electrical power supply, with voltage taking the place of pressure and electric current replacing the fluid flow. The flow can be measured and expressed in volumetric units (cubic feet per minute, liters per second, etc.) or in mass units (pounds per hour, grams per second, etc.). The pressure can be inlet pressure or discharge pressure, depending on our interest, and instead of pressure we can also display pressure difference or pressure ratio.

It is traditional in engineering literature to plot the independent variable on the horizontal axis and the dependent variable on the vertical axis. Some may wonder whether the pressure or the flow constitutes the dependent variable. When a pump begins to work, does it create a pressure difference that results in flow, or vice versa? This may perhaps be a question for philosophers. We may conclude that pressure and flow are simply two manifestations of the same phenomenon. But usually, when a given pump is tested, we control the flow rate by opening, closing, or adjusting valves. Pressure changes result as a consequence of our action, and this justifies the plot of pressure versus the flow rate. Note that in the high-vacuum technology it is traditional to reverse these variables.

Referring to Figure 4.3, if both the inlet and discharge ports of the compressor are open to a large source and sink of gas (such as the atmosphere), the flow will be the maximum obtainable and the pressures nearly equal (pressure ratio equal to 1 and pressure difference nearly zero). At the opposite end, if the inlet of a vacuum pump is completely closed, the device will develop the maximum pressure difference (and pressure ratio) but the net flow will be zero. In use, a compressor will operate somewhere between these two extremes. Note that even in a leak-free vacuum system there will usually be some flow, due to gas evolution from the chamber walls. When the discharge pressure is constant, such as in the case of mechanical rough vacuum pumps, it is of little interest, so inlet pressure can be plotted as a significant performance indicator instead of pressure ratio or difference. Depending on application, one or the other may be a more convenient choice.

Maximum achievable pressure ratio and pressure difference have definite meaning in relation to high-vacuum pumps. The example shown in Figure 3.13 can serve to demonstrate that the two should not be confused. Consider the simple arrangement of a hydraulic pump circulating a liquid through a filter. Note that the pump is analogous to a battery in a dc electric circuit, and the filter represents a resistor. Assume that the pump raises the pressure from 1 atm to 10 atm, so the pressure ratio is 10 and the pressure difference is 9 atm.

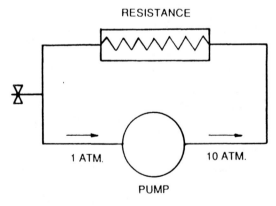

Figure 3.13 Simple pumping arrangement.

Now assume that at the valve shown on the left, we connect a high-pressure bottle, pressurize the entire circuit to 100 atm, and close the valve. This is similar to elevating the ground of an electrical circuit to a higher voltage level. The new pressures will be 101 and 110 atm; the pressure difference will remain the same. The flow rate will remain the same because neither the pump nor the density of the liquid has been substantially changed. The new pressure ratio is 1.09 instead of the 10 obtained before.

Now consider the open-loop system shown in Figure 3.14. The pump is moving the liquid from one large reservoir to another, lifting the piston shown. When the pressure difference reaches the maximum that the pump can produce, the net liquid transfer will stop (see Figure 4.3). Suppose now that we pressurize the reservoir on the right by putting some heavy weights on the piston. This would be analogous to connecting a battery with a higher voltage in reverse direction relative to the battery represented by the pump. The current flow will reverse. The pump will run backwards (if the mechanism can move backwards). Both pressure difference and pressure ratio are significant performance parameters. Their limiting values can be independently established when a pump is designed.

When the performance of a high-vacuum pump is presented in the form of a plot of volumetric flow (pumping speed) versus inlet pressure, the limiting parameters may be shown as in Figure 3.15. Regarding pressure ratios, high-vacuum pumps are very impressive. They can easily be designed to produce limiting pressure ratios over 1 million or higher (depending on the gas). Regarding the volumetric efficiency, high-vacuum pumps are rather efficient. They can have capture probabilities of 50% to, in some cases, 100% of those theoretically possible. With regard to energy efficiency they are very poor. Typically, only a few percent of used energy is converted into compression work. This is due primarily to rather extremely low mass flows associated with high-

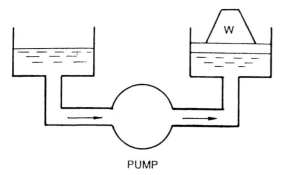

Figure 3.14 Pumping arrangement illustrating the possibility of flow reversal.

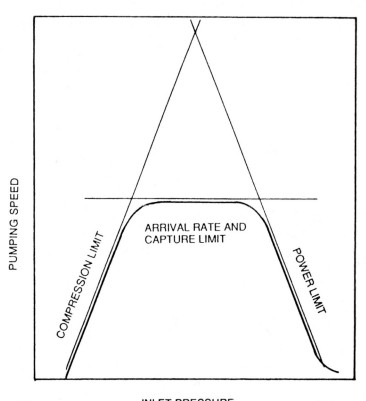

Figure 3.15 Basic limitations of pump performance.

vacuum pumping. Most of the energy used by the pump goes into friction, heating and cooling, and so on.

3.9 FLUID FLOW IN SMALL PASSAGES

In this section we describe the basic fluid flow characteristics associated with very small gas leaks. The quantitative mathematical expressions describing the rate of flow in a given passage are not always useful for precise application to engineering problems. The geometry of leaks is usually unknown, and it is difficult to define. Thus, estimates of leak rates without experimentation are usually not possible. An exception to this may be leakage by permeation, where with well-defined geometry, predictions can be made based on previous knowledge of gas and material behavior. The main purpose is to provide an appreciation of dependence of leak flow rate on outside factors. For example, sometimes the flow is proportional to the upstream pressure, sometimes to pressure difference across the leak, and sometimes to the difference of squares of pressure. Also, different gases often leak at different rates, and it is helpful to understand what governs this behavior.

In addition to the usual gas or fluid flow relationships, surface effects become important when flow passages are extremely small and when the gas is partly condensable. Therefore, adsorption and two-dimensional surface flow may become significant.

Flow rates encountered in leak detection work may range from 10^{-2} to 10^{-12} torr · L/s, and in some special cases, much lower. In addition to the difficulty of establishing the geometry of the passage, it is also difficult to establish the mode of flow. Above 0.1 torr · L/s the flow may be turbulent. In the major area of interest, 10^{-1} to 10^{-6} torr · L/s, it is usually assumed to be laminar and below 10^{-7} torr · L/s molecular. This implies that transition between laminar and molecular flows may be anywhere between 10^{-4} and 10^{-7} torr · L/s.

The significance of lack of knowledge of the flow type can be appreciated in considering the comparisons of flow rate through small leaks for different gases. It is often stated in association with leak detection that helium will flow 2.7 times faster (in throughput units) than air. This is the square root of the ratio of molecular weights. But this is true only if the flow is molecular. In laminar flow the flow rate for both gases may be nearly equal, assuming negligible adsorption, because the variation will be according to the ratio of viscosities (air 0.0169 cP and helium 0.0178 cP at 0°C).

Very small gas flow passages can easily be plugged by liquids (oil or water). If a liquid droplet enters a capillary tube, it becomes a barrier to gas flow because the diffusion of gas through the liquid occurs at a greatly reduced rate. At room temperature, it may take a very long time for the liquid to evaporate. Typical leak detection work involves the evacuation of the object that is to be tested and relying on atmospheric pressure to produce the leakage of air or a tracer gas (such as helium). However, because of the surface tension of liquids, the atmospheric pressure may be insufficient to dislodge the liquid droplet. The pressure required to dislodge the droplet depends on the size of the passage (diameter) and the surface tension ($P \sim \gamma/d$).

As an example, for a capillary with a diameter near 1×10^{-4} cm plugged with water, about 3 atm of pressure difference is required to drive the water droplet out. Since the geometry of leaks is not predictable, it is possible to have small trapped reservoirs of gas through which the tracer gas has to pass to be detected inside the vacuum chamber. The difference between the existence of molecular or diffusion flow in such reservoirs can be very significant, as illustrated in Figure 3.16, as can, in the main passage, the difference between laminar and molecular flow.

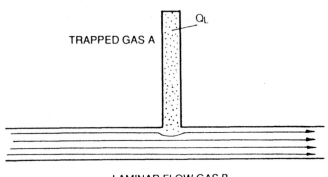

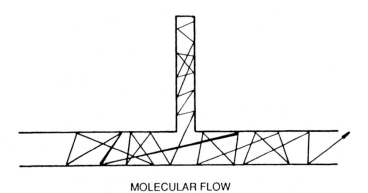

Figure 3.16 Comparison of diffusion and molecular flow patterns.

3.9.1 Delay in Flow Due to Adsorption

In very small capillaries the flow of gas may be delayed because of adsorption. The time delay of a step increase in pressure traversing a capillary is given by de Boer for molecular flow, neglecting surface migration:

$$t = \frac{L^2}{2dV} + \frac{L^2 \tau}{2d^2} \tag{3.26}$$

where t is the average time needed for a molecule to pass through the capillary of length L, d the diameter, V the molecular velocity, and τ the residence period of molecules on the surface (i.e., the sojourn time between evaporation events). The first term in (3.26) is assocaited with the cosine law of molecu-

Table 3.1 Approximate Delay Times for Transmission of Pressure Pulse Through a Capillary of Length L and Diameter d (Eq. 3.26)

	Hydrogen or helium	Nitrogen or air	Organic molecule
$L = 10$ cm, $d = 10^{-1}$ cm	3×10^{-3} s	10^{-2} s	0.5 s
$L = 10^{-1}$ cm, $d = 10^{-4}$ cm	3×10^{-4} s	10^{-3} s	50 s
$L = 10^{-2}$ cm, $d = 10^{-6}$ cm	3×10^{-4} s	6×10^{-3} s	1.5 h
$L = 10^{-3}$ cm, $d = 10^{-7}$ cm	5×10^{-5} s	5×10^{-3} s	1.5 h

Source: Ref. 10.

lar reflection; the second is associated with the adsorption time. De Boer calculated the approximate delays for a few characteristic cases (see Tables 3.1 and 3.2).

Molecules need less energy to skip on the surface than to evaporate. The nearer the gas properties approach those of condensable vapors, the more important the contribution of surface flow. In very small passages it is possible that when temperature is lowered, or the pressure increased, the total flow rate may be higher than predicted by laminar or molecular flow considerations. For a time delay in response to a step change in pressure, for molecular flow including surface migration de Boer derived

$$t = \frac{(L^2/2Vd) + (L^2\tau/2d^2)}{1 + 0.75(\tau/\tau_s)(a^2/d^2)} \tag{3.27}$$

where, in addition to Eq. (3.26), τ_s is the residence time between surface steps (skips) and a is the skipping distance.

Table 3.2 Average Transport Velocities Associated with Table 3.1 (meters per second)

1	33	5	0.2
2	3.3	1	2×10^{-5}
3	0.3	0.017	2×10^{-8}
4	0.2	0.002	1×10^{-9}

Compared to Table 3.1, the delay time for a "typical" organic molecule becomes

for $L = 10^2$ cm, $d = 10^6$ cm 0.5 h (instead of 1.5 h)
for $L = 10^3$ cm, $d = 10^7$ cm 22 s (instead of 1.5 h)
for $L = 10$ cm, $d = 10^4$ cm no effect

Surface migration is important only in very narrow passages. If the characteristic dimension (d) is near the size of a pore in a porous material (approximately 10^{-6} cm), the surface diffusion may become the dominant mechanism of gas transport. If surface migration predominates, Eq. (3.27) can be approximated by

$$t = \frac{2}{3} \tau_s \left(\frac{L}{a}\right)^2 \tag{3.28}$$

3.9.2 Surface Migration

Some liquids have a tendency to spread on solid surfaces as a thin film. This spreading is discussed in technical literature dealing, for example, with lubrication of small mechanical instruments. The spreading can be noticed sometimes as wetness on rough (sandblasted or glass-beaded) surfaces. The degree of spreading depends on the surface tension. Some low-vapor-pressure liquids that have surface tensions above 30 dyn/cm do not spread on ordinary metal surfaces. They are called "autophobic" liquids because they do not spread on their own monolayers covering a metal surface. There remains, however, the possibility of molecular surface diffusion, sometimes referred to as "creep."

The importance of surface diffusion as a flow mechanism may be estimated as follows. As noted before, the residence time on a surface site associated with surface diffusion steps is lower than the period between evaporation events. The values of heat of adsorption for surface diffusion are rarely found in literature. They range typically from 0.55 to 1.0, compared with the heat of adsorption for evaporation. The sojourn-time ratio between evaporation and surface diffusion can be used to estimate the relative importance of surface migration. It has been estimated (Ref. 11) that for a tube of 1 cm diameter, the surface diffusion can be assumed to be significant when the sojourn-time ratio is above 10^{15}.

As an example, consider the possibility of contribution of surface migration for a liquid such as DC-705, used as a pumping fluid in vapor jet pumps (Chapter 6). The respective heat of adsorption values in this case can be taken to be 28 and 16 kcal/mol. This gives a sojourn-time ratio of only 10^5. The residence time corresponding to 16 kcal/mol for surface diffusion is approximately 0.1 s at room temperature. Consider, for example, a total travel distance of 1 cm and assuming a single-step distance to be about 10 Å. A molecule has to make about 10^7 steps (traveling in the same direction!) to cross 1 cm. Multiplying 0.1 s per step times 10^7 steps, we obtain 10^6 s, or nearly 1 week.

The graph in Figure 3.17 shows the sojourn-time ratios above which the surface diffusion begins to contribute significantly to the flow rate. On the right side the sojourn-time ratio is shown depending on the difference between the energy associated with desorption and with surface diffusion (τ is the sojourn time for desorption, τ_s the sojourn time for surface steps, E_d the heat of desorption, and E_s the surface diffusion energy). The effect of absolute temperature is also indicated. The units of $E_d - E_s$ are kcal/mol.

3.9.3 Diffusion of Gases

Diffusion of one gas through another, which may be assumed to be stagnant, can be treated in analogy with heat transfer. The concentration of the diffusing gas is analogous to heat, the diffusion coefficient to thermal conductivity, and partial pressure of the diffusing gas to temperature. The diffusion coefficient is inversely proportional to pressure. Thus when pressure approaches the high-vacuum range the diffusion "conductivity" becomes very high. This can be related to the mean free path, which is also inversely proportional to pressure. General theoretical treatment of gas transport by molecular diffusion can

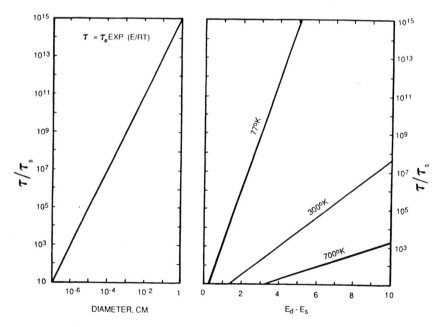

Figure 3.17 Sojourn-time ratios, above which surface diffusion is significant.

be found in books on kinetic theory of gases, chemical engineering, and mass transfer. For the one-dimensional case of two gases at constant temperature and steady state conditions, the rate of transport can be considered to be driven by the concentration gradient:

$$J = -D \, (dc/dx) \tag{3.29}$$

where J is in mols/sm^2 sec, c is in mols/cm^3, and x (the length of the diffusion path) is in cm. With these units, D (the diffusion coefficient) will be in cm^2/sec. These units are found in most tabulations. The units of D are a result of algebraic cancellations. The complete units can be stated in various ways and perhaps the most convenient is torr L/sec of flow per cm^2 of area with a gradient of torr/cm. For helium in air, at room temperature, the diffusion coefficient is approximately 0.5 cm^2/sec.

For counter diffusion of two gases at equal molal flux, and constant total pressure and temperature, the integrated equation is

$$n/t = A(D/R_0Tx)(p_1 - p_2) \tag{3.30}$$

where the flux n/t is in g · mols per sec, A the area perpendicular to the flow is in cm^2, D the diffusion coefficient is in cm^2/sec, and P and R_0 must have compatible units, for example torr and torr cm^3/K g · mol, respectively.

For the case of a gas diffusing through another "stagnant" gas the equation becomes

$$n/t = A/(D/R_0Tx)(p/p_{sm})(p_1 - p_2) \tag{3.31}$$

where p is the total pressure and p_{sm} is the log-mean pressure of the stagnant gas, which is given by

$$p_{sm} = (p_{b2} - p_{b1})/(\ln p_{b2}/p_{b1})$$

A numerical example for a helium leak in the absence of a total pressure difference is given in the leak detection chapter.

Diffusional processes are important in many engineering operations, such as drying of solids, freeze drying, humidification, and gas separation. Since our concern is related primarily to leak detection, two situations are of interest. When a tracer gas (usually helium) is introduced into the detection system (which must be kept under high vacuum) through a long flexible tube, there are three basic ways to make the connections, as shown in Figure 3.18. If the tracer gas has to pass through a long probe filled with stagnant atmospheric air, the flow by diffusion will usually be unacceptably long. If the tube connecting the leaking object (with the tracer gas) is evacuated as a part of the vacuum system of the detector, the tube will require thicker walls. If it is made of plastic materials, it may absorb the tracer gas and produce memory effects. Also, it will have a high outgassing rate, which may limit the capacity of the vacuum

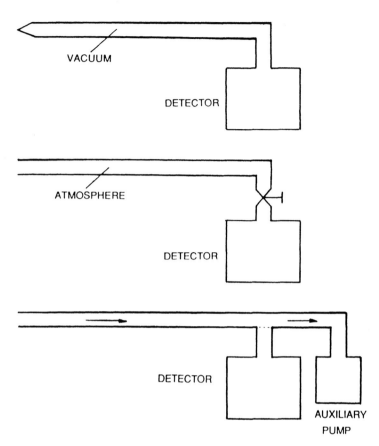

Figure 3.18 Design of tracer gas detection probes.

system. In most cases, the best results are obtained by using an auxiliary pump, as shown in the lower figure. The tracer gas is brought to the vicinity of the detector with adequate velocity to produce a response time of a few seconds. The detector can communicate with the sample stream through a capillary restriction or a semipermeable membrane, which has enhanced transmissibility for the tracer gas.

3.9.4 Permeation

Permeation is the transfer of a gas through a solid. Depending on the structure of the material and the gas, the transmission of the gas through a solid wall can occur in significant quantities in the absence of holes, cracks, or even grain boundaries. A well-known case is the penetration of helium through glass.

The basic mathematical expression of permeation is usually given as

$$q = K_p \frac{A}{L}(P_1 - P_2) \tag{3.32}$$

where q is the rate of mass flow, K_p a permeation rate constant depending on gas and the material, A the cross-sectional area (perpendicular to the flow direction), L the length of the flow path or the wall thickness, and P_1 and P_2 the partial pressures of the permeating gas on each side of the wall.

As may be expected, the permeability coefficient will depend on temperature; the relationship is often said to be

$$\log K = A - \frac{B}{T} \tag{3.33}$$

Table 3.3 Examples of Permeation Rates

Nitrogen[a]	Polyvinylidene chloride	5×10^{-11}	at 30°C
	Polyamide (nylon 6)	6×10^{-10}	
	Polyethylene	2×10^{-8} to 2×10^{-7}	
	Natural rubber	5×10^{-7}	
Helium[a]	Rubber	8×10^{-7}	
Polybutadiene acrylonitrile[a]	Nitrogen	2×10^{-8}	at 30°C
	Oxygen	1×10^{-7}	
	Freon-12	3×10^{-7}	
	Helium	5×10^{-7}	
	Hydrogen	5.5×10^{-7}	
	CO_2	7.5×10^{-7}	
	Water	1×10^{-4}	
	Benzene	3.5×10^{-2}	
Helium in glass[a]	Fused silica	6×10^{-8}	at 25°C
	Vycor	9×10^{-8}	
	Pyrex	7×10^{-9}	
	Soda lime	4×10^{-12}	
	X-ray shield	2×10^{-14}	
Hydrogen in metals[b]	Mild steel	10^{-9} to 10^{-10}	25°C
	Stainless steel	10^{-12} to 10^{-13}	25°C
	Mild steel	2×10^{-5}	400°C
	Stainless steel	8×10^{-7}	400°C
	Aluminum	6×10^{-11}	25°C
	Copper	2×10^{-14}	25°C
	Inconel, Kovar	3×10^{-13}	25°C
	Model, nickel	5×10^{-11}	25°C
	Palladium	1.3×10^{-8}	25°C

[a]Units: torr · L/s per 1 cm² of area per 1 mm of wall thickness per 1 atm of pressure difference.
[b]Units: torr · L/s per 1 mm of wall thickness per 1 cm² of area per (1 atm of pressure difference)$^{0.5}$.

where A and B are constants and T is the absolute temperature. However, for the case of helium leaking through glass, the variation with small changes in temperature (near room temperature) appears to be nearly linear with T, rather than logarithmic. There have even been cases reported where the permeation can decrease as the temperature is raised, apparently because of structural rearrangement in the material. Table 3.3 shows some examples of permeation rates.

The following general summary is given by F. J. Norton in an article on gas permeation through the vacuum envelope. For glasses, measurable permeation occurs for He, H_2, Ne, Ar, and O_2; for metals, H_2 permeates most, especially through iron (by corrosion, electrolysis, etc.); inert gases do not permeate through metals. O_2 permeates through silver, and H_2 permeates through palladium. For semiconductors, He and H_2 permeate through germanium and silicon. For polymers, almost all gases permeate through all polymers. There are many, often surprising, specificities. Some gases of higher molecular weight will permeate at a higher rate than that of helium. Water rate is likely to be high, especially with silicon rubber.

In addition to the steady-state rate of permeation, the delay encountered in reaching the steady state is significant for some engineering applications. For example, a gasket 4 by 4 mm in cross section can have a delay of 10 min for silicone rubber, 30 min for natural rubber, and 45 min for neoprene.

REFERENCES

1. A. H. Shapiro, *Shape and Flow*, Doubleday & Company, Inc., New York, 1961.
2. W. Jitschin and P. Roehl, Thermal Transpiration, *J. Vac. Sci. Technol.*, A5(3) (May/June 1987).
3. S. Dushman, *Scientific Foundations of Vacuum Technique*, John Wiley & Sons, Inc., New York, 1949.
4. F. J. Norton, in *Eighth National Symposium on Vacuum Technology Transactions*, L. E. Preuss, ed., Pergamon Press, Elmsford, N.Y., 1962.
5. A. Roth, *Vacuum Sealing Techniques*, Pergamon Press, Elmsford, N.Y., 1966.
6. D. J. Santeler and T. W. Moller, in *Second National Symposium on Vacuum Technology Transactions*, E. S. Perry and J. H. Durant, eds., Pergamon Press, Elmsford, N.Y., 1956.
7. C. E. Rogers, *Permaselective Membranes*, Marcel Dekker, Inc., New York, 1971.
8. R. J. Nunge, *Flow Through Porous Media*, American Chemical Society, Washington, D.C., 1970.
9. J. W. Marr, *Study of Dynamic and Static Seals*, CASA CR-58729 (Contract NAS 7-102) (Dec. 1963).
10. J. H. de Boer, *The Dynamical Character of Adsorption*, Clarendon Press, Oxford, 1953.
11. P. A. Redhead, J. P. Hobson, and E. V. Kornelsen, *The Physical Basis of Ultrahigh Vacuum*, Chapman & Hall, Ltd., London, 1968.

12. R. C. McMaster, ed., *Nondestructive Testing Handbook*, American Society for Nondestructive Testing and American Society for Metals, 1982.
13. M. H. Hablanian, *J. Vac. Sci. Technol.*, $A5(4)$ 2552 (1987).
14. D. J. Santeler et al., *Vacuum Technology and Space Simulation*, NASA, 1966 and Am. Inst. Phys., 1993.

4
Vacuum Systems

4.1 VACUUM CHAMBER DESIGN

Design considerations for vacuum chambers involve two distinct aspects, one structural and one related to the residual gas composition. The basic structural design is in the realm of mechanical engineering, when it is necessary to follow low-pressure vessel codes. The maximum pressure difference to which vacuum vessels are ordinarily exposed is 1 atm. However, 101,000 Pa (14.7 psi, 760 torr) pressure can produce high forces when acting over a large area. Because the pressure acting on structural shapes is usually external, buckling (structural instability) must be considered together with the usual strength issues. It is sometimes erroneously assumed that vacuum vessels are safe because they can only implode rather than explode. Brittle materials such as glass need particular attention. Once set in motion by an implosion, pieces of glass will continue flying in all directions.

It should be remembered that vacuum equipment normally is not intended for pressure work. For example, high-pressure nitrogen or argon gas containers are sometimes attached to the vacuum vessel to raise the pressure back to atmospheric with an inert gas. If atmospheric pressure is exceeded during this operation, even a small overpressure acting on a large area can produce a high force. Thus a 2-m-diameter door with a sticking gasket may suddenly swing open, causing possible injury. General safety aspects of vacuum equipment design and operation are discussed in a booklet published by the American Vacuum Society (see Ref. 19).

With regard to the internal design, there are two basic considerations: minimizing the residual gas content and providing appropriate duct sizes between the various elements of the vacuum system. This primarily involves the connections between vacuum pumps and the chamber and the connections between vacuum gauges and the area where the pressure is to be measured.

Aside from structural considerations, the materials of construction, welding and other joining techniques, surface treatment, and cleaning methods are chosen so as to minimize unwanted gas evolution. This suggests immediately that materials exposed to high vacuum should have a low vapor pressure. Zinc and cadmium, for example, should be avoided.

The design of joints, ports, gaskets, gasket grooves, and so on, must be made so as to avoid trapped gas pockets. The same general comment can be made regarding welding and brazing techniques. Figure 4.1 shows an example of such general considerations. The upper section indicated a poor design. If the internal weld has a leak, its location is impossible to find with this design.

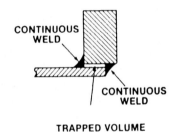

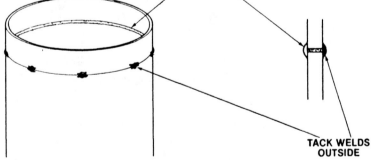

Figure 4.1 Examples of poor (top) and proper welding practice.

A better practice is shown at the bottom of the figure. For the same reason, O-ring grooves and similar seals have a narrow lateral cut, which makes it convenient to locate leaks by introducing a tracer gas into the periphery of the O-ring through the narrow cut. The few examples given here are for general guidance only; it is nearly impossible to list all situations.

The most common materials used for the construction of high-vacuum chambers are carbon steel, stainless steel, aluminum, and glass. Typically, rolled carbon steel can be used if the desired ultimate pressures are no less than about 1×10^{-7} torr. Stainless steel is generally used for high- and ultrahigh-vacuum chambers. It has significantly lower gas evolution rates than those of carbon steel and is easier to clean and keep clean. It is also relatively easy to weld and can be baked at temperatures as high as $400°C$ without severely oxidizing the external surfaces. Aluminum is easy to fabricate and has the advantage of containing less hydrogen. With proper treatment it can have very low outgassing rates, but it cannot be baked thoroughly and is therefore rarely used for ultrahigh-vacuum chambers.

Generally, the cautions associated with the design of high-vacuum chambers must be accommodated to the degree of desired vacuum level. What can easily be tolerated at 10^{-5} torr cannot be permitted at 10^{-8} torr, and so on.

Because of the requirement for extremely low external leakage and internal gas content (virtual leaks and surface outgassing), ordinary system components such as flanges, valves, etc., used in coarse vacuum and compressor industries cannot be normally utilized in high-vacuum systems. Therefore, in addition to pumps and gauges, high-vacuum equipment manufacturers offer a great variety of specialized assessories for high and ultra-high system construction such as: gate valves with or without automatic closure during power failure; valves that can have operation life of 100,000 to one million cycles; valves that can be operated at elevated temperatures; valves with removable and reparable gate carriages; valves that have only metal construction including the gaskets; valves with electropolished internal surfaces and low particle generation during operation; valves with ports arranged to provide laminar flow patterns; valves with ports for pressure (vacuum) gauges and purge gas inlets; valves with manual, pneumatic, or electric operation; valves with controllable rate of closure and gate positioning; closed loop throttling valves that operate from a predetermined setting or by a pressure monitoring device; angle valves, in-line valves, poppet valves, and butterfly valves; valves with circular and rectangular ports; precision adjustable gas leak valves; sight ports, rotary shaft and reciprocating motion feedthroughs, including ferromagnetic seals; precision positioning devices and other specialized fittings; and prefabricated high- and ultrahigh-vacuum chambers with and without assembled pumping systems. Designers of high-vacuum apparatus should become familiar with commercially available components before creating new ones.

High-vacuum technology deals with a very wide range of pressure or particle density conditions. Usually, the process of evacuation begins at atmospheric pressure and proceeds to high or ultra-high vacuum. No single pumping device can be expected to function efficiently at all pressure conditions. In addition to a multistage pumping system, the type of pumping mechanism employed is different at atmosphere and at high vacuum. Even if, in principle, a given pumping method could be used throughout the entire pressure range, such an attempt would be impractical in regard to size, weight, or cost of equipment, as illustrated in Figure 4.2.

The most common pumping arrangement for the production of high vacuum consists of a positive-displacement mechanical pump for initial evacuation, followed by a high-vacuum pump (turbomolecular, cryogenic, ion-gettering, or vapor jet pump). Usually, parallel and series arrangements are used. For initial evacuation, mechanical pumps alone are used; to obtain high vacuum, the two types of pumps are connected in series. In such a system, the gas enters the pumping train at high vacuum and is exhausted at atmospheric pressure by the final pump. In some cases, the device may be evacuated and sealed, terminating the pumping process; in others, the pumping is continuously applied to compensate for the gas evolution in the vacuum chamber. In sorption pumps, particularly in ion-getter pumps, the pumped gas is not exhausted to atmosphere.

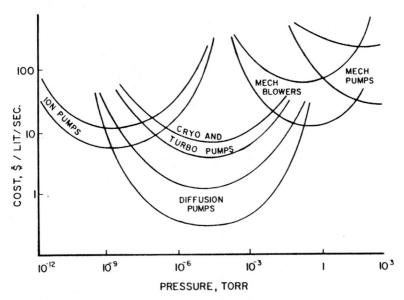

Figure 4.2 Pressure regions in which various vacuum pumps are most effective (1985 $ values).

This has the obvious advantage of isolation from the high-pressure environment and the disadvantage of limitation in gas load capacity or the necessity for periodic regeneration.

The basic performance of pumps and compressors can be associated with flow and pressure factors. With appropriate allowances for size, power, pumped fluid characteristics, and so on, all such devices generally behave according to the pressure-flow graph shown in Figure 4.3. For high-vacuum work, both high-volume flow rate and high-pressure ratio (discharge to inlet) are necessary. The flow rate is associated with the size of the pumping device. Thus for a given size, the pressure ratio needs to be increased as much as possible. Mechanical vacuum pumps, diffusion pumps, and turbo-molecular pumps produce pressure ratios of over 1 million:1. This can be compared to industrial plant air compressors, automobile engines, or aircraft compressors, for which the compression ratio does not exceed 10 to 1. Obviously, high-vacuum pumps require very special designs, and familiarity with their design, construction, operation, and maintenance is an important ingredient of success in the production and use of a high-vacuum environment. Specialized equipment, valves, gaskets, feed-throughs, sight ports, flanges, and so on, are commercially available.

4.2 EVACUATION AND PROCESS GAS PUMPING

High-vacuum pumps are used for two somewhat distinct applications. The first involves evacuation of gas (usually air) from a vessel; the other, maintenance of process pressure with a given gas load. It is difficult to choose the correct pump size with any degree of precision for both applications when the pressure required is in the high-vacuum region. In the first case, this is due to the uncertainty of surface outgassing rates, which depend on temperature, humid-

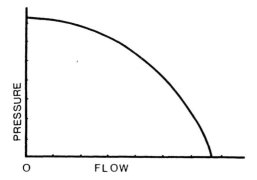

Figure 4.3 General performance characteristic of a compressor.

ity during atmospheric exposure, and the presence of gas-adsorbing deposits on the chamber walls. Water vapor is the usual cause of concern. In the second case, process gas evolution is uncertain due to variation of gas content in materials and variation in process parameters. For example, when electron beams are used for evaporation, the power distribution in the focal spot may not be exactly repeatable. This may produce a variation in temperature distribution at the surface in the vicinity of the electron beam. As a very rough rule, for evaporation, melting, welding, or sputtering of metals, at least 100 L/s of net pumping speed should be provided for every kilowatt of power used for the process if the pressure is maintained at 10^{-5} torr range.

Generally, one can hardly make an error in the direction of using too large a pump. It is common now to use high-vacuum pumps in bell-jar sized, typical process systems to produce a net system speed exceeding 1000 L/s. A few years ago, the net pumping speed at the pumping port was only 500 L/s. These differences are not important in improving system pump-down time but they do help to maintain lower process pressures.

As far as evacuation time is concerned, the difference between 500 and 1000 L/s is hardly noticeable. When outgassing is taken into account, the evacuation time is not inversely proportional to the pumping speed, as might be expected from the experience at higher pressures. However, during rapid process gas evolution or introduction, there often exists an inverse relationship between process pressure and pumping speed.

In some cases, such as sputtering, the maximum throughput becomes an important item because it may be desirable from the process point of view to have a certain flow of argon through the system. A high pumping speed for the initial evacuation of the system will be of less importance.

There are two basic ways of reducing the pressure or time needed for evacuation of a chamber: One can increase the pumping speed or decrease the outgassing rate. In some cases it may be less expensive to provide additional pumps rather than attempt to reduce gas evolution. This depends on the degree of vacuum and other process requirements. A review of the following points will be useful before specifying or designing a vacuum installation. The gas-handling capacity of the pumping system at a given pressure is usually more important than the ultimate pressure; the net pumping speed at the chamber is more significant than the nominal speed of the pump; increasing the number of pumps in a high-vacuum region does not necessarily decrease the evacuation time in the same proportion; well-chosen pumps, working fluids, baffles, and traps can reduce or eliminate contamination problems; rapid pumping requirements increase system cost; long exposure of the chamber to atmospheric conditions will increase the subsequent outgassing rate; the temperature history and distribution in the chamber before and during evacuation should not be

VACUUM SYSTEMS

neglected; and the amount of surface area present in the vacuum chamber is more significant (at high vacuum) than the volume.

Generally, under high-vacuum conditions more gas is adsorbed on the walls of the vacuum chamber than is present in the space. This depends, of course, on the size of the chamber. For small devices, the surface-to-volume ratio is higher and there is a tendency for small vacuum chambers to have a higher ratio of surface area to pumping speed. Therefore, 5-cm-diameter high-vacuum pump systems, for example, generally do not produce as low ultimate pressures as easily as do 10- or 20-cm-diameter pumping systems.

A similar comment may be made about other sources of gas. Pumping speed grows approximately with the square of the pump diameter, but the exposed area of gaskets increases by nearly a linear rate. Thus it is relatively easy to produce high vacuum in large chambers (as high as 10 m in diameter). The rough vacuum system in such cases may be the expensive part because of its dependence on volume. In laboratory systems, pumps of 15 to 20 cm diameter are usually more convenient for low ultimate pressure work, even if the high-speed requirements are not necessary.

Also, due to the size effect, it is more difficult to evacuate a small system rapidly to very high vacuum pressures rather than to evacuate a large system. In large systems, the rough vacuum evacuation period becomes a substantial part of the total time. To shorten this period, large and expensive machinery is needed—mechanical pumps and blowers. A typical evacuation process in a large chamber is shown in Figure 4.4.

After evacuation is essentially complete, and if there are no external leaks, the pressure in the chamber is given by the rate of outgassing divided by the pumping speed ($P = Q/S$). Clearly, this depends on the surface area and condition and has little to do with volume. To assess this more carefully, all materials in the chamber and the outgassing rate for each gas species should be considered separately and the partial pressures added to obtain the total ($P_1 = Q_1/S_1$, etc.).

4.3 EVACUATION TIME

One basic consideration in designing high-vacuum systems is the time required to evacuate a vessel to a given pressure. This usually ranges from a few minutes to a few hours. However, in some cases, the desired time can be a few seconds or a few days. The prediction of time and pressure relationship by theoretical or experimental methods is very difficult because of the uncertainty associated with gas evolution rates from inner surfaces of the vessel (outgassing or virtual leak).

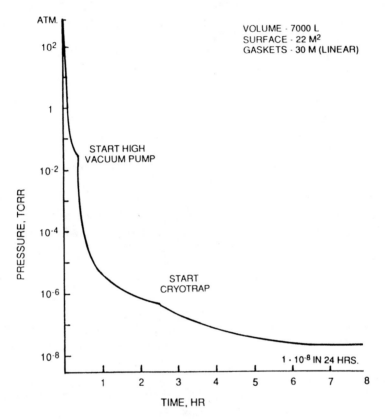

Figure 4.4 Typical evacuation progress for a large chamber.

Gas evolution from a given material depends on temperature, surface finish, previous history of exposure to a variety of atmospheric conditions, and cleaning methods. Thus, even an approximate prediction of evacuation time can be extremely difficult when outgassing becomes significant. Using well-known relationships applicable to higher pressures, the designer can predict a pressure-time curve from atmospheric pressure to about 10^{-2} torr. From steady-state (long-time) outgassing data and from characteristics of pumping devices, the final (ultimate) pressures can sometimes be predicted. The region of greatest interest often is between these two points. The following discussion attempts to develop at least a qualitative appreciation of significant effects.

4.3.1 Constant Pumping Speed and Neglecting Outgassing

In the pressure region between 760 and 10 torr, the pumping speed of the forevacuum pump is usually nearly constant. A 10% reduction at 10 torr com-

VACUUM SYSTEMS

pared to the speed near atmospheric pressure is common. Referring to Figure 4.5, the evacuation process can be represented by the relationship

$$-V\frac{dp}{dt} = Sp \tag{4.1}$$

where V is the volume of the vessel, dp/dt the rate of change of pressure with respect to time, and S the pumping speed. The physical meaning of Eq. (4.1) can be understood by noticing that the left side represents the amount of gas leaving the chamber (the minus sign indicates a pressure decrease), while the right side shows the gas entering the pumping duct. It is important to note that a solution of this differential equation will depend on the location where the pressure, P, is measured. Otherwise, to be valid, the equation must have the same pressure value on both sides of the equality sign. Thus the equation can be used as a definition of pumping speed for the given chamber, pump, and gauge location.

The evacuation time from Eq. (4.1) for constant volume and system pumping speed (as referred to gauge P_1) is

$$t = \frac{V}{S}\left(\ln \frac{P_0}{P_f}\right) \tag{4.2}$$

where P_0 is the initial pressure and P_f is the final pressure, or

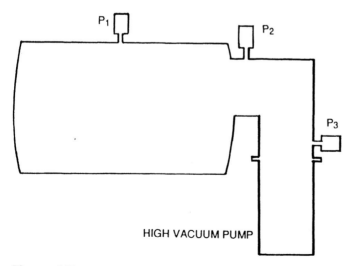

Figure 4.5 Schematic view of a vacuum chamber and a pump.

$$t = \frac{V}{S}\left(2.3 \log \frac{P_0}{P_f}\right) \tag{4.3}$$

This equation gives adequately accurate values after a few seconds from the start of evacuation and until approximately 10 torr is reached. At lower pressures, the outgassing becomes significant and the evacuation period is elongated. Generally, it is recommended that the values obtained from Eq. (4.2) be multiplied by about 1.5 between 10 and 0.5 torr, by 2 between 0.5 and 5×10^{-2} torr, and by 4 between 5×10^{-2} and 1×10^{-3} torr.

For rapid pumping of small volumes (such as load locks), the equation will not give accurate results due to complications of geometry, conductance and volume of pumping ducts, and transient effects within the ducts and the pump itself. If the desired pumping time is less than 10 s, experimental measurements are advisable.

There is more formal way to derive Equation 4.1, which may help to appreciate the meaning of pumping speed as defined in high-vacuum technology. We start with the statement of the gas law for a given amount (mass) of gas at constant temperature

$$PV = \text{constant} \tag{4.4}$$

which (at constant temperature), can be interpreted simply as a statement of conservation of mass. When a pump begins to work or, when the valve between a pump and vacuum chamber is opened, the gas volume expands and the pressure drops accordingly. We define the pumping speed, in the general case, as an instantaneous volumetric expansion of the chamber gas into the pump

$$S = dV/dt \tag{4.5}$$

A differentiation of Eq. 4.4 yields

$$p\,dV + V\,dp = 0 \tag{4.6}$$

Dividing Equation 4.6 by $p\,dt$, and rearranging, we obtain

$$dV/dt = -(V/p)dp/dt \quad \text{or} \quad dV/dt = -V(dp/p)dt \tag{4.7}$$

Assuming constant chamber volume (V) and a constant pumping speed (S), this can be rearranged, substituting S for dV/dt (Eq. 4.5) and solving for dt, as

$$dt = -(V/S)(dp/p) \tag{4.8}$$

which is the same as Equation 4.1. Since we are interested in pressure decay in time, we can integrate both sides of the last equation (for the case when outgassing is neglected), to obtain

$$t_2 - t_1 = -V/S(\ln p_2 - \ln p_1) \quad \text{or} \quad \Delta t = -V/S(\ln P_1/P_2) \tag{4.9}$$

which is the same as Equation 4.2. The ratio V/S, which has units of time, may be called the characteristic pumping time for a particular chamber and pump. In many common high-vacuum systems the value of V/S may range from 0.1 to 10 seconds.

4.3.2 Constant Throughput, Neglecting Outgassing

When constant throughput is used in Eq. (4.1), the solution becomes

$$t = \frac{V}{Q}(P_0 - P) \tag{4.10}$$

This is not an interesting case technologically, but it occurs in a narrow pressure region, 10^{-1} to 10^{-3} torr, in some high-vacuum pumps. This period is usually short, for example less than 10 s for a typical system.

4.3.3 Presence of a Leak

Equation (4.1) disregards the possibility of leaks and desorption gas loads. If $Q_\infty = Sp_\infty$ is a constant leak, we may write

$$-V\frac{dp}{dt} + Q_\infty = Sp \tag{4.11}$$

with a corresponding solution

$$t = \frac{V}{S} \ln \frac{P_0 - P_\infty}{P - P_\infty} \tag{4.12}$$

The gas load after a long pumping time, Q_∞, may be due to an actual leak, outgassing at an almost constant rate, or a finite permeability of the vessel walls.

4.3.4 Presence of Outgassing

For qualitative purposes, without implying a theoretical association with physics of desorption, the outgassing rate of a surface in high vacuum can be represented as

$$Q = Q_0 \exp\left(\frac{-t}{\tau}\right) \tag{4.13}$$

where Q_0 is an initial outgassing rate, t is the time, and τ is associated with the slope (decay) of the outgassing rate relative to time and is assumed to be constant for a reasonable period of time.

For proper machining of experimental outgassing curves, two (or more) terms can be used, such as

$$Q = \alpha e^{-at} + \beta e^{-bt}$$

where Q is the total outgassing rate and can be expressed as qA, where A is the area of the outgassing surface.

When outgassing is included, the general differential equation describing the evacuation of a vessel can be written as

$$-V\frac{dp}{dt} + Q_\infty + Q_0 \exp\left(\frac{-t}{\tau}\right) = Sp \qquad (4.14)$$

This equation is linear in p and can be put into the form

$$dp/dt + Mp = N \qquad (4.15)$$

where M and N are any functions of t.

Dividing by V and regrouping we get

$$dp/dt + (S/V)p = (Q_0/V) e^{-(t/\tau)} + Q_\infty/V \qquad (4.16)$$

To integrate the last equation, each term can be multiplied by an integrating factor $e^{-(S/V)t}$ which completes the differential on the left side. The integration yields

$$p\, e^{(S/V)t} = \{1/(S/V - 1/t)\}(Q_0/V\, e^{(S/V - 1/\tau)t} + (Q_\infty/S)\, e^{(S/V)t} + C \qquad (4.17)$$

Solving for p and inserting boundary conditions, the constant of integration can be evaluated:

when $t = \infty$, $p = Q_\infty/S$; when $t = 0$, $p = p_0$;
and $C = p_0 - Q_0/(S - V/\tau) - Q_\infty/S$ \qquad (4.18)

The final solution for pressure decay relative to time is

$$P = (P_0 - P_\infty)\exp\left(\frac{-St}{V}\right) + \frac{Q_0}{S - V/\tau}\left[\exp\left(\frac{-t}{\tau}\right) - \exp\left(\frac{-St}{V}\right)\right] + P_\infty \qquad (4.19)$$

This cannot be solved for t, but can be put into the form

$$t = \frac{V}{S} \ln \frac{(P_0 - P_\infty) - Q_0/(S - V/\tau)}{(P - P_\infty) - [Q_0/(S - V/\tau)]\exp(-t/\tau)} \qquad (4.20)$$

In many common systems, for the part of evacuation after switching to high-vacuum pumping,

VACUUM SYSTEMS

$p_\infty \ll P_0$ and $V/\tau \ll S$

Then, the solution can be simplified to

$$p - p_\infty = p_0 \, e^{(S/V)t} + (Q_0/S)(e^{-t/\tau} - e^{(S/V)t}) \tag{4.21}$$

Usually, after approximately an hour of pumping, $(S/V)t \ll t/\tau$, and the solution simplifies further to

$$p = (Q_0/S) \, e^{-t/\tau} + p_\infty \tag{4.22}$$

which can be solved for the evacuation time t

$$t = \tau \ln (Q_0/S)/(p - p_\infty) \quad \text{or} \quad t = \tau \, 2.3 \log (Q_0/S)/(p - p_\infty) \tag{4.23}$$

In a very simplified way, the entire evacuation process can be represented by

$$t = t_1 + t_2 = \alpha \, (V/S_{mp}) \ln(p_1/p_2) + \beta \, \tau \ln (p_3/p - p_\infty) \tag{4.24}$$

where S_{mp} is the speed of the roughing pump, α and β are correction factors for a deviation from an exponential relationships, p_1 is the initial pressure, p_2 is the pressure after rough pumping (cross-over pressure), τ is the time constant associated with the outgassing decay, and p_3 is the pressure established more or less immediately after switching to the high-vacuum pump, p is any pressure thereafter, and p_∞ accounts for the final ultimate pressure established due to leaks, permeation through gaskets, and limitations of pumps. The value for p_3 can be estimated by

$$p_3 = p_2 \, (S_{mp}/S_{hv}) \quad \text{or} \quad p_3 = qA/S_{hv} \tag{4.25}$$

where S_{mp} is the speed of the roughing pump, S_{hv} is the speed of the high-vacuum pump, q is the outgassing rate per unit area, and A is the total surface area exposed to vacuum.

If more than one rough pump is used (for example, when positive displacement blowers are used between the rough pump and the high-vacuum pump), then the rough pumping period must be split into two or more parts, each with its own constant pumping speed section. Also, if more than one gas is involved during the evacuation process, a summation sign could be put in front of the second term in Eq. 4.24, with different time constants for each gas. Normally, only water vapor is considered, because it is usually the major component. The value of the time constant for helium can be seconds but for water vapor it is a few hours. Thus, for helium, the first half of Eq. 4.24 can be used even in the high-vacuum region.

The value of p_3 is somewhat uncertain. It is given by Q_0/S, from Eq. (4.23). It is the initial total outgassing rate (specific outgassing rate multiplied by the total area of surfaces exposed to vacuum) divided by the pumping speed after cross-over. The time at which the p_2 value should be taken depends on the experimental conditions at which the outgassing data are obtained. F. Dylla (Ref. 23) has suggested to take zero reference point during outgassing measurements as the time when the pressure reaches 18 torr (near boiling point of water at room temperature). This is certainly much better than the absence of any clear reference, but still, does not provide a precise way of correlating outgassing data obtained by various investigators, because outgassing may depend on the period of exposure to vacuum (before measurements are taken) which, in turn, depend on the size of the roughing pump, used during outgassing measurements, compared to the size of the chamber. Also, most outgassing data are taken in laboratory-sized chambers, in the presence of water vapor as the predominant gas, which, at room temperature, provides opportunities for re-adsorption. Thus, outgassing measurements may depend on the pumping speed of the high vacuum pump and even on the geometry of the chamber.

There exist a few computer programs which, with various degrees of user-friendliness, produce pressure-time plots with sufficient approximation for design estimates and comparisons of design variations. Such programs, supplemented with correction factors from actual measurements can be useful tools for quick evaluation of a pumping system. Such programs, however, are useful for typical cases and should not be entirely trusted for unusual circumstances. For example, if the entire vacuum chamber is lined with liquid-nitrogen cooled panels (so that reabsorption is effectively prevented), the outgassing (and the evacuation rate) will be very different.

The factors V/S and τ in Eq. 4.24 are time constants which are found in any exponential decay (or rise) after a step change in input in any single lag input-process-output system. A nearly complete change in output occurs in a period of time equivalent to 4 time constants. $V/\Delta S$, where ΔS is the change in pumping speed after engaging a new pump. The typical values for $V/\Delta S$ in common vacuum systems are seconds but τ is usually measured in hours. The actual system pumping speed can be easily calculated by observing the slope of the pressure versus time evacuation curve on a log-linear graph paper

$$S = V.\Delta(\ln p)/\Delta t \qquad (4.26)$$

where V is the chamber volume and $\Delta(\ln p)/\Delta t$ is the slope of the curve. When, at the end of the rough pumping period, this slope becomes nearly horizontal, it means that the effective pumping speed is nearly zero. Then, when the high vacuum pump is engaged, the immediate pressure reduction will be in accordance with the ratio of the actual effective speed of the roughing pump (referred to the base pressure of the day, not the nominal value), and the speed of the high-vacuum pump. In many practical situations, this ratio may be between two and four orders of magnitude (see Section 4.9). Most published outgassing measurements, primarily associated with water vapor, performed after a prolonged exposure to atmosphere (an hour), show a power function relationship

$$Q = \text{constant}/t^x \tag{4.27}$$

where x typically varies from 0.5 to 1.2, and the value of the constant will depend on the chosen units of time. When this relationship is compared to practical evacuation curves, the lower value tends to apply to less carefully designed systems which may have many elastomer gaskets and mild steel walls; the higher value will apply to cleaner systems incorporating some ultrahigh-vacuum practices. For systems which have been previously degassed and exposed to atmosphere only for a short time (a minute), or to dry air or dry nitrogen, the value of x can be 2 or even higher. For approximate estimates, the value of x can be taken to be unity and the relationship written as

$$Q = Q_0(t_0/t) \quad \text{or} \quad Q = Q_1(t_1/t) \tag{4.28}$$

which gives a 45° decay line on a log-log graph (with equal-sized decade divisions). In clean systems, the pressure-time plot will reflect this relationship (after high-vacuum pressures are reached) for as long as three decades, eventually tending to a horizontal slope as ultimate system pressure is reached. In industrial systems which are pumped in a one-half hour period before the process is initiated, the relationship may last only one decade or less. The first parts of Eqs. (4.19 and 4.20) can be recognized as solutions when outgassing is disregarded.

Note that in the high-vacuum region there does not exist the simple inverse relationship between pumping speed and time of evacuation that is obtained at higher pressures. Using Eq. (4.19), it can be demonstrated that for a common 50-cm-diameter bell-jar system, the pumping speed may have to be increased six or seven times in order to cut the evacuation time in half (ca. 10^{-7} torr). However, when process outgassing occurs, the larger pump will maintain a lower process pressure, inversely proportional to speed. Figure 4.6 shows a typical relationship of pressure versus time as indicated by Eq. (4.20).

The outgassing rates of various materials can vary by many orders of magnitude. Condensable species, such as water, may exist in ordinary surfaces in amounts on the order of 10 to 100 monolayers. The behavior of such "films" is highly dependent on the temperature and binding energies involved. The outgassing rate of many substances may be considered to vary exponentially with temperature (Section 2.9). The most disturbing species are those with desorption energies from 15 to 25 kcal/mol. Below that range of values, the pumping proceeds rapidly. Above 25 kcal/mol, the presence of the material in the vacuum space is likely to be below 10^{-11} torr. The ordinary conductance

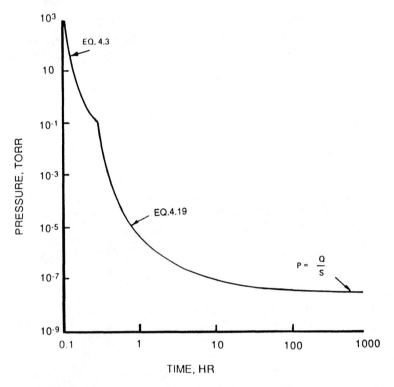

Figure 4.6 Pressure-time relationship during evacuation.

relationships cannot be applied to gases in the temperature regimes in which they are condensable.

4.3.5 Puming Virtual Leaks

J. D. Lawson published approximate calculations on the pumping of trapped volumes (Ref. 1). Five specific cases were considered (of which cases 1 and 2 are shown in Figure 4.7): 1, a small cavity connected to the vacuum in a chamber by a tube with a given conductance; 2, two cavities connected in series; 3, a chain of small cavities or a porous line; 4, a porous line open to atmosphere, initially full of air (open on both ends) exposed suddenly to vacuum on one end; and 5, the same porous line as in case 4, initially at vacuum, suddenly opened to the atmosphere on one end. To treat the effect of the trapped volume on the overall evacuation of the vacuum chamber, the geometry of the trapped volume must be known. Unfortunately, this is rarely possible. For illustrative purposes, a capillary tube may be considered. In case 1 (listed

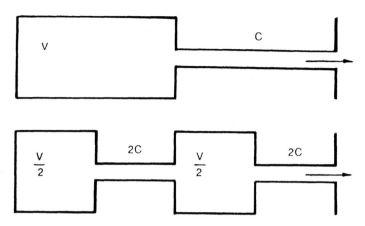

Figure 4.7 Schematic view of a virtual leak.

above), evacuation of the trapped volume proceeds as in a vacuum chamber except that conductance may be substituted for pumping speed. The assumption is made that the pressure in the chamber is very much smaller than in the trapped volume. Thus the flow rate from the cavity becomes (for molecular flow)

$$Q = P_0 C \exp\left(-\frac{Ct}{v}\right) \tag{4.29}$$

where P_0 is the initial pressure, C the conductance between the cavity and the vacuum chamber, t the time, and v the volume of the cavity. The time constant for the decay in the flow is v/C.

In a situation where the cavity may be considered as a part of a leakage passage, the overall effect on the pressure change in the vacuum system can be approximated by a combined time constant,

$$\frac{v}{C} + \frac{V}{S}. \tag{4.30}$$

where v and C are volume and conductance of the leak and V and S are volume and pumping speed of the chamber. Lawson also derives the value of the conductance, which gives the greatest flow rate after a given time t_1 (for molecular flow),

$$C = \frac{v}{t_1} \quad \text{for maximum } Q \tag{4.31}$$

and the maximum flow rate

$$Q_{max} = \frac{P_0 v}{e t_1} \tag{4.32}$$

A numerical example is given as follows. The maximum flow rate after 3 h of pumping that is obtained from 1 cm^3 of air trapped in a single cavity is

$$Q = \frac{760 \text{ torr} \times 10^{-3} \text{L}}{3 \times e \times 3600 \text{ s}} = 2.6 \times 10^{-5} \text{ torr} \cdot \text{L/s} \tag{4.33}$$

It should be appreciated that if the conductance associated with the trapped volume is comparatively large, the gas empties quickly. If the conductance is very small, the gas is removed at a very low rate. In both cases, the effect on the pressure in the vacuum system is small. If the net system pumping speed in the numerical example were to be 100 L/s, the pressure due to the flow rate from the cavity after 3 h of pumping would be

$$2.6 \times 10^{-5} \text{ torr L/s } (100 \text{ L/s}) = 2.6 \times 10^{-7} \text{ torr} \tag{4.34}$$

If the gas flow from the cavity is assumed to be viscous, the corresponding expressions for the conductance for maximum flow after a time t_1 is

$$\frac{C}{p} = \frac{2v}{P_0 t_1} \quad \text{for maximum } Q$$

and the maximum flow rate

$$Q_{max} = \frac{P_0 v}{8 t_1} \tag{4.35}$$

If, at the beginning, the flow is viscous, the cavity empties faster. Therefore, for large values of time t, the flow rate will be less than the one calculated assuming molecular flow.

For the chain of small leaks, assuming molecular flow and the same overall cavity volume and the same conductance, the time constant for the decay in the flow rate is $0.4v/C$ (2.5 times faster).

These calculations can be used as a guide for design. For example, when screws are used inside a vacuum chamber, how important is it to relieve cavities formed between the threads or inside blind holes? A small cavity may be unimportant in a system with a 10,000-L/s pumping speed operating at 10^{-4} torr but be very important at 100 L/s and 10^{-9} torr.

4.4 ULTIMATE PRESSURE

The ultimate pressure for a high-vacuum pump or system is the pressure established when sufficient time has elapsed, after which further reductions of pressure will be negligible.

The total ultimate pressure consists of partial pressure of various gases present in the system. In positive-displacement pumps such as forevacuum pumps, the pumping speed for various gases is approximately the same. However, under molecular flow conditions, in most high-vacuum pumps, each species of gas is pumped with a different pumping speed, and the ultimate pressure is given partly by the sum of the individual partial pressures obtained by dividing the gas load (throughput) of the gas by its pumping speed. The sources of gas can be in the vacuum system or in the pump itself. In addition to the normal outgassing of surfaces, lubricants, motive fluids (such as in vapor jet pumps), and reemission of pumped gases can be sources of gas at the inlet of the pump.

A clear distinction should be made between the ultimate pressure of a pump and that of a system. System gas loads often are orders of magnitude higher than the internal pump gas evolution, so that the system may never reach the ultimate pressure capability of the pump. Normally, pump performance data are given for a minimum volume and surface area at the inlet of the pump. The measuring apparatus itself is a source of gas. To obtain the real ultimate pressure capability of the pump, the contribution of all external gas sources should be subtracted. This is normally not done because of the need for much auxiliary experimentation.

In general, the inlet pressure of a pump can be expressed as a summation of effects of various gas loads and compression ratio limits,

$$P_1 = \left(\sum \frac{Q_i}{S_i}\right)_{ext} + \left(\sum \frac{Q_i}{S_i}\right)_{int} + \sum \frac{P_{2i}}{K_i} \qquad (4.36)$$

where P_1 is the inlet pressure, Q_i the gas flow rate (throughput) for a particular gas, P_{2i} the discharge partial pressure for the particular gas, and K_i the limiting compression ratio for each gas in a mixture. In practice, any one of the three terms in Eq. (4.36) can predominate. The ultimate pressure of a pump consists of the two last terms. The third term may become important at pressures approaching ultrahigh vacuum. The first term, the external sources, includes leaks, surface gas evolution (outgassing), and permeability (such as helium in the atmosphere through glass).

Ultimate pressure in a system is rarely an end in itself. If certain work is to be performed inside the vessel after evacuation, it usually results in additional gas evolution. Therefore, if a process or an experiment is to be performed at a given pressure, the ultimate system pressure should be lower than that. A convenient and very approximate rule for work that produces low gas evolution is to specify ultimate system pressure 10 times lower than the desired process environment.

To achieve pressures below 10^{-8} torr range, the entire vacuum system, including the gauge connection, must be treated with ultra-high-vacuum techniques in mind. What is usually taken to be the ultimate pressure limit of the pump is often the limit of the system. For example, the typical O-ring compression seal used to attach ionization-type pressure gauges to the vacuum chamber limits the pressure measurement to about 3×10^{-8} torr, no matter how low the actual pressure may be.

To appreciate the contribution of the first term [Eq. (4.36)], the proverbial example of the fingerprint can be used for educational purposes. To be visible by unaided eyes, the thickness of the fingerprint must be more than 100 molecular layers. If we assume the condensed area to be 1 cm^2, there are at least 10^{17} molecules comprising the fingerprint. Assume that the substances in the print are evaporating at 10^{-8} torr in a chamber with a 100-L/s pump. At atmospheric pressure, 1 L contains about 10^{22} molecules. At 10^{-8} torr, this density is reduced to 10^{11} molecules per liter or 10^{13} molecules in 100 L. Then the time required to pump the fingerprint away will be $(10^{17}/10^{13})$ 10^5 s, which, assuming no readsorption, is about 28 hours.

4.5 CONDUCTANCE CALCULATIONS

Various sections of vacuum systems can be connected by ducts. When a fluid flows through a duct as a result of an existing pressure difference, there is always a pressure drop associated with the flow. The pressure drop depends on the shape, the cross-sectional area, and the length of the duct. In addition, it depends on the flow regime, viscosity of the fluid, and surface roughness of the duct. In the design and operation of vacuum systems, knowledge of the conductances of various elements is important for utilizing pump capacities and avoiding unwanted pressure differences. Conversely, sometimes it is desirable to maintain a pressure difference between sections of the system, and an element of low conductances must then be placed between them.

For some simple geometric shapes, conductance can be calculated with precision: for example, for very thin orifices placed between large chambers in the molecular flow regime. For general engineering calculations involving complex shapes, 10% accuracy is normally adequate but not always easy to achieve.

4.5.1 Conductance in the Molecular Flow Regime

As discussed in Chapter 3, the conductance of a thin orifice placed between large chambers is given by

$$C = \frac{Av}{4} \tag{4.37}$$

where A is the cross-sectional area and v is the average molecular velocity. It is clear that the dependence on velocity indicates that the conductance will vary with the species of gas (molecular weight) and temperature. For Eq. (4.37) to be accurate, a few conditions must be present. The orifice must be thin compared to its size. In other words, its length should ne near zero. In practice, this means that the thickness of the orifice plate should be 20 times smaller than the size of the orifice and perhaps have a knife edge at the periphery. The orifice must be much larger than the molecular dimensions. The chambers on both sides of the orifice must be at least 20 times larger than the orifice. In addition, it must be understood that the value of the orifice conductance used in flow calculations is associated with pressures existing in the large chambers (rather than in the vicinity of the orifice). Note also that for a thin orifice, the area should be considered rather than the shape. In other words, it does not matter whether the orifice is circular, square, or any other shape as long as the conditions stated above are observed.

The value of orifice conductance for air at room temperature (22°C) is then $11.6A$ liters per second if A is in square centimeters. The values for other gases will be (as discussed in Section 2.4) proportional to $(T/M)^{0.5}$.

For long ducts of uniform cross section, the conductance is given by

$$C = \frac{4A^2 v}{3LF} \tag{4.38}$$

where A is the cross-sectional area, v the average molecular velocity, L the length, and F the periphery. For long tubes of circular cross section, this becomes

$$C = \frac{2\pi R^3 v}{3L} \tag{4.39}$$

and for air at room temperature the value is

$$C = \frac{12.1 d^3}{L} \tag{4.40}$$

if conductance is in liters per second and d and L are in centimeters. For a long rectangular duct with sides a and b (constant cross section),

$$C = \frac{k^2 a^2 b^2 v}{3(a + b)L} \tag{4.41}$$

where k is a factor depending on a/b ratio.

For air at room temperature,

$$C = \frac{31 k a^2 b^2}{(a + b)L} \tag{4.42}$$

where the value of k is as given in Table 4.1. When $a \gg b$, the rectangular shapes become a slot and Eq. (4.42) reduces to

$$C = \frac{31 a b^2}{L} \quad \text{units are L/s and cm} \tag{4.43}$$

For a long tube of annular cross section and air at room temperature,

$$C = 12.1 (d_1 - d_2)^2 (d_1 + d_2) \frac{k}{L} \tag{4.44}$$

where, again, dimensions are in centimeters, conductances is in liters per second, and k is as given in Table 4.2 (d is the diameter). When the ratio d_1/d_2 is close to 1, Eq. (4.43) can be used taking d for a and $(d_1 - d_2)$ for b. Note that Eq. (4.40) for air can be written as follows:

Table 4.1 Shape Factor k for Eq. (4.42) Depending on the a/b Ratio for Rectangular Ducts

a/b	k
1	1.11
0.667	1.13
0.5	1.15
0.333	1.20
0.2	1.30
0.125	1.40
0.1	1.44

Table 4.2 Correction Factor k for Eq. (4.44) for Tubes of Annular Cross Section

d_1/d_2	k
0	1
0.26	1.07
0.5	1.15
0.7	1.25
0.87	1.43
0.97	1.68

$$C = 12.1 d^2 \frac{d}{L} \tag{4.45}$$

This does not change anything arithmetically but suggests conceptually that conductance depends on the cross-sectional area (d^2) and the diameter-to-length ratio. It should immediately be apparent that if a very thin wall is placed inside a duct, dividing the duct into two sections but leaving the total area the same, the conductance will be reduced because the new d/L ratio is reduced.

Equations (4.38) to (4.45) are applicable only for long ducts. The length of the duct should be about 20 times longer than the cross-sectional dimension to make these equations directly applicable. In these formulas, the end effects are neglected. It may be noted immediately that if L (the length) approaches zero, the conductance values become infinite. These equations would then give erroneously large values when the length of the duct approaches cross-sectional dimension. Correction factors, or more correct equations must be used for short ducts.

The *pressure distribution* in a long tube with a steady-state distributed gas load is also of interest in some practical cases. Figure 4.8 shows a long tube with a closed end attached to a pump or any other duct of the pumping system. We would like to determine the maximum pressure at the end of the tube assuming uniform outgassing rate from all surfaces. The pressure at the entrance to the tube will be the total outgassing rate divided by speed.

$$P_1 = \frac{Q}{S} = \frac{qA}{S} = \frac{q\pi DL}{S} \tag{4.46}$$

where q is the outgassing rate per unit area. The outgassing of the circle at the end of the tube is neglected, but its effect can be added to the final result sepa-

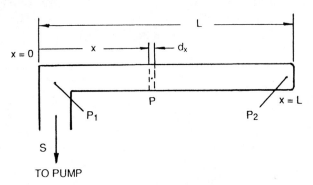

Figure 4.8 Long tubular chamber with multiple pumps.

rately. This problem can be solved by integrating the outgassing rate of a differential element and the corresponding conductance, giving

$$P_2 - P_1 = \frac{qA}{2C} = \frac{Q}{2C} \tag{4.47}$$

where C is the conductance of the entire tube. Note that the same result would be obtained if the entire outgassing load were concentrated at the midpoint of the tube. The conductance between the inlet and the midpoint is $2C$. The pressure distributions are parabolic; that is, the pressure at any distance X from the entrance to the pipe is

$$P_x = q\pi D\left(\frac{-X^2}{2CL} + \frac{X}{C} + \frac{L}{S}\right) \tag{4.48}$$

with the maximum occurring at the end of the tube.

Note that if the pressure is reduced at the entrance to the tube, the pressure difference will remain the same [Eq. (4.47)]. The pressure difference depends only on the outgassing rate and the conductance. This solution can be extended by symmetry to the case of a very long tube with many pumps placed at intervals along the tube. In that case, L should be taken to be half the distance between pumps. The maximum permissible pressure in the tube can then be obtained by appropriate design specifications for P_1 and C. A somewhat different approach to the problem of pumping long ducts has been published by C. Henriot.

Conductances of short tubes can be obtained either by using the conductance of an orifice of the same diameter and applying a correction based on the L/D ratio or starting with the long tube and applying a different correction also according to the L/D ratio. Here the first method will be used. Using Clausing's solution for the probability of a molecule entering the tube at one end and escaping at the other, the conductance of the tube can be compared to the conductance of an orifice of the same diameter.

$$C = \frac{\alpha A v}{4} \qquad (4.49)$$

For air at 22°C this will be [see Eq. (4.38)]

$$C = 11.6\alpha A \qquad (4.50)$$

if C is in liters per second and A in square centimeters. The correction factor can be found from Figure 4.9.

When dealing with more complex shapes, the simplified formulas discussed previously cannot be used. To obtain accurate values, a higher level of mathematical or experimental modeling is required. Two common shapes will be noted here: elbows and conical ducts. Some additional shapes can be found in Ref. 5 (see Figures 4.10 and 12). A 90° elbow does not represent an additional impedance to molecular flow if the axial length is used in calculations. Some textbooks recommend adding a length to the axial length equivalent to 1.33 (θ/180°) D, where θ is the angle of the elbow and D the diameter. However, this correction has not been found to be necessary by experimental and theoretical evidence (Ref. 5) at least for 90° elbows.

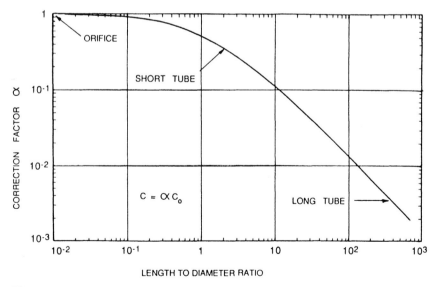

Figure 4.9 Correction factor for conductances of short tubes.

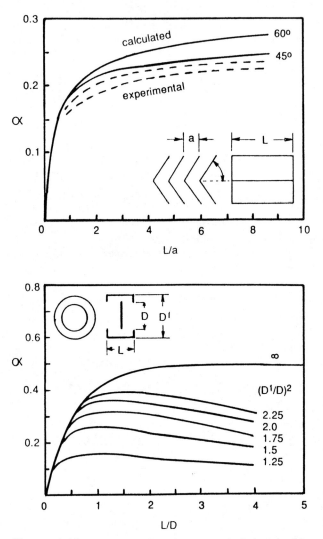

Figure 4.10 Examples of conductance calculations by Monte Carlo method. (From Ref. 5.)

Conical ducts are important because they often serve as transition pieces between conductance elements having different diameters. Transmission probabilities for truncated conical ducts are shown in Figure 4.11. In this case, the probability is associated with the inlet diameter. In other words, it is based on a comparison with an orifice having the same diameter as the R_1 circle.

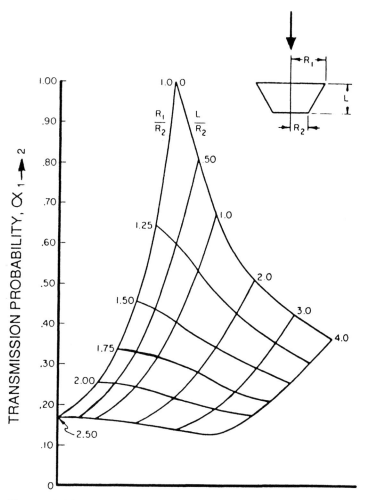

Figure 4.11 Transmission probabilities for conical ducts. (From Ref. 4.)

There are several ways to determine the conductance of a duct with a complex geometric structure. First, the particular complex shape may be separated into component shapes for which simple solutions exist, and the results summarized as conductances in series. In some cases, the solutions may be obtained by employing classical differential calculus. In more complex systems, finite element analysis or Monte Carlo computations can be used. The latter consists of following individual histories of a molecule entering a given duct and assigning to it random interactions at the walls of the duct according to the laws of reflection. By following histories of many thousands of molecules, integrated statistical results are obtained. The technique may be laborious for complex three-dimensional shapes, but it has been successfully applied in many important cases.

When the system does not lend itself to convenient theoretical analysis, experimental measurements of models can be very useful. The value of conductance can be obtained from the transmission probability, which depends only on the geometric shape of the structure and is independent of its size. All that is necessary in such measurements is to compare the conductance of the model to that of an orifice of known size. The experimental setup is shown in Figure 4.12. A set of models (baffles) is placed on a rotating disk and each indexed into the position above the high-vacuum pump for measurement. The apparatus can be calibrated by placing one or more orifices of known diameter in place of models. To maintain the calibration conditions, one of the models on the disk should be an orifice.

To assign a transmission probability to a duct that has different entrance and exit diameters, a decision must be made as to which diameter is to be used as the base. Although conductance of a duct is always equal in both directions, the transmission probability is not. In other words,

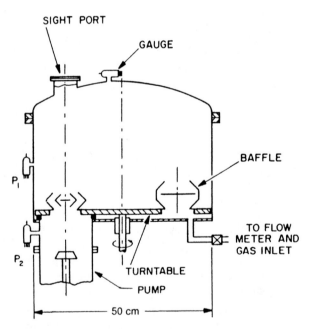

Figure 4.12 System for measuring conductance.

VACUUM SYSTEMS

$$\alpha_1 A_2 = \alpha_2 A_1 \tag{4.51}$$

where α is the transmission probability and A is the area of entrance or exit. For example, if the shape under consideration is a conical section, α_1 can be taken to be the transmission probability when the larger diameter (A_2) is the entrance and similarly, α_2 and A_1 in the opposite direction. In the test setup of Figure 4.12, the reference orifice was made equal to the exit area of the model and the transmission probability obtained directly from pressure measurements.

$$\alpha = \frac{C_m}{C_0} = \frac{(P_1 - P_2)_0}{P_1 - P_2} \tag{4.52}$$

where C_m is the conductance of the model and subscript zero is for the reference orifice. The conductance then can be obtained from

$$C_m = \alpha C A \tag{4.53}$$

where C is 11.6 L/s · cm² for air at room temperature. Transmission probabilities of various shapes obtained with the system shown in Figure 4.12 are given in Figures 4.13 to 4.16. The experimental accuracy of these values is between ± 5 and $\pm 10\%$, as judged from the correspondence with the expected orifice conductances and the scatter of pressure measurement data.

4.5.2 Combining Conductances in Series

The combination of conductance elements that are connected in a parallel arrangement does not represent a problem. The conductance values are simply added. When conductance elements are connected in series, the general rule is to add the resistance values (the reciprocals of conductances)

$$\frac{1}{C_n} = \frac{1}{C_1} + \frac{1}{C_2} \tag{4.54}$$

where C_n is the net conductance and C_1 and C_2 are the conductances of two elements connected in series. This relationship holds as long as end effects are not significant, for example, in the case of long ducts or tubes.

In the case of short ducts, it should be recalled that conductance is defined as

$$C = \frac{Q}{P_1 - P_2} \tag{4.55}$$

where the two pressures are measured in large chambers located at the two ends of the duct. The presence of such chambers between each element will assure that the conductance values are not affected by the geometrical matching of the

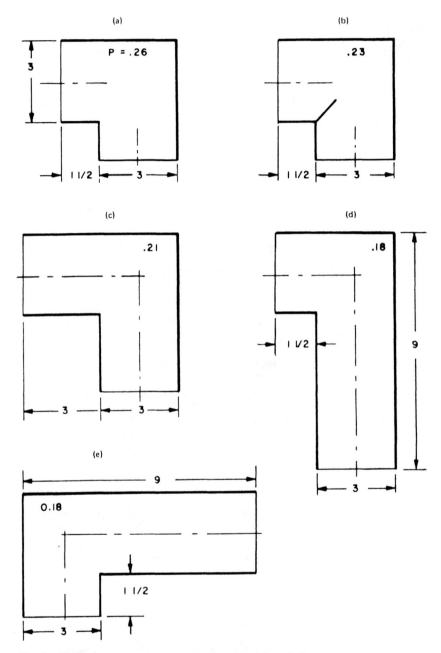

Figure 4.13 Examples of transmission probability of elbows.

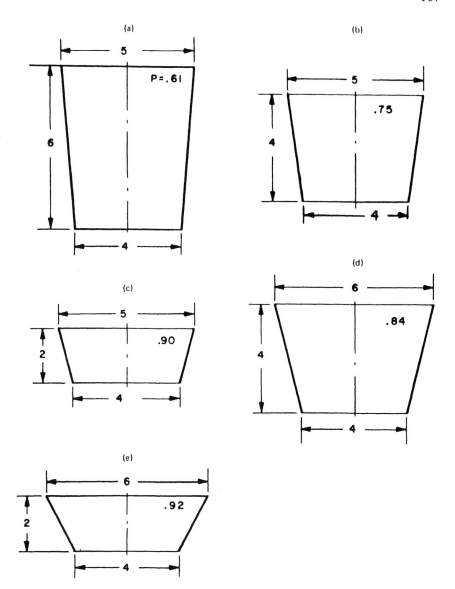

Figure 4.14 Examples of transmission probabilities of conical ducts.

adjacent element. However, in practice such chambers are usually absent and significant errors can be made using Eq. (4.54) without correction.

Using the example of connecting two short tubes of equal diameter, Oatley analyzed the effect of omitting a large chamber between the two items and produced a correction as follows:

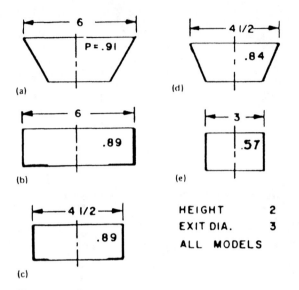

Figure 4.15 Transmission probabilities of some common shapes.

$$\frac{1}{\alpha_{1-2}} = \frac{1}{\alpha_1} + \frac{1}{\alpha_2} - 1 \tag{4.56}$$

Transmission probability α is defined as C/C_0, the ratio of actual conductance to the conductance of the orifice of the same diameter. In case of conical ducts, the conductance in both direction is the same but the transmission probability is not (i.e., $\alpha_{1-2}A_2 = \alpha_{2-1}A_1$).

If Eq. (4.56) is rewritten in terms of conductances, it will become

$$\frac{1}{C_c} = \frac{1}{C_1} + \frac{1}{C_2} - \frac{1}{C_0} \tag{4.57}$$

Using an example of two elements, each having a transmission probability of 0.3, we would obtain $\alpha_{1-2} = 0.15$ by direct reciprocal addition and 0.17 by Eq. (4.57). This would give a nearly 12% error. For three elements in series, the correction should be

$$\frac{1}{\alpha_{1-3}} = \frac{1}{\alpha_1} + \frac{1}{\alpha_2} + \frac{1}{\alpha_3} - 2 \tag{4.58}$$

and so on.

Additional difficulties will be encountered with abrupt changes in cross section. Consider the case shown in Figure 4.17, which shows three pipe sections

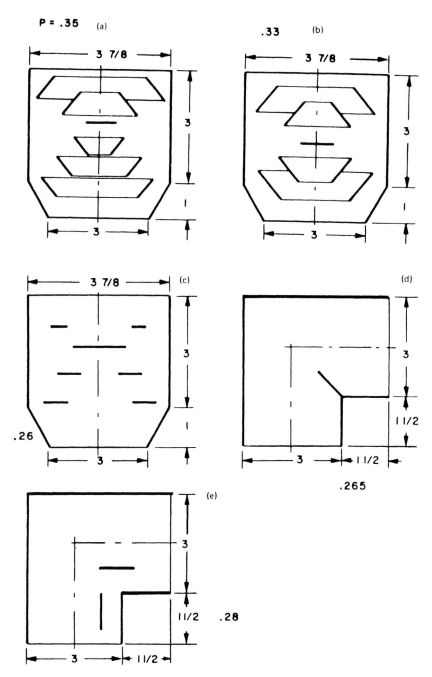

Figure 4.16 Examples of transmission probabilites of baffles.

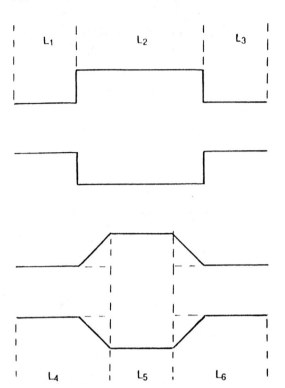

Figure 4.17 Approximations in calculating conductances of ducts with abrupt changes in dimensions.

in series. To estimate the net conductance of this section without resorting to a higher level of mathematical analysis, we may appreciate that in analogy to flow of electricity or distribution of stresses in elements of the same geometry, the corners of the larger tube do not provide much additional conductance compared to the modified lower figure. Thus, instead of adding the lengths L_1, L_2, and L_3, we can, for the first approximation, add conducatnces of lengths L_4, L_5, and L_6. A better approximation will be obtained by adding the conductances of sections L_1, L_5, and L_3 and the two conical sections. Then, because there are no large chambers between the elements, Oatley's method should be used. The equivalence of shallow conical sections, and short tubes with an aperture, has been confirmed experimentally. For example, two geometries shown in Figure 4.15a and b produced nearly identical transmission probability (based on the smaller diameter), using the apparatus shown in Figure 4.12.

The case shown in Figure 4.17 is relatively simple. Ducts of more complex geometry can be analyzed to establish an equivalent case consisting of recognizable elements for which conductances (or transmission probabilities) are known. It is often not an easy task. To produce a reasonable equivalent system, both the internal geometry and the transition between adjacent components must be considered. An example of an incompatible arrangement is shown in Figure 4.18.

Two identical baffles (of commercially available design) are arranged in series. In Figure 4.18a the arrangement is compatible, permitting the use of basic conductance value of the baffle in Eq. (4.54) (or a modified corrected form of it). The arrangement in Figure 4.18b, which is used only as an illustrative example, has nearly zero conductance. To make an estimate of conductance in this incompatible case, one should completely disregard the actual individual conductance of the baffles and treat the narrow space between them as a rectangular slot.

Much of the difficulty that seems to persist in the addition of series conductances is due to the tendency of using concepts of flow in high vacuum taken from simple electrical analogy. In molecular flow and for short duct geometries, simplified treatment is not directly applicable. Consider, for example, the problem of specifying the conductance of a baffle. There is no single number that can be assigned to a given baffle and used in all situations because the value will depend on entrance and exit conditions (i.e., the geometry of adjacent elements). There may be three alternatives to treat this problem:

1. Insert the baffle betwee two large chambers having pressure P_1 and P_2 and specify conductance as $C = Q/(P_1 - P_2)$, which is the classical method.

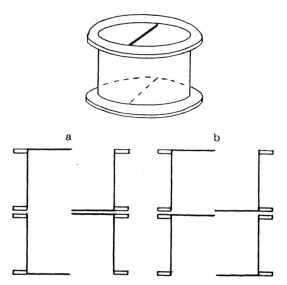

Figure 4.18 Example of compatibility of two adjacent elements.

2. Insert the baffle between tubes of the same diameter as baffle entrance and exit, measure pressures very near the entrance and exit, and use the same formula as above.
3. Measure the net speed of the baffle and pump, or any two elements in combination, without any reference to classical conductance values.

Perhaps a system could be developed permitting direct use of the simple reciprocal equation for routine engineering computations not requiring accuracy better than 10%. This may necessitate a dual set of values for conductance elements, corresponding to cases 1 and 2 above. Case 1 obviously has very basic physical meaning but is not easily applied without end-effect corrections. Cases 1 and 2, however, will both give unique conductance values for a given component.

The values obtained in case 2 may permit direct use of Eq. (4.54) for compatible components. Case 3 has been used in commercial situations because of lack of a better method. After measuring the net speed of a baffle with a high-vacuum pump using the standard speed measuring dome, the value of "nominal conductance" is derived from Eq. (4.54). Tihs procedure has the obvious advantage of direct measurement but does not provide a unique value for specifying the conductance. But for a system of well-matched components, Eq. (4.57). can also give reasonably good approximations in the latter case.

In cases involving short elbows joined at various angles and including various baffle structures inside, we cannot expect great accuracy. In many practical cases, it is impossible to divide the structure into logical parts of simple geometry to assign transmission probability values to each part for even an approximate analysis. The simplest approach may be direct measurement with models of the entire system. In extreme cases where compatibility is absent, gross errors can occur through careless application of simplified equations.

The reduction of pumping speed of a high-vacuum pump due to the presence of inlet ducts can be established using the following expression:

$$\frac{1}{S_n} = \frac{1}{S_p} + \frac{1}{C} \tag{4.59}$$

where S_p is the speed of the pump, C the conductance of the inlet duct (or baffle, or valve), and S_n the net pumping speed. This expression, often quoted in textbooks, contains an error if the conductance of a baffle is defined in the classical way as associated with large chambers on both ends. Santeler recommends a correction similar to Oatley's method (Ref. 9). The correction consists of subtracting the pressure drop associated with the exit of the baffle (i.e., the difference in pressure inside the tube, its exit, and the pressure measured in the large chamber). Thus the first correction is obtained as

$$\frac{1}{C'} = \frac{1}{C} + \frac{1}{C_0} \qquad (4.60)$$

and then

$$\frac{1}{S_n} = \frac{1}{S} + \frac{1}{C'} \qquad (4.61)$$

4.5.3 Conductance in the Transition–Viscous Flow Region

In the molecular flow range, the value of the conductance is independent of pressure. Since collisions between molecules are insignificant, the conductance depends only on molecular velocity and the geometry of the duct. At higher pressures, the value of the conductance depends on pressure. Resistance to mass flow (per given pressure difference) is reduced with rising pressure because fewer molecules interact with the stationary walls and the flow rate depends on the viscosity of the gas. In the viscous flow domain, the conductance values for long tubes can be obtained from

$$C = \frac{0.0327 P_{av} D^4}{\eta L} \qquad (4.62)$$

where C is in liters per second, P_{av} is the average pressure in torr, D the diameter in centimeters, η the viscosity in poise, and L the length in centimeters. For air at room temperature, this becomes

$$C = \frac{179 P_{av} D^4}{L} \qquad (4.63)$$

with units of liters per second, torr, and centimeters.

Figure 4.19 gives some examples of conductance values in the range of pressures comprising the transition between molecular and viscous flow domains. A more generalized graph is presented in Figure 4.20, where the ratio of conductance is plotted versus the product of average pressure and the diameter for air at room temperature. Equation (4.63) does not give accurate results at pressures above 10 torr, when pressure ratios are high. Conductance values, as shown in Figures 4.19 and 4.20, do not grow indefinitely. When bulk velocities reach values higher than half the speed of sound, the flow becomes choked (see Eq. 3.21).

Turbulent flows can also occur in vacuum systems during the initial evacuation and in flow of atmospheric air through air-release valves. However, the conductance of the pipes connecting the rough vacuum pumps to the chamber

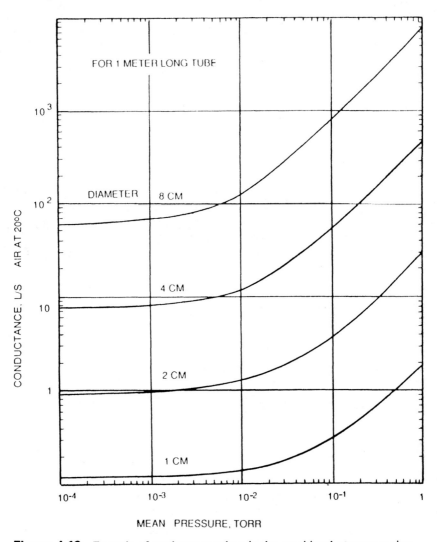

Figure 4.19 Example of conductance values in the transitional pressure regime.

is normally so high compared to the speed of the pump that the net pumping speed is not significantly affected by the presence of turbulence. In other words, if the piping is correctly dimensioned for the laminar flow regime, it will be sufficient even if turbulent flow occurs. In unusual cases, textbooks on gas flow and pneumatic pressure loss tables should be consulted. When pressure drops are high (higher than 20% of inlet pressure), the pipe should be divided in sections and then the separately computed resistances added.

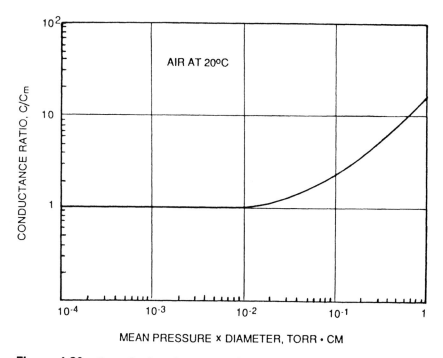

Figure 4.20 Generalized conductance relationship in the transitional pressure range.

Turbulent and choked flows can occur in systems with multiple, differentially pumped chambers when a continuous thin web substrate is passed through narrow slits from atmospheric air into a high-vacuum process chamber. Another example is the passage of an electron beam from high vacuum into atmospheric pressure for electron beam welding (in argon). Such cases are treated, for example, by A. H. Shapiro (Ref. 17), B. W. Schumacher (Ref. 16), and P. Shapiro (Ref. 15).

4.6 OUTGASSING EFFECTS

4.6.1 General Considerations

Most pumping devices in their normal range of operation tend to have a more-or-less constant pumping speed. When the volumetric flow rate of a gas is constant, the mass flow depends on density. During evaluation of a vacuum chamber, as the pressure (and density) is lowered, fewer and fewer gas molecules are removed per unit time. Therefore, to achieve lower pressures, any unwanted source of gas must be minimized.

There are several unintended sources of gas in vacuum chambers in addition to the adsorption effects discussed in Chapter 2. First, there may be leaks through the walls either in a concentrated capillary flow or by permeation in porous materials. Then there are the so-called virtual leaks. These are gas sources inside the chamber without any penetration through the walls.

These could be cavities or porous regions in internal welds, captured gas in plugged holes or under screws in blind holes, or even gas left between tightly clamped plates with fine surface finishes. Often, the source of addition gas evolution is the process performed in the vacuum system. For example, during evaporation of thin films, a thick deposit of the evaporated material eventually accumulates on the walls of the chamber. Usually, this deposit is porous and it adsorbs large quantities of water vapor when exposed to atmosphere. For this reason, the walls of the vacuum chamber should not be cold when it is open to air. To prevent condensation, the chamber should be a few degrees warmer than room temperature. In systems that contain water cooling, the cooling of the walls should be interrupted a few minutes before the system is exposed to air and cooling should be restarted a few minutes after the evacuation.

The amount of gas evolving from metals increases rapidly when the metal is heated. In vacuum furnaces, electron beam welding, and similar equipment, the vacuum system must be designed to maintain a desired pressure at the elevated temperature and not when it is cold. As noted before, the following very approximate rule may serve as an illustration. If, after the initial evacuation, common metals in a vacuum chamber are suddenly subjected to a thermal energy input, at least a 1000-L/s pumping speed may be required for each kilowatt of energy to maintain 1×10^{-6} torr pressure.

4.6.2 Water Vapor in Vacuum Systems

Water vapor in the vacuum system tends to accumulate in the colder areas, depending on location of its sources and sinks and rates of desorption and sorption (or outgassing and pumping speeds). Therefore, the temperatures of various parts of the vacuum system should be considered carefully. In principle, the behavior is similar to common occurrences of the fogging of windows of automobiles or homes, mirrors in bathrooms, and the cold-water pipes in cold basemenets during changes of temperature and humidity conditions. The major difference is that instead of dealing with bulk liquid condensation, we have a high-vacuum system with thin films of dense absorbed vapor that may display properties different from the liquid. For example, a water molecule will usually adhere to the metal surface more strongly than to its own liquid.

Generally, areas near the pump should be kept colder than the system. A liquid-nitrogen-cooled trap above a vapor jet pump or a cryopump will produce this requirement naturally. Pump-down procedures, baking, and system shutdown sequences should also be arranged carefully to prevent unnecessary contamination of the vacuum system with vapors previously pumped and adsorbed. In systems that are frequently exposed to the atmosphere, it is generally recommended not to cool the chamber walls below room temperature. Even the vapor jet pump cooling water should not be started until rough pumping is completed, to prevent condensation from the water vapor in the air, before it is removed by the rough vacuum pump.

In ultrahigh-vacuum systems, the water vapor content is reduced by pumping the water during baking. The remaining major constituent is then hydrogen. In unbaked high-vacuum systems, the major residual gas content is due to water vapor (as much as 95%). One way to reduce the water vapor density is to use as high a pumping speed as possible. However, if a process sensitive to the presence of water vapor is conducted near its sources, the high pumping speed may not be entirely helpful. Even if the process area is surrounded by liquid-nitrogen-cooled surfaces, providing 100% sticking efficiency for water vapor molecules, the density of the vapor will be given by teh rate of outgassing or desorption from the uncooled surfaces. This example indicates the possibility of controlling the presence of water vapor by careful consideration of vacuum system design, not only by providing a high pumping speed, but also by controlling the sources of the vapor.

4.7 PUMPING SYSTEM DESIGN

High-vacuum pumps are used, together with mechanical pumps, in applications where system operating pressures of 10^{-3} torr and below are desired. The physical arrangement of system components depends on the characteristics of the process to be carried out, such as the pressure level, cycle time, cleanliness, and so on. To some extent the availability and compatibility of components influences system design. In some instances, the economic aspects of component selection may determine system layout. The following paragraphs illustrate briefly the most common component arrangements, referred to as valved and unvalved systems, and outline their major respective advantages and disadvantages. A recommended operating procedure to prevent misoperation and work chamber contamination is outlined for each type.

To furnish maximum effective pumping speed at the processing chamber, it is generally desirable to make the interconnecting ducting between the chamber and the pump inlet as large in diameter and as short in length as practical. The amount of baffling and trapping required depends on the desired level of cleanliness in the chamber, the necessary reduction of the inherent backstreaming characteristics of the pump, and the migration of the pumping, sealing, or lubricating fluids. The effects of this phenomenon, which is discussed later, can be reduced significantly by correct component selection and operating procedures. Baffles, traps, and valving must be selected for minimum obstruction to gas flow, without sacrifice of efficiency of their specific function.

The size of roughing and foreline manifolding is governed by the capacity of the mechanical pump, the length of the line, and the lowest-pressure region in which it is expected to function effectively. Additionally, the size of the foreline is influenced by the forepressure and backing requirements of the high-vacuum pump under full-load operating conditions.

Several design considerations for vacuum systems cannot easily be put into a single set of requirements. A practice that can be easily tolerated in one case must be completely forbidden in another. The requirements span over three or four orders of magnitude in pressure range and surface area. For a small device that must be evacuated rapidly to higher vacuum with a system of limited pumping speed, any trapped air pockets must be carefully eliminated. For example, a screw or bolt in a threaded hole can be a source of virtual gas leakage. It is good practice to provide an evacuation passage for the trapped gas. A hole can be drilled through the bolt, threads can be filed flat on one side, or the blind holes can generally be avoided. However, all this will be of no significance in a system that has a pumping speed of 50,000 L/s and is operated only at 10^{-5} torr. Obviously, vacuum level, pumping speed, size of chamber, and the desired time of evacuation must be considered to provide a balanced design.

In large vacuum systems, structural distortions during evacuation may be high enough to be troublesome. To preserve precise alignments of optical equipment, mechanical drives, and so on, it may be necessary to support the various components independent of the external vacuum vessel.

As noted earlier, for a 2.5-m-diameter door used in an industrial coating system or furnace, the total force due to 1 atm pressure difference in nearly 50 tons. This is certainly sufficient to seal an elastomer gasket without the use of any clamping devices. However, if the chamber is "air-released" with bottled gas, it is advisable to prevent an internal overpressure. A small overpressure may suddenly open the door if the rubber gasket has a tendency to stick. A spring-loaded catch can be used to prevent this.

Another common malfunction due to distortion occurs in large high-vacuum valves. The pneumatic pistons used to actuate such valves are usually not strong enough to provide the vacuum seal. The pressure from the atmospheric side is relied upon to provide the tight closure. Thus such valves cannot be opened against the force of atmospheric pressure. A malfunction can occur during the roughing cycle. As the atmospheric pressure in the chamber is reduced, the valve seat may lift slightly and begin to leak. If the valve is used above a high-vacuum pump system, the leak may be large enough to overload the pump and cause backstreaming, stalling, and pumping fluid-loss problems.

4.7.1 Pump Choice

The selection of high-vacuum pumps is not as straightforward as the selection of mechanical pumps. The considerations for reaching a given pressure in a short time may not prove to be an adequate measure of the system's capability to handle steady gas load characteristics; the pumping curves of the high-vacuum pump will indicate the proper components. More frequently, gas loads are not known and selection of the pump must be based on experience with similar applications.

Vacuum-equipment suppliers generally will offer performance based on pump-down characteristics of an empty chamber. This can be misleading if it is construed to be a measure of pumping efficiency under full-load conditions. The vacuum equipment designers must relate performance to a set of conditions that are known and can be defined and duplicated.

The selection of high-vacuum pumps should be concerned with all factors involved in the system. These factors normally consist of pumping speed, tolerable forepressure, backstreaming rates, pressure at which maximum speed is reached, protection devices, ease of maintenance, backing pump capacity, roughing time, ultimate pressure of the roughing pump, baffles or cold traps employed, valve actuation, and so on.

Unless these factors are carefully weighed in the system design, the system may not perform satisfactorily as a production system. It is to the advantage of both the supplier and the user to consider these factors carefully before finalizing system design.

It is obvious from the previous discussion that pump overload should be avoided in system operation. As a general rule, high-vacuum pumps should not be used at inlet pressures above their maximum throughput capacity for prolonged periods. In a properly designed system with a conventional roughing and high-vacuum valve sequence, the period of operation in an overload condition should be measured in seconds. As a rule of thumb, it may be said that if this period exceeds half a minute, the pump is being misoperated.

It is clear, then, that in applications requiring only 10^{-4} torr process pressure, the mechanical pump roughing period is normally much longer than the high-vacuum-pump portion of the pumpdown cycle, or at least it should be for economic reasons.

4.7.2 Use of High-Vacuum Valves

For applications involving rapid recycling, a fully valved pumping system is essential. This type of system is shown schematically in Figure 4.21. It permits isolation of the high-vacuum pump for the work chamber at the conclusion of a pumpdown and prior to air admittance. The pump can therefore remain at operating condition during periods when the chamber is at atmosphere and

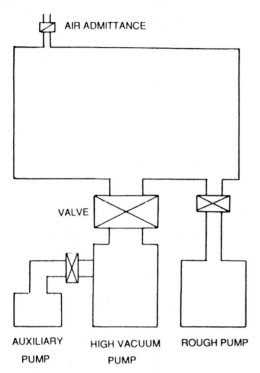

Figure 4.21 Schematic view of a typical vacuum system.

during the rough pumping portion of the cycle. The length of these periods may indicate the need for a separate holding pump. The main isolation valve also permits continuous operation of the cryobaffle between it and the high-vacuum-pump inlet. Neither this nor rapid cycling can be realized without the valving indicated, in view of the shutdown and startup time lapse inherent in high-vacuum-pump operation, and the cool-down and reheat time of the cryobaffle. Judicious operation of the main valve at the changeover phase from roughing to high-vacuum pumping can significantly reduce the backstreaming of contaminants to the work chamber.

Leak testing and leak hunting are considerably easier in valved systems, and repair procedures are also generally less time consuming. However, the following disadvantages can be noted. Valved systems are initially more expensive, especially when large valves are involved. In addition, the use of valves inevitably adds to the system complexity and generally results in lower effective pumping speed at the chamber.

Valveless vacuum pumping systems are generally considered for applications where the length of the pumpdown is of less importance, and process or test cycles are of extended duration. They may be found on baked ultrahigh-vacuum chambers using large high-vacuum pumps, due to lack of availability and/or prohibitive cost of valves compatible with the proposed operating pressure levels. The valveless system generally offers higher effective pumping speed at the chamber and, in view of its prolonged operation at very low throughputs, lends itself to the use of a smaller holding backing pump. This pump is sized to handle the continuous throughput of the high-vacuum pump at low inlet pressure, 10^{-6} torr and below, and allows economy of operation by permitting isolation and shutoff of the considerably larger main roughing and backing pump. The valveless system, however, also has a number of disadvantages. More complex operating procedures are necessary to ensure maximum cleanliness in the work chamber and minimum contamination from the pumps.

To minimize chamber contamination, it is recommended that at least one of the traps be kept operative while the vacuum chamber is being baked. When the inlet ducts are at elevated temperatures, the rate of contaminants migration into the system is accelerated. At temperatures near 200°C the oil vapors are not condensed in the chamber and are subsequently pumped by the high-vacuum pump. But a very thin (invisible) film of oil will condense in the chamber during the cool-down. This film formation is minimized by continuous operation of at least one of the traps.

4.8 OPERATION OF HIGH-VACUUM SYSTEMS

4.8.1 General Considerations

Operation of high-vacuum systems requires certain care and attention to several items. General cleanliness can be extremely important, especially in small systems. It should be remembered that a small droplet of a substance that slowly evaporates, producing a pressure of 10^{-6} torr, may take many days to evaporate and be pumped away if the available pumping speed is 100 L/s. Humidity and temperature can be important in view of the constant presence of water vapor in the atmosphere. When vacuum systems are backfilled or air-released and then repumped, it will make a significant difference whether the air was dry or humid. The time of exposure is also significant. If necessary, the backfilling can be done with nitrogen or argon. For short exposures (30 min or less) this appears to reduce the amount of water vapor adsorption in the vacuum system.

It is extremely important to develop good habits in sequences of valve operations, especially in systems with manual valves. Graphic panels showing valve locations and functions are very useful. A single wrong operation may require costly maintenance procedures. It is well to pause a few seconds before operating any valves in a high-vacuum system unless a routine procedure is followed. Automatic process control sequences have been widely used in recent years to avoid operational errors.

The performance of displacement-type high-vacuum pumps is usually displayed in the form of a plot of pumping speed versus inlet pressure, as shown in Figure 4.22. Note that the graph consists of four distinct sections. In the first section on the left, the speed is seen to decrease due to the limit of maximum pressure ratio obtainable with the pump. This will vary with different gases, and in most applications only light gases such as helium and hydrogen will show this effect. The constant-speed section results from the constant arrival rate of gas into the pump at molecular flow conditions. In this section, the amount of gas being pumped is lower than the maximum mass flow capacity of the pump and the pump speed efficiency is also constant. The section marked "overload" is a constant throughput section which indicates that the maximum mass flow capacity of the pump has been reached. This information is important for knowing when to switch from roughing to the high-vacuum pump in the process of evacuating a chamber. Note that it makes little difference whether the inlet pressure is at 1 or 10 mtorr as far as the amount of pumped gas is concerned, because here the mass flow is nearly constant.

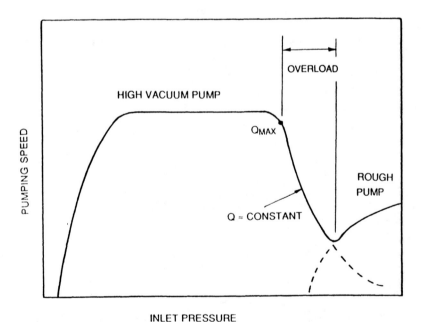

Figure 4.22 Typical performance of a high-vacuum pump (pumping speed versus inlet pressure).

In the last section on the right, the high-vacuum pump performance is dependent on the size of the backing pump. At this condition, the upper stages of a vapor jet pump, for example, are not doing any pumping, and most of the pumping is done by the mechanical pump. Strictly speaking, a high-vacuum pump should not be used above the critical point shown in Figure 4.22. If process pressure demands high pressure, the pump should be throttled to limit the throughput to the maximum permissible.

This performance feature can be illustrated much more clearly by displaying throughput of the pump instead of speed and by reversing the traditional position of inlet pressure in the horizontal axis of the graph. This is shown in Figure 4.23. This graph is somewhat idealized. The actual pumps will not produce a completely vertical midsection as shown. However, it should be clear

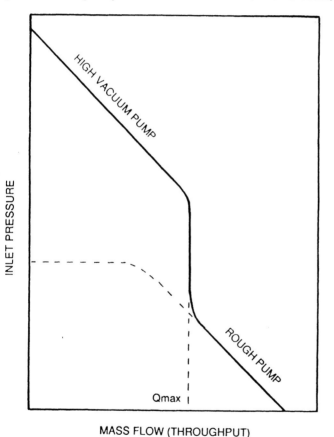

Figure 4.23 Typical performance of a high vacuum pump (inlet pressure, increasing downward, versus mass flow).

that to the left of the vertical line, representing the maximum mass flow rate for the high-vacuum pump, it is performing the primary pumping. To the right of the vertical section, only the mechanical pump is pumping; the high-vacuum pump is idling.

Thus, the correct condition for switching from roughing to high-vacuum pumping is clear. It is when the amount of gas from the chamber is below the maximum throughput of the high-vacuum pump! How to recognize this condition is the subject of the following section.

4.8.2 Switching from Rough to High-Vacuum Pumps

During the evacuation of a vessel, the question arises regarding the proper time to switch from rough pumping to the high-vacuum pump. In other words, when should the high-vacuum valve be opened? No general answer can be given because it depends on the volume and the gas load of the chamber.

Ideally, the answer should be: when the gas flow from the high vacuum system is below the maximum throughput of the high-vacuum pump. This would imply an inlet pressure near 1 mtorr. However, it is not necessary to obtain this pressure with mechanical roughing pumps or even with mechanical booster pumps (Roots type). In practice, the transfer from roughing to high-vacuum pumping is normally made between 50 and 150 mtorr. Below this pressure region, the mechanical pumps rapidly lose their pumping effectiveness and the possibility of mechanical pump oil backstreaming increases. Although the throughput of the high-vacuum pump is nearly constant in the region of inlet pressure of 1 to 100 mtorr, the initial surge of air into the pump when the high-vacuum valve is opened may overload the high-vacuum pump temporarily.

The general recommendation is to keep the period of inlet pressure exceeding approximately 1 mtorr (0.5 mtorr for large pumps) very short—a few seconds, if possible. Consider the following example. For constant throughput, the evacuation time (typically between 1 and 100 mtorr) is

$$t = \frac{V(P_1 - P_2)}{Q} \tag{4.64}$$

For a common 45-cm-diameter bell jar, the volume is about 120 L and using, for example, a 15-cm-diameter vapor jet pump with a maximum throughput of about 3 torr · L/s, we obtain

$$t = \frac{(120 \ L)(0.1 \ \text{torr})}{3 \ \text{torr} \cdot \text{L/s}} = 4 \ \text{s} \tag{4.65}$$

If the high-vacuum valve were to open slowly to admit the gas into the diffusion pump at the maximum throughput rate, it would take only 4 s to reach the stable pumping region, from point B to point A as illustrated in Figure 4.24.

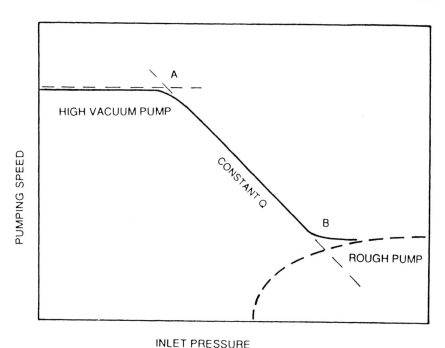

Figure 4.24 Crossover region between rough and high-vacuum pumps.

Because points A and B lie on the same gas load line (constant throughput), either one may be called the "crossover pressure." To avoid this confusion, we should not speak of crossover pressure but only of crossover throughput (or crossover gas load!). This maximum gas load limitation for a given high-vacuum pump exists regardless of the size of backing pump. As a further illustration, consider the system shown in Figure 4.25. In principle, a 100-L/s high-vacuum pump can be connected in series to a 100-L/s mechanical pump (212 ft^3/min), obtaining a constant pumping speed from atmosphere to the high-vacuum region.

In this system, points A and B coincide. The high-vacuum pump must still be isolated from excessive gas loads to prevent increased backstreaming, stalling, pumping fluid loss, or even complete destructive failure. To the right of point A the mechanical pump is doing the pumping; to the left, the high-vacuum pump. The crossover point is still point B. If we persist in thinking in terms of crossover pressure rather than crossover throughput, we arrive at an incongruous situation where a smaller mechanical pump would seemingly tolerate a higher crossover pressure.

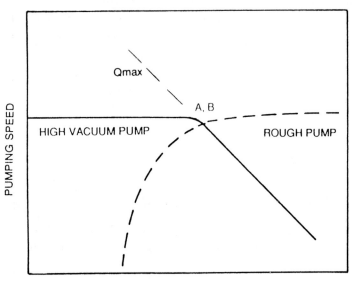

Figure 4.25 Crossover between rough and high-vacuum pumps of equal speed.

4.8.3 Measure Gas Load

Most manufacturers of high-vacuum systems specify a pressure value for switching to the high-vacuum pump or for opening the high-vacuum valve. Such pressure value can only be specified for a given system or a class of systems with fairly well known basic design, such as conventional bell-jar coaters. In general, the pressure will depend on the volume of the system, its process gas load, and the maximum throughput of the pump. Thus, a single pressure value cannot be given for all cases.

There are several ways to recognize the condition of exceeding the maximum throughput. Depending on installation, one or the other method may be more convenient to use.

1. The simplest way of detecting a high-vacuum pump overload is to observe the time it takes to reach about 0.5 mtorr (5×10^{-4} torr) after the high-vacuum valve is opened. This period in a well-designed and properly used system should be only a few seconds. If the high-vacuum pump is not overloaded, the transient-time response should be: system volume divided by the pumping speed. Otherwise, pump fluid loss, backstreaming, stalling, and system contamination may occur. Worse still, simply no significant pumping will be done. Nothing will be gained by switching too soon.

2. The more definite method consists of a rate-of-pressure-rise measurement under process conditions. After evacuation to a given pressure, say 0.1 torr,

the roughing valve should be closed and the pressure rise inside the fully loaded chamber measured. The rate of rise will be measured in units such as torr per second. Multiplied by the volume of the chamber (or the entire isolated volume), this will give the gas load (or throughput) in torr · L/s. This measurement should be repeated at various pressures until the gas load becomes lower than the maximum specified throughput of the high-vacuum pump.

3. Perhaps the simplest way to establish the gas load at the time the high-vacuum valve is opened is to relate it to the pumping speed of the rough vacuum pump. The gas evolution from the vacuum chamber remains essentially the same immediately preceding or immediately following the opening of the valve. Thus the mass flow rate (throughput) into the roughing port and into the high-vacuum-pump inlet will be the same: $P_1 S_1 = P_2 S_2$.

Then it is only necessary to estmiate the net speed of the roughing pump (S_1) and to measure the pressure in the chamber (P_1) just before the high-vacuum valve is opened. The gas load from the chamber must be less than the maxmium throughput specified for the given high-vacuum pump. If it is not, the roughing process should be continued to a lower pressure. After the high-vacuum valve is opened, the chamber pressure should drop to a new valve according to the ratio of net pumping speeds S_2/S_1.

4.8.4 Operation Near Overload

When designing a vacuum system for a given process, it is always advisable to provide safety factors in the selection of pump size. There are some uncertainties associated with pump performances, pressure and flow measurements, and external leakage. But by far the most important uncertainty is due to outgassing. The outgassing rates given in technical literature can easily vary by a factor of 10, depending on the history of a surface, its treatment, humidity, temperature, and the period of exposure to vacuum.

Some vacuum systems, such as furnaces and paper or plastic web coating plants, often operate near the maximum throughput point of high-vacuum pumps. In addition, such systems often have porous materials or accumulated coatings on the walls which can absorb large quantities of water. During evacuation, after near-high-vacuum conditions are reached, the outgassing rate does not change with pressure but depends mostly on time (and temperature).

It is generally wiser to continue pumping with the roughing pump until the outgassing is reduced below the maximum permissible rather than cross over to the high-vacuum pump too soon and operate it in the overloaded condition. Backstreaming during such periods can be reduced by a gas purge at about 0.1 torr.

It often happens that a process chamber that has been pumped to a given pressure in 15 min "suddenly" requires two or three times as much time. On such occasions, typically, the operators begin a frantic search for causes and measure pump temperatures and nozzle throat dimensions, while the problem is a relatively small increase in outgassing. This possibility is illustrated in Figure 4.26, where it is shown how widely different evacuation times can occur due to a small change in gas loads.

4.8.5 High-Vacuum Valve Control

Usually, in a high-vacuum system that is expected to be pumped to 10^{-6} torr in less than half an hour, the outgassing will be negligible (compared to the maximum throughput of the pump) during the initial period after the high-vacuum valve is opened. In practice, due to lack of precise valve control, the period between the time when the valve is opened and the time when inlet pressure of 1 mtorr is reached can be shorter than that computed using Eq. (4.10). This is due to the expansion of air across the high-vacuum valve into the higher-vacuum space downstream. This downstream space (part of the valve, trap, baffle, upper part of the pump) can be significant compared to the chamber volume. On the other hand, when outgassing is severe, the pump may be overloaded for longer periods of time. The same is true if the pump is too small for the chamber size. This can be encountered in furnaces having porous insulating materials and in coaters where large drums of thin-film plastic are present in the vacuum chamber. In such cases it may be better to elongate the rough pumping period before returning it to the high-vacuum pump. When

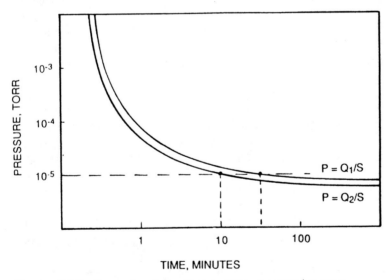

Figure 4.26 Evacuation time with two different outgassing rates.

throughput is nearly constant, the pumping time will be the same regardless of whether the high-vacuum valve is opened only slightly or is fully opened. In general, the high-vacuum valves should be opened slowly—very slowly at the beginning. The motion of ordinary pneumatic valves can be controlled to some extent with the air inlet and exhaust adjustments. Special valve controls can be made either to maintain approximate constant throughput during initial opening or to have two-position interrupted operation. When the valve is almost closed, it serves as a baffle. This prevents the possibility of pumping vapor (in case of vapor jet pumps) getting into the chamber due to the turbulent flow of air entering the pump.

4.9 ESTIMATION OF PRESSURE–TIME PROGRESS DURING EVACUATION

With modern calculators, it is not very difficult to use Eq. 4.19 for direct calculations of pressure values during the process of evacuation. But nature often defies simple mathematical expression. The actual outgassing measurements indicate that the decay rate (τ) is not constant over the entire range of interest. In addition, the initial outgassing values are uncertain. For example, data given in technical literature shows outgassing values for metals varying by as much as 10 times whether the surface is said to be covered with heavy oxide (rusty), "normal," or "exceptionally clean." The outgassing values for elastomers are also uncertain because their gas (water) content depends on the exact conditions of manufacture (ingredients, temperature, time of exposure to humid air, etc.).

The increase of the time constant, τ may be due to the different binding energies involved in adsorption of water on surfaces. At the beginning, the surface layers may evaporate rather quickly, especially after the chamber pressure passes approximately 18 torr. However, at this pressure, the amount of outgassing is still small compared to the space gas leaving the chamber. Normally, no sudden change is noticeable in the pressure-time curve near 18 torr. Therefore, it is perhaps a good compromise to assume the onset of significant outgassing (in typical vacuum chambers) to be between 1 and 0.1 torr. The higher values are applicable for smaller systems.

Outgassing values are often reported on log-log graphs, to cover a larger span of time. Such graphs are somewhat confusing for engineering purposes because it is difficult to extrapolate to "zero" time. For estimating evacuation time, the outgassing rates should be shown as in Figure 4.27, which attempts to establish the "zero" time. The choice of this time should not be made arbitrarily. It can be referred to the onset of molecular flow in the chamber or some other related concept, such as a ratio of volume gas and surface gas.

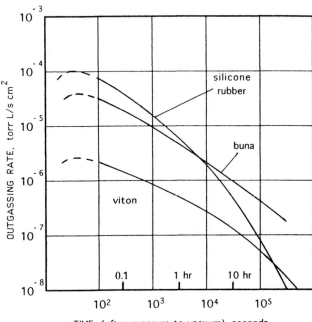

Figure 4.27 Outgassing rates for elastomer gaskets depending on time of exposure to high vacuum. Actual values can vary as much as ± factor of three from these rates.

4.9.1 Initial Evacuation (Roughing)

The following outlined procedure for esitmating evacuation time should be used in conjunction with full understanding of vacuum pump characteristics and limitations (see Section 4.8.4 on overload). The rough pumping period can be defined in two different ways. First, as the time during which the roughing pump was used, and, second, as the period after which the space gas had been essentially removed and the predominant gas flow is from the surface. In the past, the former designation was more convenient, but with the advent of hybrid turbomolecular pumps the distinction between roughing and high-vacuum pumping has been blurred in regard to the type of pump used for the initial evacuation and the following high-vacuum pumping. Thus, for the purposes of this discussion, the second definition should be preferred.

Step 1. Rough pumping from atmospheric pressure. Roughing pumps, generally, tend to have constant pumping speeds. If not, their speed versus pressure curves can be divided into approximately constant speed sections. Then Eqs. 4.2 or 4.3 can be applied to these separate segments to estimate the time period. The same holds for the successive use of different pumps, such as Roots blowers. Figure 4.28 can be used for quick estimates. It shows the values of

VACUUM SYSTEMS

pumping time for $V/S = 1$. The values obtained from this figure can be multiplied by the V/S value for any chamber to obtain the time for reaching a given pressure. Any consistent units can be used, as indicated on the graph. Note the deviation from the logarithmic curve due to the onset of outgassing (or reduction of effective pumping speed) near 1 torr. The values of the deviation, associated with the correction factor alpha in Eq. 4.24, are for typical (nonultrahigh-vacuum) medium-sized chambers. They depend on surface-to-volume ratios, and should be higher (and deviation begin sooner) for smaller chambers, and lower for larger chambers.

The slope of the evacuation curve in Figure 4.28 can be viewed as an indication of pumping speed value. Note that at 0.076 torr the time correction factor is near 2 but the slope for the curve A (sand blasted chamber walls) is about

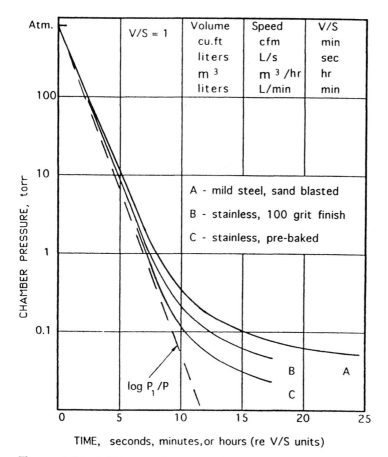

Figure 4.28 Initial evacuation process for typical metal systems with some elastomer seals for $V/S = 1$, shown on a log-linear graph (any consistent set of units can be used).

10 times lower. If, for example, at 0.076 torr another pump is engaged, the lower slope should be used for obtaining the ratio of pumping speeds of the new pump compared to the previous one. It is also good to remember that the values of deviation from a logarithmic curve are associated with the usual presence of water vapor in the vacuum chamber. The curve for helium would be nearly a straight line unless it is dissolved in pump oil or retained in some other way.

Step 2. Switching to a high-vacuum pump (cross-over). Usually, when the roughing valve is closed and the valve to the high-vacuum pump is opened, the pressure drops rapidly because of the increase in pumping speed, assuming that the high-vacuum pump is not overloaded (see overload chapter). This sudden drop should be regarded as the end of the roughing period (the last "gulp" of the volume gas), just before the surface gas evolution becomes predominant. In other words, the nearly vertical part of the typical evacuation time curve can be associated with roughing and the nearly horizontal one with high-vacuum pumping. These time periods are associated with the two terms of Eq. 4.24. Both of these periods can be divided into a few sections which may be assumed to have a constant speed.

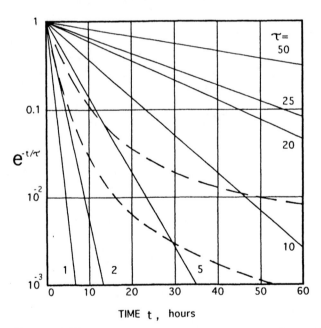

Figure 4.29 Outgassing curves (dashed lines) superimposed on a grid of values of τ (Eq. 4.23). The upper curve is stainless steel, mechanically polished; the lower curve is stainless steel, chemically cleaned.

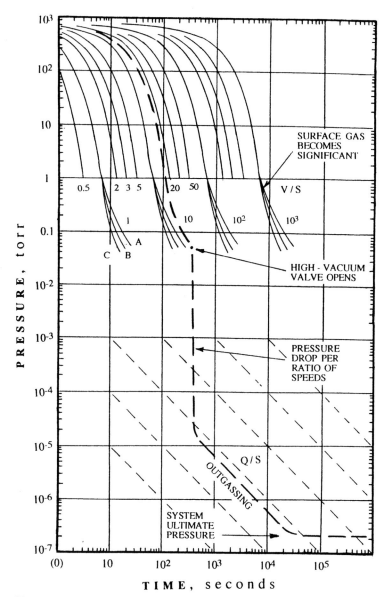

Figure 4.30 Evacuation process, from atmosphere to the high-vacuum region, shown on a log-log graph.

Step 3. High vacuum pumping. As noted before, τ in Eq. 4.24 is not constant. To help estimate correction values (for β), Figure 4.29 shows outgassing curves (essentially for water vapor) superimposed on the values of $e^{-t/\tau}$. The two dashed curves represent a mechanically polished and a chemically cleaned stainless steel surface. When the primary outgassing species is water vapor, τ may start with a value of 0.5 hours and grow to 4 or 5 hours after 10 hours of pumping, as indicated by the matching slopes in Figure 4.29. The values for pressure obtained in these estimates must be considered to be $(p - p_\infty)$. Usually, after about 24 hours of pumping, the pressure changes become too small to be significant because the system reaches a condition when either leaks, or permeation through gaskets, or the limitation of pumps, produce the final ultimate pressure. At this point inlet pressure is given by the residual gas flow divided by the pumping speed.

The above procedure may be convenient when the pressure-evacuation time relationship is plotted on a log-linear graph. Then, the entire curve (beginning from atmospheric pressure) can be constructed from an outline provided by a few straight lines on the graph, including the final horizontal line when the ultimate system pressure is reached.

The alternate procedure may be used as indicated in Figure 4.30, where the log-log plotting produces straight lines in the high-vacuum section, following the relationship for $Q = \text{constant}/t$ for the outgassing function. An example for using the graph is shown. The deviations from the exponential lines in the roughing section are the same as in Figure 4.28. The drop in pressure, after the valve the high-vacuum pump is opened, is calculated from the ratio of actual net system pumping speeds prevailing at the time. The net pumping speed of the roughing pump can be obtained as indicated by Eq. 4.26.

REFERENCES

1. J. D. Lawson, *J. Sci. Instrum.*, *43*, 565 (1966).
2. G. P. Brown, A. DiNardo, G. K. Cheng, and T. K. Sherwood, *J. Appl. Phys.*, *17*(10), 802 (1946).
3. R. P. Benedict, *Fundamentals of Pipe Flow,* John Wiley & Sons, Inc., New York, 1980.
4. J. Pinson and A. Peck, in *Transactions of the Ninth Vacuum Symposium of the American Vacuum Society*, G. K. Bancroft, ed., Macmillan Publishing Company, New York, 1962.
5. L. Levenson, N. Milleron, and D. H. Davis, in *Seventh National Symposium on Vacuum Technology Transactions*, C. R. Meissner, ed., Pergamon Press, Elmsford, N.Y., 1961.
6. C. W. Oatley, *Br. J. Appl. Phys.*, *8*, 15 (1957).
7. D. H. Holkeboer, D. W. Jones, F. Pagano, and D. J. Santeler, *Vacuum Engineering,* Boston Technical Publishers, Cambridge, Mass., 1967.

8. J. O. Ballance, in *1965 Transactions of the Third International Vacuum Congress*, H. Adam, ed., Pergamon Press, Elmsford, N.Y., 1965, p. 85.
9. D. J. Santeler, *J. Vac. Sci. Technol.*, *A5*(4), 2472 (1987).
10. M. Wutz, *Vak. Tech.*, *14*, 5 (1965).
11. Y. N. Lyubitov, *Molecular Flow in Vessels*, W. H. Furry and J. S. Wood, Transls., *Consultants Bureau*, New York, 1967.
12. G. L. Saksaganskii, *Molecular Flow in Complex Vacuum Systems*, Gordon and Breach, Science Publishers, Inc., New York, 1988 (Atomisdat 1980).
13. M. H. Hablanian, *J. Vac. Sci. Technol.*, 7, 237 (1970).
14. R. G. Livesey, *Vacuum*, *32*(10/11), 651 (1982).
15. P. Shapiro, in *Eighth National Symposium on Vacuum Technology Transactions*, L. E. Preuss, ed., Pergamon Press, Elmsford, N.Y., 1962, p. 1186.
16. B. W. Schumacher, in *Eighth National Symposium on Vacuum Technology Transactions*, L. E. Preuss, ed., Pergamon Press, Elmsford, N.Y., 1962, p. 1192.
17. A. H. Shapiro, *Vacuum*, *13*(3), 83 (1963).
18. J. A. Freeman, *Microelectron. Manuf. Test* (October 1985).
19. L. C. Beavis, V. J. Harwood, M. T. Thomas, *Vacuum Hazards Manual*, Amer. Vac. Soc., New York, 1979.
20. R. J. Elsey, *Vacuum 25*, 7, 299, (1975). and *Vacuum 25, 8*, 347, (1975).
21. G. P. Brown et al., *J. Appl. Phys.*, *17*, 10, 802, (1946).
22. P. A. Redhead, *J. Vac. Sci. Technol.*, *A13*, 467, (1995).
23. H. F. Dylla, *J. Vac. Sci. Technol.*, *A11*, 5, 2029, (1993).

5
Coarse Vacuum Pumps

5.1 INTRODUCTION

The term "coarse vacuum" or "rough vacuum" usually refers to the pressure region between atmospheric pressure and 1×10^{-3} torr. This is a somewhat artificial barrier. Some rough vacuum pumps are capable of achieving 10^{-4} torr and even lower pressures.

There are many different types of pumping devices which can be employed in this pressure region, although only a few are commonly used. Depending on conditions such as the volume of the vacuum system or the vacuum vessel, the amount of evolved process gas, and the required time of evacuation, an unusual combination of pumps can be used to evacuate a chamber to a given pressure or to maintain a pressure at a given gas load. Depending on the pumping characteristics of various pumps, many different parallel- and series-connected pumping arrangements can be used.

Generally, there are the following major categories of pumps:

Positive displacement, where the gas is transported by means of pistons, gears, vanes, rotating cams, and so on;

Momentum transfer, where the pumped gas is entrained in and removed by an energetic stream of a pumping fluid or directed by fast-moving blades;

Entrainment, where the gas is absorbed on the surface of a suitable material that may or may not be refrigerated.

Among so-called mechanical vacuum pumps used in vacuum technology, there are reciprocating piston pumps and a variety of rotary pumps. The latter include vane, rotary piston, cam, lobe, and screw-type pumps. Momentum transfer pumps include liquid, gas, or vapor jet pumps. Centrifugal and axial compressors are not used for coarse vacuum applications, but this exclusion is not intrinsic and is due to practical limitations.

Each pumping device has certain advantages of performance, cost, and process compatibility which make its use preferable in a particular application. This general concept of practicality is illustrated in Figure 5.1, where approximate costs for a variety of pumping speeds is shown relative to the inlet pressure. Similar plots can be made based on size, weight, and other required characteristics.

The basic performance aspects of rough vacuum pumps are the pumping speed (or volumetric capacity) and the ultimate pressure (or the best achieved vacuum), but there are many other secondary aspects that must be considered to obtain a reliable performance. A list of such properties follows.

Pumping speed (displacement, volumetric capacity)
Ultimate pressure (highest-obtainable vacuum)
Weight, size, noise, vibration
Water vapor pumping (with and without gas ballast)

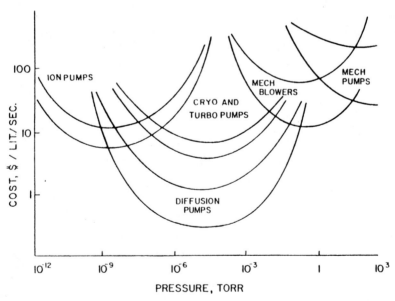

Figure 5.1 Pressure regions in which particular pumps are most effective.

Inlet pressure during gas ballast
Recovery time after pumping a certain gas
Pressure and speed stability
Retention of vacuum and oil when stopped
Temperature (oil change period)
Oil flow rate, size of reservoir (oil change period)
Corrosion resistance
Maintenance, life, cost
Power consumption (per unit speed or per unit mass flow)
Sensitivity to particulate contaminants
Degree of backstreaming
Presence or absence of oil inside pumping spaces
Appearance

Depending on requirements, careful review of these properties will help in establishing a choice between, for example, vane and rotary piston pumps or pumps offered by different manufacturers.

The performance of all compressors, including ordinary ventilation fans, can be described by two basic parameters: the amount of flow and the achieved pressure. This is represented in Fig. 5.2 in a normalized way using nondimensional flow and pressure (head) coefficients, where Δp is the pressure difference, ρ is the density, N is the rotational speed, D is the diameter, and S is the volumetric flow rate.

Generally, for common designs with more or less optimized efficiency, most devices can be placed on the curve in Fig. 5.2. Fans will be at the right side (low pressure, high flow) and reciprocating piston pumps on the left side (high pressure, low flow).

Pumps of different construction can be used more effectively in certain inlet pressure regimes. They can be compared by cost per unit pumping speed,

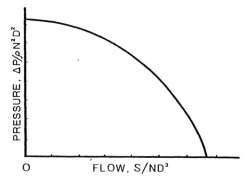

Figure 5.2 Nondimensional performance parameters of pumping devices.

or per unit mass flow, or size, weight, and energy usage per unit pumping speed or throughput. However, almost regardless of cost and other factors, certain pumps can be chosen for their unique advantages such as a low degree of contamination of the working environment.

Only twenty-five years ago, high-vacuum pumping was primarily accomplished by vapor jet pumps and oil-sealed mechanical pumps, sometimes assisted by residual gettering. Now, there are three more fully developed methods of pumping available to high vacuum technologists: turbo-molecular, ion-getter, and cryopumps.

Some of these pumps have unique features useful for particular applications. Some have extensive overlap in their performance. Thus, a user should be familiar with the properties of each pump to make an informed selection. Some general and specific comments made in the following sections may help this process.

From the chosen coordinates of Fig. 5.2, it can be seen that the pressure difference is proportional to the density and the peripheral velocity of an impeller squared $(ND)^2$, and flow is proportional to peripheral velocity (ND) and the area (D^2). Then, the power associated with raising a certain flow rate to a higher pressure level is proportional to $S \cdot \Delta P$, which is density, peripheral velocity cubed, and the area. Therefore, in viscous flow, at the same pumped gas conditions, if the diameter of the impeller is doubled at the same rotary speed (rpm), then the flow rate will increase 8 times, the pressure 4 times, and the power consumption 32 times. This must be considered in designing large pumping systems. To save power, it is often necessary to use different types of pumps which are more suitable for certain pressure ranges and at reduced volume of highly compressed gas.

The selection of various coarse and fine (high-vacuum) pumps for pumping in different pressure ranges generally reflects the positions seen in Fig. 5.3. Typical pressure performance of pumps presently used in high-vacuum technology is shown in Fig. 5.3. If electro-magnetic pumps, such as sputter-ion pumps, were included in the same graph, the decline of their pumping speed would occur near 1×10^{-5} torr.

A variety of pumping mechanisms is available from the gas moving and compressor technologies. Some of these are more adaptable for the achievement of higher pressures and some for higher flow rates. For high-vacuum pumps, both high compression ratios and flow rates are desirable. For coarse vacuum pumps, the flow rate requirements are more modest but high compression ratios are essential. Practical useful ranges of basic pumping devices are illustrated in Fig. 5.4. This graph should be viewed in normalized way (i.e., the devices should be considered as if they were, for example, of equal cost, or equal size, or equal weight). In other words, viewed in the sense of an index of practicality.

COARSE VACUUM PUMPS

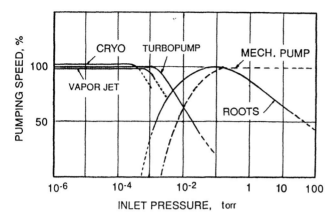

Figure 5.3 Practical range of flow rates and compression associated with pump types.

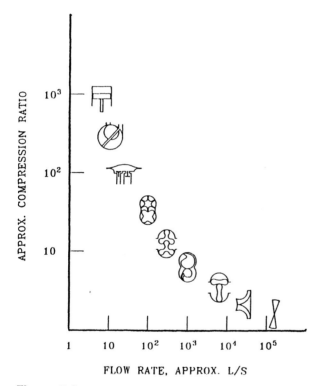

Figure 5.4 Comparison of various pumping mechanisms in regard to their pressure-flow characteristics (from left to right: piston, vane, diaphragm, screw, three-lobe Roots, hook and claw, two-lobe Roots, centrifugal, and axial).

5.2 ROTARY VANE PUMPS

Vane pumps used in the high-vacuum industry are made in two basic varieties. One has a sliding spring-loaded vane in the stator. The vane remains in contact with an eccentric cam that may be called a "rotary piston," and the distances between the cam and the stator walls are kept at a practical minimum (see Section 5.8.1).

The other basic type has two or three (or more) sliding vanes placed in the rotor. The vanes are pushed toward the stator wall either by springs or centrifugal forces, as demonstrated in Figure 5.5. To make high-pressure ratios achievable, the principal requirement of these pumps is that the inlet and discharge always remain separated. This separation is achieved as seen in Figure 5.5 by placing the rotor in close proximity to the stator, the near contact line being placed between the inlet and discharge, and spring loading the blades so that they remain in contact with the stator. In addition to the seals shown in the figure, oil is used to seal the end surfaces of the vanes and along the entire length of the two vanes, as well as across the ends of the rotor.

The rotors in such pumps are placed inside the stator so that the gaps along the cylindrical "touching" line and the two ends have a spacing of about 0.02 mm for pumps with a pumping speed of 3 to 5 L/s. Oil is used in these pumps for a variety of purposes: to effect a seal between the inlet and discharge areas, to lubricate, to fill the space under the discharge valve, to serve as a heat transfer medium and keep the rotor temperature within acceptable range, to flush particulate matter out of the pump, and sometimes also to serve as a hydraulic fluid for operation of auxiliary valves.

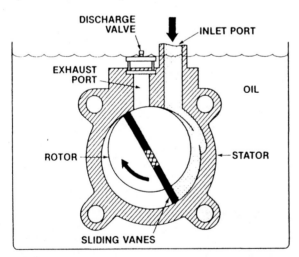

Figure 5.5 Basic mechanism of a sliding vane pump.

Vane pumps are rather simple in the basic concept and have been used for pumping liquids at least as early as 500 years ago. However, for high-vacuum application, which began about 90 years ago, a remarkable degree of effectiveness has been achieved. A single-stage oil-sealed rotary vane pump can produce inlet pressures near 0.010 torr for air while discharging to atmosphere. This represents a compression ratio of 10^5. Modern two-stage pumps, Figure 5.6 can produce at the inlet a partial pressure of air below 1×10^{-5} torr while discharging to the atmosphere, thus achieving a pressure ratio of almost 10^8.

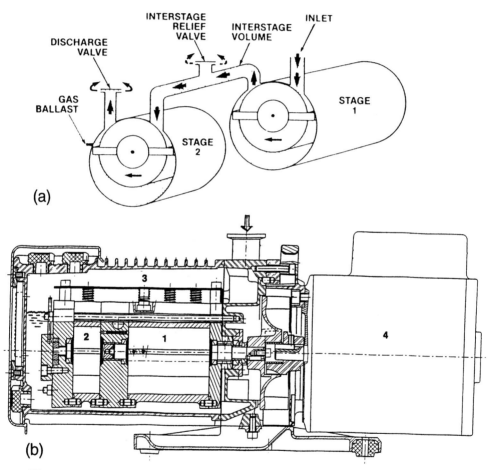

Figure 5.6 (a) Two-stage pump. (b) Cross section of a two-stage, oil-sealed, rotary-vane pump; (1) inlet stage, (2) second stage, (3) relief valve, (4) motor. (Courtesy of Alcatel Vacuum Technology.)

In two-stage pumps, special arrangements are made to seal and lubricate the upstream stage with degassed oil and maintain the proper amount of oil circulation. Oil is continuously supplied to the pumping space and ejected together with the pumped gas in both stages. Typically, in pumps with a speed of 3 to 5 L/s the oil circulation rate is over 1 cm^3/s, which is sufficient to cool the rotor.

In modern pumps, an arrangement is made to interrupt the oil flow automatically when the motor stops. This prevents filling the pumping mechanism with oil and also prevents air leakage into the pump and subsequently into the vacuum system. Such "anti-suck-back" devices may be small valves at the oil inlet hole kept open by a variety of centrifugal switches attached to the pump rotor. Also, automatically operated inlet valves and solenoid valves are used.

Pumps of this type are generally very reliable, requiring only occasional oil changes. However, although the oil is the primary ingredient of successful performance, it also represents the source of the two major weaknesses: the possibility of backstreaming of oil vapors into the vacuum system and the destruction of the oil by corrosive gases and solvents.

There are advantages and disadvantages in all these methods of preventing oil from entering the pump when the pump is stopped as well as in having two or three vanes and other details of construction. Preventing the filling of the pumping volumes with oil is important. Otherwise, when the pump is restarted, it will be subjected to high forces in the attempt to compress the liquid. Most vane pumps have a depression near the exit port along the parts of the cylindrical surface to allow the oil to flow back around the vanes at the very end of the compression stroke.

5.3 PERFORMANCE CHARACTERISTICS

The traditional display of basic performances of a vacuum pump is a plot of volumetric pumping speed versus inlet pressure. This practice is the reverse of the common presentations used in other mechanical engineering fields. The discharge pressure is neglected because it is usually the "constant" atmospheric pressure. Unfortunately, this reversed traditional display has been carried over to performance presentations of pumping trains (connected in series) and also to high-vacuum pumps. In addition, the relative neglect of mass flow has often produced considerable conceptual confusion in understanding pump performance.

The common performance curves are shown in Figure 5.7 for single- and double-stage pumps. In applications associated with high-vacuum technology, double-stage pumps are used much more frequently than single-stage pumps. In following sections the discussion will generally be limited to double-stage pumps. The main reason for the use of two-stage pumps is the achievement of

COARSE VACUUM PUMPS

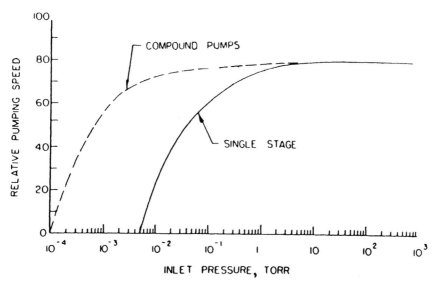

Figure 5.7 Pumping speed at various inlet pressures.

lower pressure, but in addition, the low inlet pressure can be maintained for a longer period because only degassed oil is usually permitted to enter the first stage. Modern designs of double-stage pumps have direct-coupled motors and both stages arranged on the same axis. The ratio of volumetric capacity of the first and second stages typically varies from 1:1 to as high as 20:1. This variation is due to purely practical considerations. In very small pumps the similarity of parts is more important than the advantages in size, weight, cost, and power requirements obtained by using a smaller second stage.

The basic performance characteristics shown in Figure 5.7 are the almost constant pumping speed near atmospheric pressure, a gradual reduction, and the final drop toward so-called "ultimate" pressure, where the speed is shown to be zero (however, see Section 5.7). Specifications for rough vacuum pumps, often called simply "mechanical vacuum pumps," occasionally include the term "displacement," which, unfortunately, does not have a clear definition in textbooks or even in standard documents of various engineering societies. The figure for displacement is somewhat higher than the pumping speed, indicating that the volumetric efficiency of the pump is less than 100%.

The displacement is sometimes defined as a purely geometric or kinematic factor (i.e., the geometric volume swept by the pumping mechanism during one revolution of the shaft or one piston stroke). Others associate the displacement with the amount of gas at inlet pressure that enters the pump during one revolution, presumably at very low rotational speed. The two definitions are not identical because in vacuum pumps there is often an internal expansion of the gas after it has been isolated from the inlet before it is finally compressed. Knowledge of the displacement figure, however defined, is of no value to the user, who should only be concerned about the actual pumping speed at normal rotational velocity (rpm).

Some pumps show irregularities in the pumping speed curve compared to Figure 5.7. Sometimes there is a stepped reduction in speed between 1 and 10 torr, sometimes a dip in speed near 0.1 torr, and sometimes the reduction in speed toward the lower pressures is much steeper than in Figure 5.7. These variations may be genuine aspects of design due to stage matching, amount of oil, effectiveness of sealing in various passages, and so on. It is also possible that deviations from the typical curve are due to the measurement technique and changes in the type of pressure gauges used in different pressure ranges. Usually, positive-displacement pumps are reliable and predictable in regard to the pumping speed performance unless there is a specific mechanical malfunction, for example, if a vane is not free to move (broken springs or swelling and sticking). An insufficient supply of oil to the first stage will be noticeable as a high ultimate pressure rather than as a low pumping speed.

There exist recommended practices and standards for measuring pumping speed of vacuum pumps used in high-vacuum industry such as by American Vacuum Society, European, and International Standards Organization. The basic measurement consists of measuring the flow rate at a set of selected inlet pressures (for the given gas) and deriving the pumping speed as $S = Q/P$. An example of actual measurements is shown in Figure 5.8 for two oil-sealed dual-stage rotary-vane mechanical pumps. The pumping speeds measured with a standard vacuum chamber do not necessarily coincide with the actual pumping speed existing in the vacuum chamber. Despite the relatively high conductance of connecting ducts at the higher pressure (in viscous flow), pumping speed is reduced because of the pressure difference across the duct. The actual system speed can be easily established by using the slope of the evacuation curve at its straight section and by referring to Eq. 4.2. The evacuation progress for a 100 liter chamber of ordinary steel (pressure versus time) for the two pumps of Fig. 5.8, is shown in Figure 5.9. The small discontinuity near 1 torr is due to the change of the pressure measuring instrument.

It should be understood that the ultimate pressure near 0.1 mtorr is obtained only with a pump in good condition (new shaft seals, little wear, new degassed oil, etc.) and in relatively clean applications. Otherwise, if the pump is subjected to high pressure, high temperature, reactive gases, or too much water vapor and solvents, the ultimate pressure will rise into the 10 mtorr range.

A significant characteristic of the performance is the power required to operate the pump. The peak power occurs typically at inlet pressures between 1/2 and 3/4 atm. Power is proportional to both mass flow rate and the pressure difference created by the pump (Fig. 5.2). There are several advantages

COARSE VACUUM PUMPS

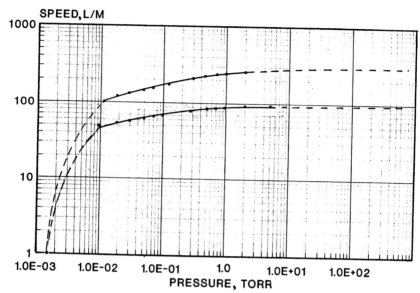

Figure 5.8 Pumping speed performance obtained using the recommended practice (American Vacuum Society) for two dual-stage oil-sealed vane pumps.

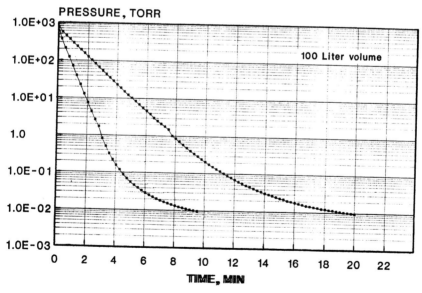

Figure 5.9 Evacuation time for a 100 liter volume (mild steel chamber) using the two pumps of Figure 5.8; the discontinuity near 1 torr is due to the change of the pressure gauge.

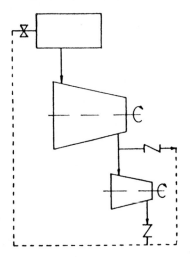

Figure 5.10 Bypass valve arrangement reducing amount of compression.

in keeping the power requirement down. To minimize power, almost all double-stage pumps have intermediate exhaust valves placed between the two stages. Thus at the higher inlet pressures most of the compressed gas is exhausted through the intermediate valve rather than being subjected to useless additional compression in the second stage. This is illustrated schematically in Figure 5.10.

5.4 ULTIMATE PRESSURE

The concept of ultimate pressure does not have a simple clear definition. In general, it is the lowest pressure obtainable after a long period, perhaps 10 to 20 h. There are several difficulties associated with this concept. First, the ultimate pressure of a pump is the sum of at least three components: external gas loads, internal gas loads, and compression limits:

$$P_{ult} = \frac{Q_e}{S} + \frac{Q_i}{S} + \frac{P_d}{k} \tag{5.1}$$

where Q_e is the external gas load, Q_i the internal gas load, P_d the discharge pressure, S the pumping speed, and k the limiting compression ratio of the pump. At pressures approaching high vacuum, each of these three terms consist of a summation of gas loads and compression limits for each gas present in the system.

In other words, Q_e, Q_i, S, P_d, and k may be different for each gas species. For rough vacuum pumps, when the inlet port is closed, the first term is usually negligible. However, for very small pumps the gas load from the measurement system may become significant.

COARSE VACUUM PUMPS

When the pump performance is measured, metered amounts of gas are introduced into the pump controlled by a gas inlet valve and the resulting pressure is recorded. To establish the ultimate pressure, the gas inlet valve is closed. But there remains some residual gas supply from the outgassing in the inlet ducts, gauges, seals, and so on. In other words, in high-vacuum conditions the inlet valve is never truly closed; there is no true zero-flow condition.

For oil-sealed pumps, the second term in Eq. (5.1) is an important part of the total ultimate pressure. Even in a clean application, where the oil does not react with external agents, the oil molecules will fracture due to mechanical shear and localized high-temperature points at contacting surfaces. The rotor of the vane pump, for example, is tightly fitted at the line of contact with the stator. Actually, this is a short cylindrical surface along the entire length of the rotor where the rotor and stator curvatures are equal. This special shape introduced into the stator is made to widen the contact area and thereby increase the impedance between the discharge and inlet spaces in the pumping mechanism. The rotor is also only about 0.003 cm shorter than the stator in small pumps. This implies that the ends of the rotor must be closely perpendicular to the axis of rotation. Small errors in machining, axial forces from shaft coupling, and even warping of the stator during assembly can produce areas at the contact surfaces where boundary lubrication conditions may exist instead of desired hydrodynamic lubrication.

The oil vapors present at the inlet of oil-sealed mechanical vacuum pumps are not due to the equilibrium vapor pressure of the fresh oil but to molecular fractions of that oil, which include light hydrocarbon gases. This can be demonstrated using partial pressure measuring instruments (such as mass spectrometers) and observing the time of arrival of hydrocarbons at the detector when a valve to the pump is opened.

An additional difficulty in establishing the ultimate pressure is due to difficulties with pressure measurements. Until only a few years ago the usual gauges used for testing rough vacuum pumps were McLeod gauges. Because the gas is compressed during the measurement, McLeod gauges do not measure condensible vapors accurately. To eliminate the ambiguity, sometimes liquid-nitrogen-cooled traps are placed between the pump and the gauge. This traps oil vapors (except the lightest hydrocarbon gases) but also traps water vapor, which presumably represents a part in the performance of the pump. At the very low end of the pressure scale, ionization gauges can be used, but they quickly become contaminated by the oil vapors. Diaphragm gauges such as capacitance manometers are useful, but certain precautions have to be taken to measure pressures near or below 1 mtorr.

Further illustration of the difficulty associated with the concept of ultimate pressure can be derived by measuring the performance of the mechanical pump using a gas other than air. A good example is helium because it is used as a

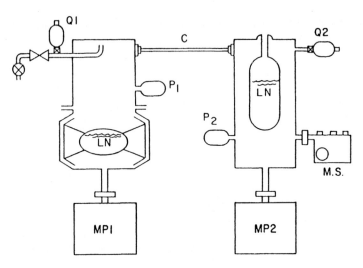

Figure 5.11 System for measuring pumping speed at very low pressures.

tracer gas in leak detection work and as a carrier gas in some analytical instruments. The apparatus shown in Figure 5.11 may be used to demonstrate that the pumping effects do not cease at pressures below the conventionally displayed ultimate pressure. In addition to the liquid-nitrogen traps, the entire vacuum system was assembled with the reduction of outgassing in mind. Using a known conductance of the tube connecting the two pumps, the helium leaks of known flow rate and a mass spectrometer for measuring partial pressure of helium, pumping speeds can be measured at very low partial pressures of helium. The results shown in Figure 5.12 and other evidence obtained from the leak detection technology indicate that the pumping speed is retained at pressures as low as 10^{-10} torr as long as the amount of the gas in question is low at the discharge end of the pump (i.e., as long as the limiting compression ratio is not exceeded).

At higher pressures, helium and air mixtures are pumped by mechanical pumps in the normal manner. Helium is absorbed by the oil and, after a large dose of helium, the pump may produce an elevated background in the leak detector. A sudden ingestion of a few atmospheric cubic centimeters of helium can produce a noticeable background that may need an hour to decay completely. The worst situation occurs if a part pressurized with helium develops a gross leak during evacuation of the vacuum chamber. Following such an event, it is often impossible to find small leaks until the background signal is reduced to an acceptable level.

All mechanical pumps are not identical in their ability to purge themselves from a gross helium exposure. Table 5.1 shows typical results of comparisons

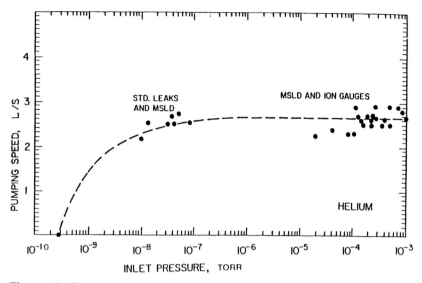

Figure 5.12 Pumping speed for helium at very low concentrations.

made with a fixed test procedure. A simple test system was devised to introduce 100 atmospheric cubic centimeters of gas into the pump. Pure helium and 10% helium mixture in air were used. The leak detector was a contra-flow type (Varian 925-40). Because of manual valves, the initial timing is not accurate, but the general behavior can be clearly observed using the arbitrary unit scale on the leak rate meter.

The information from Table 5.1 is reproduced in Figure 5.13 without showing the pressure bursts. The 10% helium test is more realistic and most pumps will recover reasonably well after about a minute (e.g., pump A, as shown), but 100% helium tests cause bursts in some cases even after 15 min of waiting (pump D). The differences depend on design features, existence of unvented volumes, size of the oil reservoir, and arrangement of oil flow in the pump. The major reason for the lingering helium presence is the absorption of helium in the oil reservoir. This can be demonstrated in a simple experiment by introducing fresh oil directly into the pump. It should be noted that with a fresh oil supply, not only the cleanup of residual helium proceeds at a much higher rate, but the reversals, instabilities, and bursts (typical in 100% helium tests) are completely absent. The indicated differences in behavior are significant not only for the selection of components but also in regard to their interaction under varying operational conditions, valve arrangements and sequencing, and electronic controls.

As noted in Eq. (5.1), the ultimate pressure is affected by internal gas loads. These can be due to leaks, permeability of gaskets, outgassing of surfaces at

Table 5.1 Helium Decay in Mechanical Pumps After a Gross Leak

Time (min)	Pump A (8.5 ft³/min)		Pump B (8.5 ft³/min)	Pump C (5 ft³/min)	Pump D (5 ft³/min)
	10% He	100% He	100% He	100% He	100% He
0	400	4400	3500	5600	6200
1	7	150	100	280	280
2	6	130	60	250	86
3	2	70	50–75	200	200–2000
4	5	120	60	190	78–84
5	6	110	60–70	190	78–84
7	3	64	56–72	140	78–84
10	3	40	42–58	130	60
15	(0)	30	38–52	78	35–1000
20	(0)	19	35–45	54	24–28

COARSE VACUUM PUMPS

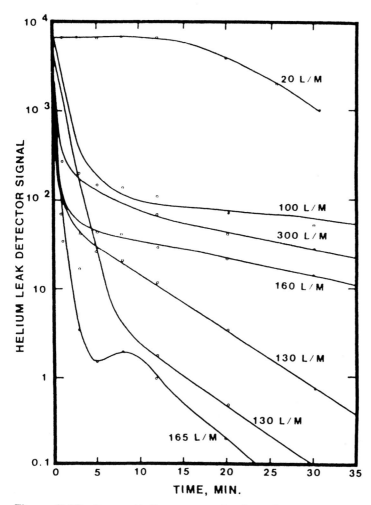

Figure 5.13 Decay of helium concentration for various pumps after pumping a given amount of helium.

the inlet of the pump, and the gases dissolved in the oil. Therefore, memory effects from a previously pumped gas may linger for periods ranging from a few minutes to a few hours. A practical example of this behavior can be seen in pumps used in helium leak detectors (Chapter 13).

5.5 CONDENSABLE VAPORS AND GAS BALLAST

Coarse vacuum pumps that have high compression ratios have a certain difficulty in pumping condensable vapors. Normally, it is the water that causes the problem, but in some applications used in the semiconductor industry, certain process gases can also condense inside the pump toward the end of the compression stroke. Excessive amounts of condensed water vapor produce the typical milky appearance of the pump oil. In this condition the oil does not have sufficient lubricating properties and, due to the presence of the water vapor at the inlet, the ultimate pressure capabilities of the pump deteriorate. An effective solution for this problem was proposed by W. Gaede, who developed several early vacuum pumps at the beginning of the century. It consists of deliberately introducing dry gas (ordinary air will suffice) into the pumping chamber at a later stage of compression. This lowers the partial pressure of water vapor in the mixture of air and water and prevents condensation at the prevailing temperature. Note that if the pump temperature were to be maintained somewhat about 100°C, condensation would not occur. The addition of air has the effect of increasing the pressure at the exhaust side of the pumping chamber. The discharge valve will then open sooner than it would open if the gas ballast were absent. With the gas ballast, the partial pressure of water vapor in the pumping chamber does not reach its condensation level and the vapor is discharged together with the air before condensation can occur.

Usually, the air is introduced through a manually adjustable valve. In double-stage pumps, which are ballasted only in the second stage, fixed valves are often used. The amount of gas ballast air is typically 2 to 5% of pumping speed in single-stage pumps and 10% in double-stage pumps. The volume rate of air is taken at standard conditions and expressed as a percentage of pumping speed (or displacement). The effect of using the gas ballast is an increased inlet pressure, as shown in Figure 5.14. Depending on process conditions, pumps can be operated with the gas ballast activated continuously or used only when required to clean the oil between process cycles.

5.6 OILS AND BACKSTREAMING

The primary requirements for mechanical pump oils are adequate lubrication properties, reasonable viscosity range, low vapor pressure, and resistance to degradation when used with aggressive gases. There is no single oil that can be recommended for all possible vacuum applications. Oils used in mechanical pumps range from mineral oil (not very different from automobile engine oils) to special perfluorinated fluids that cost 500 times as much.

There are several classes of oils. Mineral or hydrocarbon oils have an average molecular weight of 280 to 440 and vapor pressures of 10^{-4} to 10^{-5} torr. They are widely used for general-purpose applications but cannot be used with oxygen and chemically active or corrosive gases. With oxygen, at a pressure higher than about 1 torr, there exists a danger of explosion. With other gases that react chemically with the oil, such as chlorine and fluorine, the oil quickly

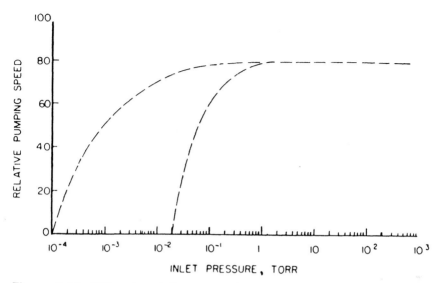

Figure 5.14 Effect of gas ballast on pumping speed.

deteriorates. Its viscosity increases, resulting in plugging of oil passages and consequently in lack of lubrication and high wear.

To improve operation with oxygen and corrosive gases, specially distilled fluids have been used which are generically classified as perfluoroethers and polychlorofluoroethylenes. Their molecular weights range from 800 to 3000 and vapor pressures from 10^{-4} to 10^{-5} torr. Pumps are usually specially prepared for use with hazardous gases. Some of their parts may be coated with corrosion-resistant paints or films and arrangements are made for providing inert gas flushing. Nitrogen, for example, can be introduced either through the gas ballast valve or at the discharge. The oil reservoirs of vacuum pumps are usually exposed to the atmosphere as well as to the exhaust gases. There is a possibility of accumulating explosive mixtures and the creation of a corrosive environment due to the presence of water vapor in the air.

In addition to the possibility of damaging the oil due to pumped gases, the presence of the oil in the pump produces the possibility of migration or backstreaming into the vacuum chamber. Many high-vacuum applications are extremely sensitive to the presence of even minute amounts of hydrocarbon contamination. Examples of these are surface study instruments and various microcircuit fabrication processes. Backstreaming becomes significant in typical high-vacuum installations whenever inlet pressures of less than 0.5 torr are reached. At this pressure, molecular or transition flow conditions are reached in pipes or ducts a few centimeters in diameter (and a few meters long). In the molecular flow range the oil molecules are free to flow in the upstream direction because they do not collide with pumped gas molecules.

The rate of backstreaming at the vacuum chamber depends on pump design, the oil used, the dimensions of the inlet ducts (the longer and narrower, the better, but consider concomitant loss of pumping speed), and the pump and duct temperatures. However, the rate is usually high enough that the variation of a factor of 10 due to the foregoing conditions does not represent a satisfactory range for a selection. Typical measurements indicate backstreaming rates that correspond to oil film deposition at the rate of one monolayer per second (ca. 10^{-5} g/cm^2·min).

There are several ways to reduce this. First is the use of liquid-nitrogen refrigeration. The simplest is to form a U-shaped bend in the duct and dip it into a container filled with liquid nitrogen. This will stop all hydrocarbons with the exception of the lightest, containing only one or two carbon atoms. A more effective method is cryosorption, combining liquid-nitrogen refrigeration with zeolite trapping. Zeolites are porous minerals known as molecular sieves. They have a very large internal surface because their pore sizes are on the order of atomic dimensions (5 to 15 Å). There exists a variety of trap designs, some with a single pass, some with a dual pass through a bed of zeolite pellets, some heated for degassing and then cooled to room temperature, some cooled to liquid-nitrogen temperature. A typical trap operated at room temperature is shown in Figure 5.15. Properly designed and adequately large, such traps do not significantly reduce the pumping speed, at least not in the functional pressure range of mechanical pumps. A comparison of evacuation time achieved with or without zeolites in the trap is shown in Figure 5.16.

Zeolite traps can be very effective when properly maintained, as illustrated in Figure 5.17, which shows the results of a residual gas analysis obtained using a mass spectrometer. The mass spectrum in Figure 5.14a shows a typical grouping of peaks (see Chapter 11) associated with hydrocarbons which are usually 14 mass numbers apart (CH$_2$ group additions). With zeolite in the trap, an improvement can be seen in Figure 5.17b.

In addition to traps used to protect the vacuum system from a possible contamination from the pump, sometimes it is necessary to protect the pump from process-related residual materials that can harm the pump. System designers must evaluate the possibility and estimate the quantity of particles and the amount of condensate that may enter the pump. The damage can be mechanical (due to interference with moving parts), or chemical (due to, for example, a reaction between the residual material and the lubricating oils). Vacuum pumps are not designed to pump liquids. Therefore, if more liquid is produced than can be accommodated by gas purges, a suitable condenser must be employed before the pump and the accumulated liquid periodically drained. Large two- or three-stage traps can be used to capture most of the particles before they

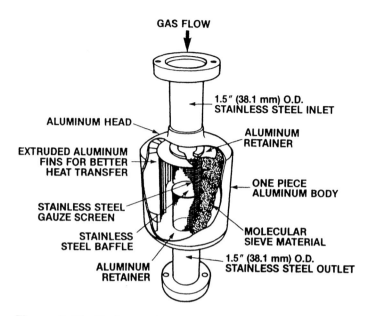

Figure 5.15 Typical zeolite trap.

enter the pump. Such a trap is shown in Figure 5.18. Multiple filter cartridges filled with suitable materials for various processes can be used in the same trap body. The trap shown in the figure has 10 cartridges (5 in each of two rows). The gas first enters the outside cylindrical space, then passes through the filter toward the hollow cylinder inside the cartridges, then enters a header (underneath the middle wall), then passes through the second row of filter cartridges. The choice of the trap size depends on the expected service, but when in doubt, it is always better to opt for the larger trap.

5.7 OPERATION AND MAINTENANCE

Generally, mechanical pumps are rugged, reliable devices. In many cases, all that is needed is an occasional change of oil. In clean applications once or twice per year may be sufficient. Belt-driven pumps which operate at low rotational speeds may last 10 or 20 years under favorable conditions, modern high-speed pumps perhaps 5 to 10 years of continuous operation.

The situation is different in applications where highly reactive gases are present. Oil changes may be required as frequently as once per week, and parts of the pump may even be destroyed once a month. Use of corrosion-resistant fluids is helpful. In recent years, auxiliary devices have been developed to filter the oil, to remove acid-forming compounds, and otherwise to monitor the

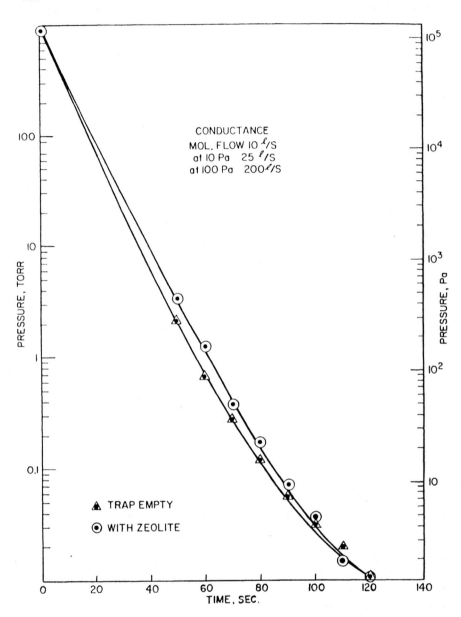

Figure 5.16 Effect on pumping speed with a trap of adequate size.

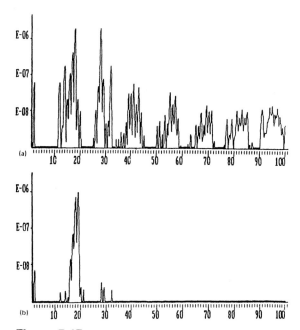

Figure 5.17 Residual gas spectrum at the pump inlet: (a) without trap; (b) with trap.

oil condition while the pump is in operation. For pumping reactive corrosive and generally hazardous gases, it is most important to follow the manufacturer's operating instructions and general recommendations for pumping hazardous gases [e.g., American Vacuum Society publications (Refs. 6 and 7)].

Some precautions should be obvious. For example, if a pump filled with oil is brought inside from a cold winter environment, it should be allowed to warm up before starting. Otherwise, the motor may not be able to turn. Also, before changing the oil the pump should be operated until it reaches normal operating temperature. Otherwise, the drainage will be incomplete.

Large pumps should be fully or partly air-released before they are stopped. Otherwise, the pump may be filled with oil due to the pressure difference. This means that a valve should be installed in the line because it may be desirable to keep the vacuum system under vacuum. When large pumps are started, it is generally better to start them with the inlet valve open; otherwise, the vacuum achieved with the first stroke produces a torque requirement on the motor before it has time to accelerate to a normal speed.

Also, some pumps during startup may eject a slug of liquid oil into the inlet line. A T-shaped duct with the blind leg facing the pump inlet may be used to prevent the oil from reaching the vacuum system.

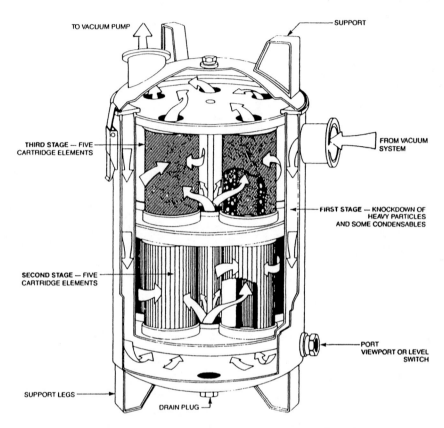

Figure 5.18 A two-tier, ten-cartridge trap for use upstream of rough vacuum pumps (including Roots-type). (Courtesy of Mass.-Vac.)

Mechanical vacuum pumps are usually operated continuously, but in some applications this is not required. When possible, interrupted operation will help in maintaining low temperature, conserve oil, and reduce backstreaming. For example, if a rough vacuum system requires a pressure of 200 mtorr or lower, the pump can be operated until 100 mtorr is reached and then stopped. A valve in the line can be closed at this time and opened again and the pump restarted when the pressure reaches 200 mtorr. Depending on the volume, outgassing rate, size of the pump, and so on, the period of pumping could be a few minutes and the idle period perhaps an hour. A simple automatic system with a solenoid valve can control the alternating sequence using a pressure gauge with pressure set points.

Generally, if backstreaming is significant, it is best not to leave the pump operating at or near the ultimate pressure. Either the inlet valve should be closed or a deliberate gas leak should be arranged to keep the inlet line in viscous flow pressure range.

When Roots blowers are used in series with mechanical pumps, considerations of protective gas bleeds should not be excluded. An unprotected system with a Roots blower may have backstreaming rates two or three times lower than that of the pump itself.

Air or inert gas bleeds are also helpful in preventing undesirable gas concentrations inside the pump and also inside the oil reservoir of the pump. At very low pressures the oil reservoir may not be adequately vented and the presence of atmospheric moisture at the discharge areas of the pump may promote acid formation with accumulated pumped gases.

The performance of oil-sealed mechanical pumps depends on the temperature of operation. Elevated temperatures affect the viscosity of the oil and its lubricating properties. At higher temperatures the rate of the physical degradation of the oil will be increased. Not only will more light oil fractions be produced, but they will migrate at a higher rate if the ducts are at an elevated temperature. It is therefore significant to consider the ambient temperature surrounding the pump. If it is placed inside cabinets, adequate air cooling should be provided.

The rate of degradation depends on the oil and pump design. For example, for typical hydrocarbon-type oils used in vane pumps, backstreaming may double when the ambient temperature increases from 20°C to 25°C, and quadruple at 30°C. Some oils, such as perfluorinated oils (Fomblin or Krytox), will have significantly lesser increases between 20 and 40°C.

This temperature sensitivity is apparent when the same pump is operated at 50 to 60 Hz electrical service. Because of the convenience of using induction motors, the selected operating nominal rpm will be 900, 1200, 1800, and 3600 at 60 Hz and correspondingly at 750, 1000, 1500, and 3000 at 50 Hz. Standard induction motors are available at particular power ratings. Therefore, the choice of the motor power can become a controlling parameter for optimizing the pump performance in regard to its size and pumping speed. If during the design phase, the pump is optimized at 60 Hz and 1800-rpm operation and then taken to a country that has 50-Hz service, the practical choice is to operate it at 1500 rpm.

The situation is somewhat different in the reversed case. A pump optimized for 1500 rpm when operated at 60 Hz may be operated at either 1800 or 1200 rpm. Due to competitive pressures, the temptation is to operate at 1800 rpm, which produces higher temperatures, faster oil degradation, and higher backstreaming rates. Fortunately, in cabinet systems, cooling fans will also run faster and this may alleviate the problem. If backstreaming is important, additional cooling may be advisable. Generally, when zeolite traps are used, they should be placed in the coldest available environment. It is difficult to give generally valid comparisons, but it may be expected that zeolite traps will reduce backstreaming rates by about 10 times and, again, the rate may quadruple if the trap temperature increases from 20°C to 30°C.

When the performance of a coarse vacuum pump is assessed from the performance of the system, in addition to the ultimate pressure (5.7), pumping speed may be of concern. At pressures above about 1 torr, pumping speed can be directly related to the evacuation time (Chapter 4). There are, however, some difficulties. To use the evacuation relationship, the net pumping speed at the chamber must be known. This requires an exact calculation of the pressure drop in the inlet lines. In addition, the pumping speed is not strictly constant but usually decreases with decreasing pressures. To reconcile the pumping speed measured by the American Vacuum Society practices with the speed obtained from the rate of evacuation, the pump must be mounted directly at the chamber using a conical duct (widening toward the chamber) to prevent turbulence at the pump entry. Calculations of line conductances using conventional equations in textbooks on vacuum technology usually produce higher net speed values than those obtained in practice.

Calculated values also produce optimistic expectations when pumping times are extremely short, for example, when pumping load locks of very small volume or during industrial leak detection operations when the entire test cycle may be only 10 s. The dynamic or transient behavior of the atmospheric air inrush into a pump, which is at vacuum, does not correspond to the equations derived at nearly steady-state conditions. Inrushing air may momentarily slow down the pump, blow out oil seals, and change valve action. Large single-stage pumps in such service appear to hesitate less when they have lubrication forced by an auxiliary oil pump. An additional effect is due to the volume of the inlet ducts, which have to be reevacuated after each air exposure. Thus when pumping time is only 5 s, experimental measurements for a given system are highly advisable.

5.8 OTHER COARSE VACUUM PUMPS

5.8.1 Rotary Piston Pumps

Vane pumps are made in the range of sizes from 0.5 to 25 L/s (30 to 1500 L/min). For higher pumping speeds, there is another well-established design, shown in Figure 5.19 in cross section. This rotary piston or plunger pump is similar in principle to cam pumps with a single sliding vane in the stator. The shaft and the eccentric cam are keyed together. As the cam rotates inside the rotary piston, it forces the piston to orbit inside the stator, nearly touching it. The piston slide is rectangular in shape, forming a single body with the piston along its entire length. This piston extension slides inside a cylindrical slide pin

COARSE VACUUM PUMPS

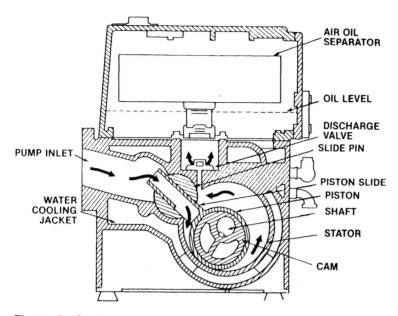

Figure 5.19 Rotary piston pump.

that rocks back and forth as the piston orbits. The arrows in Figure 5.19 indicate the path of pumped gas flow. The inlet and discharge are always separated by the "contact" point between the piston and the stator as the contact point moves around the periphery.

Pumps of this type are usually made in a single-stage version, producing ultimate pressures near 10 mtorr. They are usually machines with less precision than are vane pumps and use oil having a higher viscosity than the vane pump. Sometimes they are made in three separate pistons on the same shaft to achieve better balancing. If lower pressures are required, a smaller pump with its own motor (or connected by belts to the main motor) is arranged in series with the main pump.

Rotary piston pumps are made in sizes up to 100 L/s pumping speed and in dual pistons on a common shaft up to 200 L/s (12,000 L/min).

5.8.2 Roots Blower

When higher pumping speeds are needed, particularly at pressures near 1 torr, Roots blowers are used as a first stage connected in series to other vacuum pumps. Figure 5.20 shows the basic mechanism and the pumping progression. Roots blowers are similar to gear pumps but with only two lobes. To rotate the two rotors in proper angular alignment, synchronizing gears are used in a sepa-

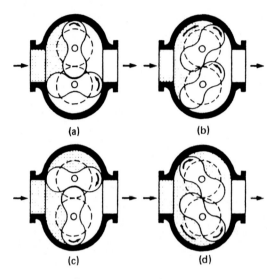

Figure 5.20 Outline of a Roots blower.

rate chamber located beside the pumping chamber together with the bearings. A similar chamber on the other side contains bearings. Shaft seals on both sides isolate the auxiliary chambers from the pumping chamber.

Roots blowers can be used at pressures near atmosphere (and even for compression service), but then they need very large motors, intercoolers to remove the heat of compression, and silencers at the exhaust because of the pulsating exhaust pressure as the rotor delivers the gas to the exit port (see the upper rotor in Figure 5.19c). In high-vacuum practice the blowers have much smaller motors and are used for pumping only below 10 to 15 torr. During the evacuation of a chamber, usually another mechanical pump is used to pump the chamber through a bypass line and lower the pressure to near 10 torr. Then the blower is started. A typical pumping speed performance is shown in Figure 5.21, where it can be seen that not only is the pumping speed much higher compared to the backing mechanical pump, but the ultimate pressure is also lower. Near atmospheric pressure, Roots blowers typically produce only a pressure ratio of about 3, but at the high-vacuum range, this ratio increases to 40 or 50.

To simplify the starting procedure as well as to provide optimum pumping speed utilization, some blowers have a fluid coupling to the motor. This way the blower can be started at atmospheric pressure and gradually reach its normal rotational speed without overheating and without using large motors. Also, to simplify pumping procedures, some blowers have a built-in bypass valve

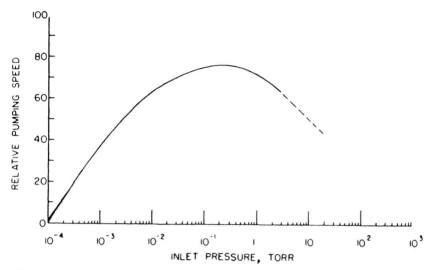

Figure 5.21 Pumping speed with a Roots blower.

between the inlet and discharge regions which is set to open at a prescribed maximum pressure difference. Such blowers can also be started at higher pressures. The limitation of the pressure difference is required to reduce bending of the shafts and other mechanical difficulties. In some installations two Roots blowers can be used in series, providing ultimate pressures in the 10^{-5} torr range.

An example of the bypass valve arrangement (Stokes) is shown in Figure 5.22. The force required to open the valve can be adjusted by the choice of a suitable spring so that the operation is automatic. The permissible maximum pressure difference across the rotors is maintained even if the motor is large enough to turn the machine at high inlet pressures and when the backing pump is too small to remove the gas delivered by the blower and keep the blower discharge pressure adequately low.

In transient situations (for example, if the blower is used for rapid evacuation of a relatively small volume starting at atmospheric pressure), it is possible that the shaft seals in the blower may leak outward and pressurize the side chambers containing bearings and gears. At the end of the evacuation process, this excess pressure may drive the lubricating oil from the sides through the shaft seals into the inner pumping space. In such cases, small pressure-equalizing tubes are used to connect the side chambers to the discharge side of the blower.

Depending on the volume of the vacuum chamber and evacuation time requirements, a variety of Roots-type blowers and backing pumps can be used.

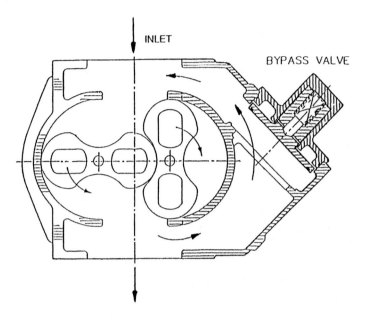

Figure 5.22 A Roots-type blower with a pressure-actuated bypass valve.

For small pumps, the pumping speeds of the blower and the backing pump may be equal (staging ratio 1:1), but this is impractical in large installations because of the requirement of very large motors and the additional expense of large backing pumps. A typical staging arrangement is shown in Fig. 5.23. Often the blower is started by a signal from a pressure switch at inlet vacuum chamber pressure between 10 and 20 torr, as indicated by the dashed vertical line in the figure. Depending on the relative size of the vacuum chamber, the pumping speed of the blower, and the staging ratio of the blower/backing pump train, the blower can be started at a higher pressure. It will reach its full rotational speed in time to be in a safe pressure range as illustrated by the shaded area in Figure 5.23. Because of the possibility of many variations, especially when faced with unusual applications, it is prudent to review the selections with a competent system manufacturer before a final design is established.

5.8.3 Ejectors

Ejectors are devices that produce a pumping action by transferring energy from one fluid to another. They are often called Venturi pumps (Figure 5.24). The motive fluid can be air, steam, or oil vapor (Chapter 6). Pressurized air accelerates entering the narrow section. This, at first, may appear counterintuitive,

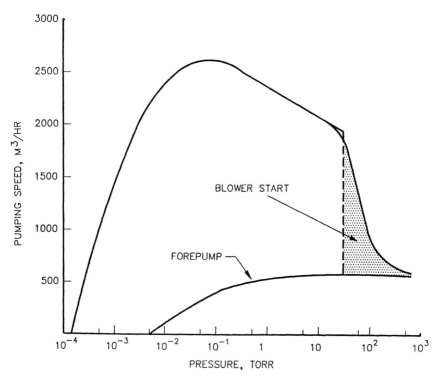

Figure 5.23 A typical inlet pressure region for starting Roots-type pumps.

but assuming a noncompressible fluid and considering preservation of mass flow, it must be apparent that the velocity must increase in a narrowing duct and decrease again at a greater cross section. Considering the total energy in the moving fluid, following Bernoulli's principle, it will be seen that pressure is low when velocity is high, and vice versa. In other words, the potential energy associated with high pressure is exchanged with the kinetic energy associated with the high velocity, but the total energy remains constant.

Typically, in air ejectors the motive fluid, the pressurized air starts at 60 to 100 psi (4 to 7 atm) and is finally discharged to atmosphere. The pressure in the narrow section is typically 10 to 15 times lower than atmosphere. Thus a small air ejector with a pumping speed of a few liters per second can evacuate a bell-jar chamber to about 60 torr in a few minutes.

Multistaged ejectors can be also used, but they require excessive amounts of pressurized air for operation. For example, when large amounts of water vapor must be pumped, five- or six-stage steam ejectors can be used to achieve pressures below 0.1 torr.

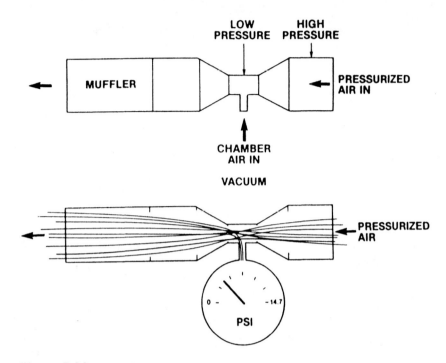

Figure 5.24 Venturi pump.

5.8.4 Liquid Ring Pumps

A liquid ring pump consists of a bladed rotating impeller placed eccentrically inside a cylindrical housing (Figure 5.25). The impeller blades almost touch the housing at the top. An amount of liquid (often water) is kept in the pump such that during rotation the liquid is forced against the cylindrical housing by centrifugal forces, and it provides a barrier between the inlet and discharge ports. The ports are placed at the end plate.

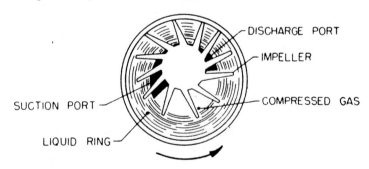

Figure 5.25 Liquid ring pump.

COARSE VACUUM PUMPS

Such pumps can be useful for pumping water vapor or other vapors when a compatible liquid of reasonable viscosity is used in the pump. These pumps typically produce 10 to 30 torr ultimate pressure, depending on water temperature. This ultimate pressure can be lowered by using an air ejector in series with the liquid ring pump. The motive air comes directly from the atmosphere and is discharged into the pump. About half of the pumping speed is sacrificed, but the pressure is reduced to a few torr.

Liquid ring pumps are rarely used in the high-vacuum industry because their ultimate pressure is poor and they are inefficient from the point of view of power requirements. About 90% of power is dissipated due to friction losses.

5.8.5 Sorption Pumps

Degassed zeolite pellets (Section 5.6) cooled to liquid-nitrogen temperature can hold large quantities of condensible gases, the amounts by weight being comparable to the weight of the zeolite. This is a very simple system that has been used successfully for evacuating vacuum chambers, which are subsequently pumped by utlra-high-vacuum pumps (Figure 5.26). The operation proceeds as follows. Both bottles filled with zeolite pellets are prechilled for about 20 min. Then the valve at one of the pumps is opened. An inrush of condensible gases will carry all other gases in a mixture into the pump. Before the noncondensible gases can migrate back into the system, the valve is closed and the second valve is opened. The ultimate pressure after such an operation may be as low as 10 mtorr, depending on the amount of gases pumped previously, but with careful operation using three pumps, pressures of a few millitorr can be obtained.

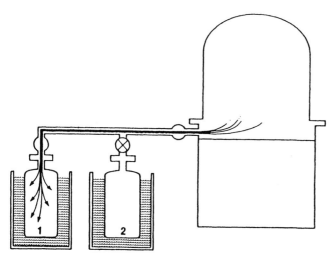

Figure 5.26 Vacuum system with sorption pumps.

An automatic pumping system of this kind (Varian–Megasorb) can hold over 1.5 million torr·liters of air and has a pumping speed comparable to that of medium-sized mechanical pumps.

The vapor pressures of common gases are shown in Figure 8.1. At the temperature of liquid nitrogen, which happens to be a relatively convenient and inexpensive method of cooling, nitrogen and oxygen are not condensed. However, the addition of zeolite and the process of physisorption have the effect of lowering the temprature about 30°C. Then only helium, hydrogen, and neon are not pumped. Thus if the sequencing of two or three sorption pumps is not carried out properly, the vacuum system may contain primarily these light gases at the end of evacuation.

5.9 OIL-FREE VACUUM PUMPS

The demand for a higher purity vacuum environment has precipitated a number of attempts to develop pumps that do not contain any liquids, either as lubricants or as motive fluids. Although it is possible with careful design and operation to prevent hydrocarbon contamination from oil-sealed forepumps and vapor jet pumps, the mere presence of oil or oil-like substances in the pump provides the opportunity for backmigration, especially as a result of a malfunction. The problems of maintaining inlet traps, and the chemical reactions between the oil and aggressive gases used in many vacuum processes became annoying enough to seek other methods.

Oil backstreaming occurs when molecular flow conditions are approached. In typical mechanical pump installations, this occurs at pressures below about 0.1 torr depending on the length and the diameter of the duct, its temperature, and the type of oil. Typical values (from Ref. 9,11) are shown in Figure 5.27. These amounts are not negligible and many applications require inlet traps, gas purges, or both, to prevent oil vapors from reaching the work chamber.

Many recently developed processes used by the semiconductor industry (plasma etching) employ and produce corrosive gases as well as solid by-products which must be handled by a vacuum system. Thus, the pump must be protected from the accumulation of particles, which may be abrasive, the deterioration and depletion of the oil must be controlled, and the dangerously reactive and toxic by-products must pass safely through the pump (to be later disposed of properly). This implies external oil filtration, inlet traps, venting or purging to prevent accumulation, monitoring the condition of these accessory devices, etc., as well as the use of expensive, less reactive oils. The associated need for maintenance, which can be troublesome and very costly, led to the development of entirely oil-free pumps or, at least, pumps in which oil is not intended by design to be present in the pumping spaces of the pump.

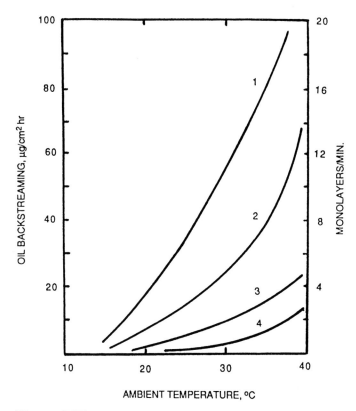

Figure 5.27 Typical values of oil backstreaming. 1-pump unprotected; 2-with a roots pump; 3-mechanical pump with a trap; 4-pump with roots and trap.

During the last ten years a variety of partly or fully oil-free pumps have been introduced and described in technical journals. It is practically feasible to produce pressure ratios near 100,000 in a self-contained single device discharging to the atmosphere without using any conventional lubricants in the vacuum environment. Thus a pressure near 10 mtorr can be produced, which is entirely sufficient for the initiation of high-vacuum pumping devices. Most of the lesser known mechanisms are discussed briefly in the next section.

5.9.1 Rotary Compressors

Roots blowers (Section 5.8.2) are not the only positive-displacement pumps (or compressors) that can be used for high-vacuum service. There are a variety of similar devices that can be adapted for vacuum service. For example, instead of only two lobes, it is entirely practical to use three (or more) lobes in the rotors. This tends to produce a higher pressure ratio at the expense of the reduction in volumetric capacity (pumping speed). Also the shapes of the lobes do not have to be identical in both rotors. A variety of intermeshing cams can be employed. For example, by using multistaged three-lobed Roots blowers, pressures reaching into the low-millitorr range can be obtained while discharging into atmospheric air. Three to five stages separated by walls can be arranged on the same shaft in a compact housing. Because the rotor synchronizing gears as well as the bearings must be lubricated, the freedom from oil vapors in such blowers depends on the integrity of the seals separating the pumping chamber from the gear and bearing sections.

Another essentially similar device is the claw compressor, which is sometimes combined with Roots blowers in a series arrangement, for example, a four-stage arrangement on a common shaft of two Roots lobes followed by two stages of claw compressor lobes.

The claw compressor produces comparatively higher compression ratios but less pumping speed. The inlet and discharge are at the end of the unit rather than arranged laterally as in Roots pumps. The rotors (Figure 5.28) are cylindrical through about two-thirds of the cylinder but have a depression followed by a protruding claw. The rotors almost touch at their cylindrical surfaces, producing a seal in the center. The claw, which almost touches the stator surface, enters the depression in the mating rotor, and vice versa, Figure 5.29. In three stages this pump can produce an inlet pressure of 50 mtorr (7 Pa) while discharging to atmosphere. The pumping speed peaks at about 1.5 torr (200 Pa) and is reduced at atmospheric pressure to about 60% of its peak. At atmospheric pressure, a 90-m^3/hr (25-L/s) pump requires a power input of nearly 8.2 kW (11 hp).

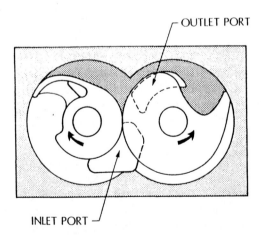

Figure 5.28 Claw compressor rotors.

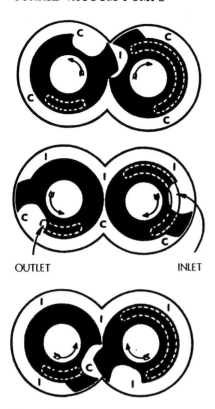

Figure 5.29 Pumping mechanism of synchronized claw-type rotors.

As in Roots pumps, the two rotors are synchronized by timing gears that are lubricated by oil. There are seals between the gearbox and the pump as well as external shaft and bearing seals. The gearbox is evacuated to prevent leakage and the driving of lubricants into the pumping space. Pumps of this type are operated at relatively high rotational speeds (to obtain higher pumping speeds). This necessitates water cooling due to heat produced at bearings, seals, and the gas compression at the high-pressure stage. Typical performances of five-stage Roots-type pumps of four different sizes are shown in Figure 5.30.

Pumps of this general class are made mainly for process gas pumping associated with the semiconductor industry (by Alcatel, Edwards, Leybold, Stokes, and others). An example of the arrangement of rotors for a four-stage vertical claw pump is shown in Figure 5.31 (Leybold). The cross section schematic of this pump and an example of interstage pressure distribution are shown in Figure 5.32 (with a gas at the inlet at 0.75 torr). When high pumping speeds are required, the multi-stage Roots or claw pumps are often used with a sepa-

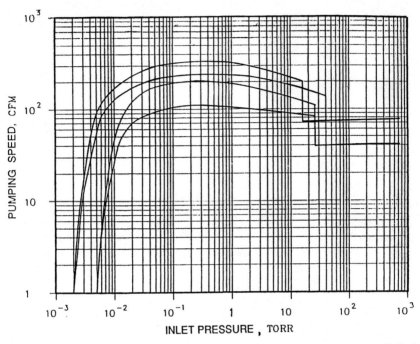

Figure 5.30 Pumping speeds of four different five-stage Roots-type pumps (Stokes).

rate full-sized Roots pump connected in series and arranged in a combined system structure.

5.9.2 Particle Formation

Semiconductor materials processing steps often involve corrosive gases that can produce particles, resulting from chemical reactions, and internal condensation as a result of internal compression. To counteract the formation of internal solids and liquid materials inside the pump, special gas purge and temperature control arrangements are used. Generally, particles tend to agglomerate at the higher pressures because of higher collision frequencies. Also, the last stage of the pump may introduce water vapor from the atmosphere into the pump during the exhaust, unless the exhaust lines are protected by an additional dry air (or nitrogen) venting system.

The bulk velocity of gas flow in pumps of high compression is naturally reduced in the downstream stages. A gas purge increases this velocity, which may help to flush small particles out of the pump. The situation with larger particles is not entirely clear, not only because larger particles have higher terminal velocities (due to gravity), but also because the force of impact of particles on

Figure 5.31 Rotor arrangement of four-stage claw type pumps (Leybold).

a solid surface is proportional to $\rho \cdot V^2$ (ρ, density and V, velocity), which may increase the possibility of sticking. Therefore, it may be necessary to introduce purge gas already in the second stage of the pump to prevent further particle agglomeration. When purge gas amounts and locations are properly arranged, the effect on the ultimate pressure is minimal (see Figure 5.33).

The selection of gas purges, the amounts of purge gas flows required in various stages of the pump, the timing of purge sequences, and the temperature control must be made with the knowledge and experience of process requirements. The same can be said about maintenance and periodic cleaning schedules. The arrangement of exhaust lines should be made carefully so that the accumulated or deposited materials (liquid and solid) do not find their way into the pump. Even the simple admonition given in Figure 5.34 can be very helpful.

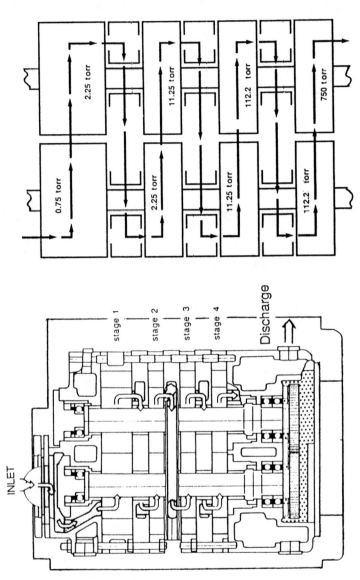

Figure 5.32 Staging arrangement and an example of pressure distribution for a four-stage claw pump (Leybold).

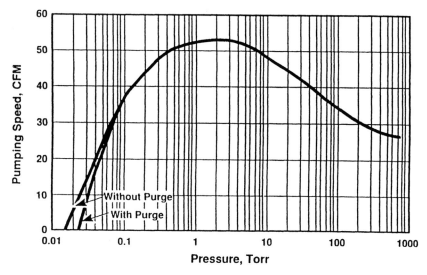

Figure 5.33 Ultimate pressure effect of purge gas flow for a four-stage claw pump (Leybold).

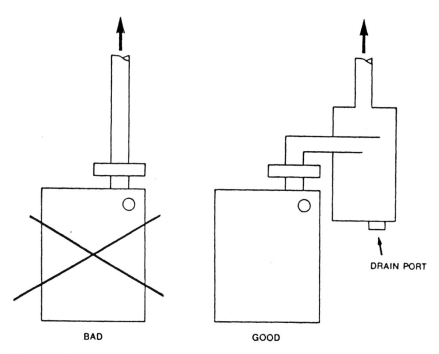

Figure 5.34 Preferred exhaust duct arrangement for rough vacuum pumps (Ref. 11).

This subject, however, is better served in the realm of system-, plant-, and safety engineering rather than vacuum technology.

Many microcircuit manufacturing processes must also be protected from the presence of particles in the vacuum chamber, which may deposit on the process substrates. Particles can be introduced into the vacuum system in many ways, for example, from atmosphere during an air release. One item that received some attention is the effect of high gas velocities due to high pumping speeds during the initial evacuation.

High gas densities and velocities existing at the beginning of evacuation can dislodge particles, which may fall on or be brought to the substrate surfaces. This observation may lead to the ironic claims that small pumps are better than large pumps. This however, is a rather complex subject. The movements of particles can depend on many things including the pumping system and vacuum chamber design.

The important parameters that can influence the flow patterns inside the vacuum chamber are:

Density or pressure
Mass flow rate
Velocity (magnitude and direction)
Nature of flow (turbulent or laminar)
Temperature (aerosol formation)
Flow pattern (geometry)
Existence of convection currents
Local conditions at surfaces (characteristics of boundary layers)

Some of these parameters are interrelated and can be expressed in terms of ratios such as, for example, the Reynolds number (ratio of inertial to viscous forces in the fluid stream). Some parameters may be of significance in the independent and absolute sense (threshold magnitude).

From the data in the technical literature, some examples compare fast and slow roughing. In some cases, the ratio between particle counts favors slow pumping by a small factor, but in other cases it is as much as a factor of 100. Slow venting has been noted to be more important than slow pumping and the suggestion has been made to vent through multiple orifices to achieve the same venting rate.

For load-lock pumping (Ref. 40), it was demonstrated that in one case the particle contamination was unchanged when the pumping port orifice area was changed by a factor of 10.

The idea of multiple ports can also be applied to pumping. Multiple "orifices" are simple to make by interposing a perforated shield between the chamber and the roughing port. The geometry of these perforations is important. The orifices can be shaped as smooth nozzles, which greatly reduces the velocity at the center of the stream.

It is common in discussions of slow (soft) rough pumping to specify the requirement for the Reynolds numbers for the flow conditions in the gas. When Reynolds numbers are mentioned in conjunction with attempts to delineate the conditions for laminar and turbulent flow, however, it is not clear which characteristic dimensions are used. In the same chamber, there may be several regions of interest having different Reynolds numbers. A heated wire may have a characteristic dimension of 0.020 cm, the roughing port 3 to 10 cm, the chamber diameter 50 to 100 cm, etc. Local conditions are important for both the creation and deposition of particles, and even the benefit of having laminar flow is not always clear. For example, using tubes of 1 cm diameter and an air flow with Reynolds numbers of 5,000 to 10,000, P. Rossi notes that experimental and theoretical data indicate that the least amount of deposition is obtained *with turbulent flow* (Ref. 41). The experiments were conducted at rather low velocities, 1 to 50 m/s. There were four main mechanisms of deposition identified: diffusion (Brownian motion), which is the principal mechanism for particles smaller than 0.1 micrometers; sedimentation; inertial deposition (for particles greater than 1 micrometer); and electrostatic attraction. It was observed that, for flow with Reynolds numbers above 10,000, turbulence was very high and impaction of particles occurred. The corresponding velocity was near 40 m/s. Other authors (Ref. 35), also indicated that Brownian motion is a significant factor in the deposition of submicron particles.

At a partial vacuum in a chamber, particles fall (precipitate) due to gravity, depending on their size and density, and the pressure (or density) of the remaining gas (see Figure 5.35). In some cases it may not be beneficial to elongate the evacuation process. The gravitational velocity may be significant if the rough pumping period becomes too long. Timing can be important for sweeping away particles.

At higher pressures, submicron particles are only slightly influenced by gravity. Ultrafine particles follow the mass flow of gas and at higher velocities, deposits can be formed (Ref. 36). The actual values depend on particle size and gas properties.

It would appear that one of the more important parameters is the velocity, rather than the Reynolds numbers as such. The dislodgement and deposition of particles must be associated with forces of interaction between the gas flow and the surfaces. The mechanical forces at least should be proportional to the density of the gas and the velocity. The prominent effect of the velocity should be apparent.

Generally, the gas velocities in typical bell-jar-sized vacuum chambers are not very high during the initial evacuation. For example, using a 15 cm (7 L/s) roughing pump and 1.5 inch (3.8 cm) roughing line, and dividing the volu-

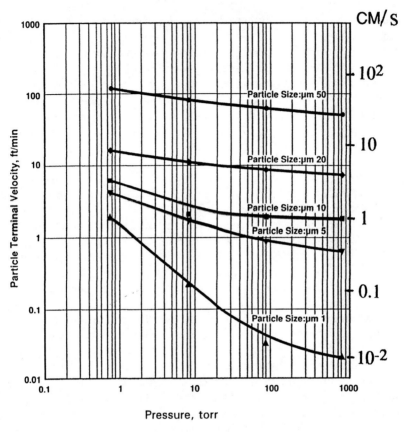

Figure 5.35 Terminal velocities of falling particles of different sizes at various pressures.

metric flow over the cross sectional area, we obtain a velocity of about 6 m/s. This can hardly be considered a high velocity in regard to particle dislodgement and redeposition. Inside the vacuum chamber, where the areas are much higher, the velocities would, of course, be much lower.

In a small chamber (such as a load lock), however, the velocity may be very high because of the sudden communication of the chamber with the duct, which has already been evacuated. The transient pumping speed in this case may be ten times higher than the speed of the pump (for a few seconds). Compromise solutions may be obtained by locating the valves further downstream, by designing the roughing line geometry to prevent turbulence, and to reduce velocities (for example, multiple ports). One may also open the load-lock valves slowly.

One should distinguish between real turbulence (high Reynolds numbers) and the "turbulence" in the sense of basically laminar circulatory patterns of eddies which can develop as a result of convection currents or simply due to interplay of velocity and geometry of the flow path.

Particle agglomeration can occur even when calculated Reynolds numbers indicate laminar flow. Strong convection currents can develop in vacuum chambers at pressures near 25 torr. Depending on geometry, convection currents can form eddies which can be sites of particle production due to stagnation conditions in the center of rotation.

Aerosol formation during rapid evacuation of vacuum chambers has been experimentally demonstrated, usually in the presence of water vapor (Ref. 28,29). This occurs due to cooling during expansion, for example, temperature dipping to near $-50\,°C$ for a 48 liter chamber (with 55% relative humidity) between 100 and 1 torr and 5 to 15 seconds after pumping start (characteristic pumping constant $V/S = 2.7$ second). A corresponding sudden increase of particle concentration began at 500 torr, peaked near 300 to 400 torr, and was eliminated near 1 torr. To prevent the formation of water aerosols, the usual recommendation is to purge the vacuum chamber with dry nitrogen and use slow pumping at the start. Also, the purge gas may be preheated and filtered.

If sources of particle formation and causes of their deposition can be identified and the situation corrected, there may be no need to accept the rather significant sacrifice of slow rough pumping, which can significantly elongate the evacuation process (Ref. 43).

5.9.3 Gas Mixture Effects

In high-vacuum work the pumped gas is usually a mixture, such as atmospheric air. Truly oil-free pumps do not add significant amounts of internal gas evolution to the main flow. In a pressure range where internal leakage is insignificant, positive displacement pumps produce approximately the same pumping speed and the same compression ratio for noncondensable gases. When the pumps exhaust to atmospheric pressure, however, the ultimate partial pressure for a given gas depends on the discharge pressure of that gas (Eq. 5.1). Therefore, the usual pumping speed curve can be represented as in Figure 5.36 for a pump that has a compression ratio of 40,000. A corresponding evacuation curve of a reasonably sized chamber would appear as shown in Figure 5.37. The curvature for the water vapor line is produced by the gradually reducing outgassing rate.

Despite the lower compression ratios, in comparison to oil-sealed pumps, the dry pumps have the advantage of not requiring time for the sealing oil to degas after each exposure to a higher pressure. This is reflected in Figure 5.38,

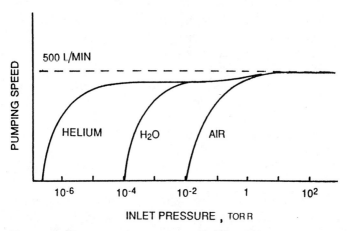

Figure 5.36 Pumping speed for different gases, depending on their usual presence at the exhaust.

where a smaller dry pump takes more than twice the time to reach the pressure where turbopump is started but does not increase the overall evacuation period. This is because the discharge pressure of the turbopump have less lighter gases which affect the pressure ratio of the turbopump.

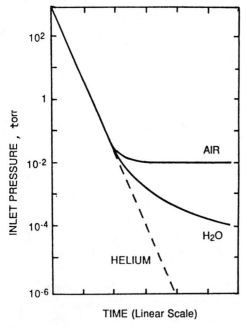

Figure 5.37 Evacuation curves corresponding to Figure 5.36.

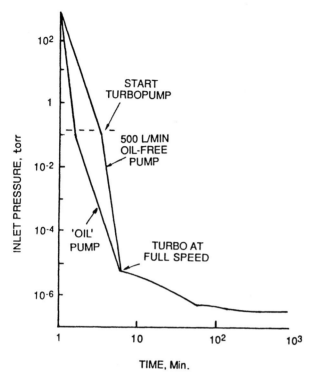

Figure 5.38 An example of evacuation time with an oil-sealed and an oil-free pump.

5.9.4 Screw-Type Pumps

Another device is the screw compressor, in which the two meshing rotors look like helical gears. The two rotors almost touch, and the pumping direction is axial. If the rotors are long enough (longer than the pitch of the screw), the inlet and discharge are always separated. Screw compressors are used as air compressors and can be used for vacuum service. Their performance will be similar to claw devices and perhaps have no significant advantages. They are more difficult to multistage because the rotors are long, but they have high pressure ratios (Figure 5.39).

There are several different shapes of rotors that may be used in screw-type pumps (see Figure 5.39). The screw pump rotors do not need to be rotated at the same rotary speed (rpm). Like gears, they can be synchronized to rotate according to the number of protrusions and grooves. For example, in Figure 5.39 the top rotor has five helical protrusions but the lower rotor has only four helical grooves. In Figure 5.40, the cross section schematic shows a typical arrangement of rotors supported on both sides. The synchronizing gears are on

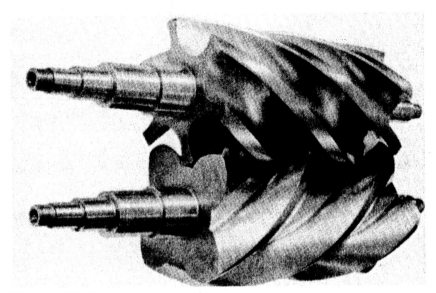

Figure 5.39 Screw compressor rotors.

the left size and it can be seen that they have different diameters. Two additional rotor shapes are shown in Figure 5.41. The upper pair is supported on both sides while the other is cantilever supported, having two sets of bearings at the discharge side of the pump Figure 5.42. Manufacturers of screw pumps often claim that they are single stage pumps. Actually, it is more sensible to

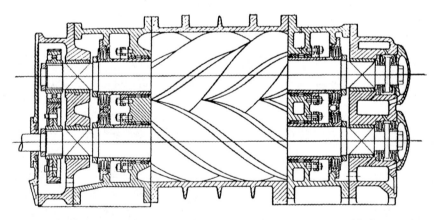

Figure 5.40 Schematic cross section of a screw-type compressor (timing gears at left).

COARSE VACUUM PUMPS

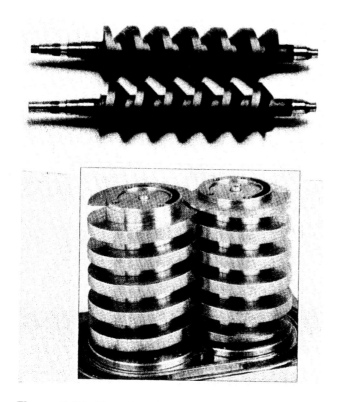

Figure 5.41 Examples of rotors for screw-type pumps (Busch; Kashiyama).

consider each completion of the pitch of the screw as a stage. Then, they have nearly the same general compression performance per stage as other multistaged dry blowers.

Screw pumps have an advantage of certain mechanical simplicity because separating walls between stages are absent. The pumping speed versus pressure performance of screw pumps resembles that of any mechanical pump, with the speed usually being constant, starting at atmospheric pressure and decaying at the ultimate presssure, which is near 10^{-2} torr.

5.9.5 Reciprocating Piston Pumps

The major advantage of rotary pumps is that they can be operated at high rotational speeds, giving high pumping speed for a given size and cost. Because in an oil-free pump the rotors are difficult to cool, the burden shifts to the

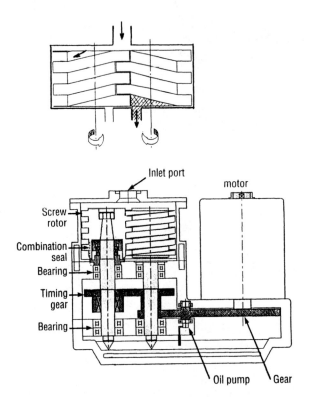

Figure 5.42 Screw type pump showing two sets of bearings at the discharge side of the pump.

design of bearings and shaft seals. Even if the rotating parts do not touch each other or the stator, heat will develop in bearings and seals. Heat will also develop due to gas compression, and it has to be removed by cooling the shaft, which may involve mechanical complications. The reciprocating devices are easier to cool because the pistons and cylinders are exposed to ambient air on one side.

A four-stage reciprocating piston pump with an overall compression ratio of 50,000 was introduced ten years ago. Its basic design features are given in Figure 5.43. It consists of four pistons of equal size, two connected in parallel to obtain a higher speed. The last piston is double-sided (i.e., it has two stages of compression). All pistons are stepped and have an atmospheric lip seal at the end of the small-diameter section (Figure 5.44). The back sides of the

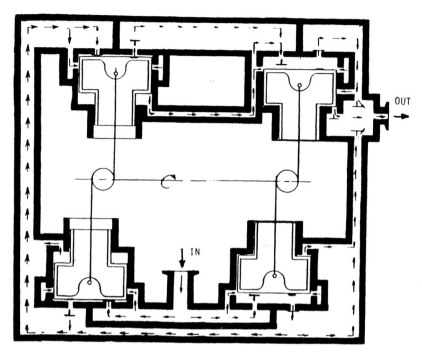

Figure 5.43 Four-stage reciprocal piston pump arrangement.

pistons (except the final exhaust piston) are pumped by the next stage to reduce the effect of leakage through the lip seals. All pistons have a dual-valve system. At higher pressures the valves function as in ordinary compressors, but as the pumped gas becomes rarefied, the inlet occurs through narrow slots in the cylinder placed at the end of the piston travel; the exhaust valves are then opened by small elastomer bumpers or lifters. Both cylindrical surfaces are lined by a polymer materials having a low coefficient of friction and a very low wear rate. The shape of the pumping speed curve and the base pressure are essentially identical to those for single-stage oil-sealed mechanical pumps. Such pumps produce final pressures near 10 mtorr, producing a compression ratio near 100,000. This limit is only due to practical considerations. At least another order of magnitude can be achieved by lengthening the stroke, increasing the rotational speed, and improving the atmospheric seals. With a compression ratio of 10^6, such a pump can equal the performance of double-stage oil-sealed mechanical pumps in most applications.

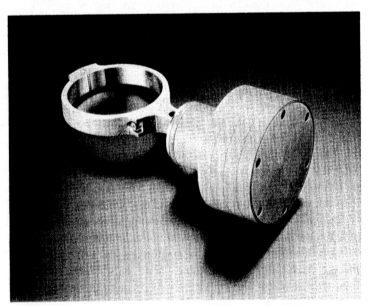

Figure 5.44 General view and piston assembly of the four-stage (four-piston) pump.

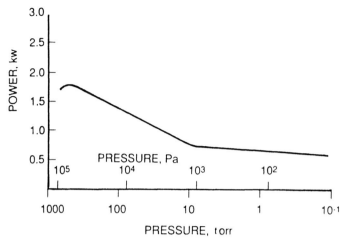

Figure 5.45 Power consumption at various inlet pressures for a 500 L/min pump.

Despite the absence of oil, the temperature of the pump is lower than that of the oil-sealed pumps. Only about 200 W is dissipated in the pump at low inlet pressure. Figure 5.45 indicates that the choice of the motor depends on the desired application and the need to keep temperatures low to avoid the complications of water cooling. Because a 1-hp motor was chosen, it is evident from Figure 5.45 that for continuous operation, the inlet pressure should be limited to 30 torr. Similarly, the motor size also limits the maximum size of the vacuum chamber that can be evacuated. Figure 5.46 shows the evacuation progress up to a 1000-L chamber, which is the recommended maximum. In determining the permissible inlet pressure in continuous (or prolonged) operation, the temperature of the internal parts must also be considered. Figure 5.47 shows a temperature history of the internal valve during the evacuation of the 650-L chambers. This temperature history, during the start of the pump, demonstrates the importance of distinguishing between transient and steady-state operations and the timing of evacuation as it relates to the size of the vacuum chamber. For vacuum pumps, temperatures can also be too low when there is the possibility of internal condensation. Most pumps have a threshold for quantity of pumped water vapor, above which condensation may occur. An example of the amount of gas purge required to prevent internal condensation for the piston pump is shown in Figure 5.48.

In devices that rely on the performance of sliding surfaces either at shaft seals or piston liners, tribological conditions at sliding surfaces have to be chosen

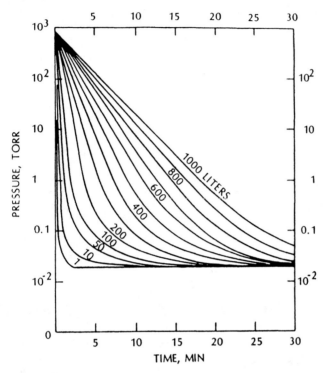

Figure 5.46 Evacuation time for various volumes (500 L/min pump).

carefully. It is in this context that the advantages and disadvantages of rotary and reciprocal devices become apparent.

5.9.6 Scroll pumps

A scroll compressor is essentially a positive displacement device. In a way, it is related to rotary piston compressors but the pumping action is produced by an orbiting motion rather than a rotation. The basic mechanism was known as early as the 1920s, but only in the last 20 years has it been adapted for use in refrigeration compressors. Its advantages are ease of multi-staging, absence of valves, and relatively smooth action. The arrangement of pumping mechanisms is shown in Figure 5.49.

The scroll mechanism can be placed somewhere between the screw pump and the diaphragm pump. It generally has a relatively high pressure ratio but low volumetric displacement. It consists of two involute spiral scrolls attached

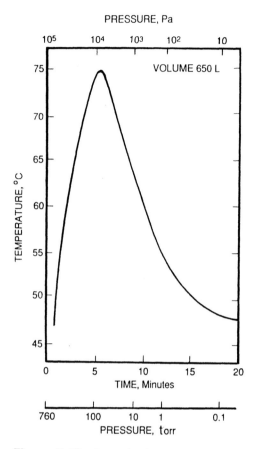

Figure 5.47 Internal valve temperature during evacuation.

to disks and placed into each other in such a way that one orbits relative to the other (see Figure 5.49). The orbiting motion forms crescent-shaped spaces into which the pumped gas enters and becomes isolated from adjacent spaces as soon as the entrance lip of the scroll closely approaches the other scroll surface, as shown in Figure 5.50. Usually, one scroll remains stationary (attached to the pump body), while the other is orbited by an eccentric crank on the motor shaft. The scrolls do not touch but are kept close to each other to prevent backward gas leakages. To create an orbiting motion (and prevent touching), the scrolls are kept apart by either a special cross-grooved coupling, or by two or three additional cranks with the same orbiting distance. For some special designs,

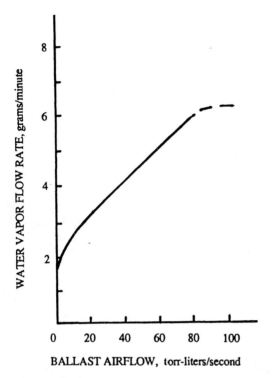

Figure 5.48 Water vapor tolerance depending on the amount of purge gas (500 L/min pump).

both scrolls may be rotated but one displaced (relative to the other) to create an orbiting motion (co-rotating scrolls). As the orbiting motion continues, the gas is moved toward the center of the scroll while the crescent-shaped volumes are gradually reduced in size compressing the gas, which is then discharged at the center. Pumping speed and achieved compression level depend on the orbiting distance and the number of involute wraps. Pumps designed for high-vacuum applications achieve compression ratios between 10 and 20 per wrap, depending on the quality of sealing between the scrolls. To reduce backward leakage, the free edges of both scrolls have a groove into which a "gasket" of low friction material is placed and backed by an elastic pad. This sealing surface is allowed to touch the bottom surfaces of the opposing scroll.

The scroll pumping space can be arranged singly or dually (180° apart), depending on the position of the last external wrap. This is a matter of the de-

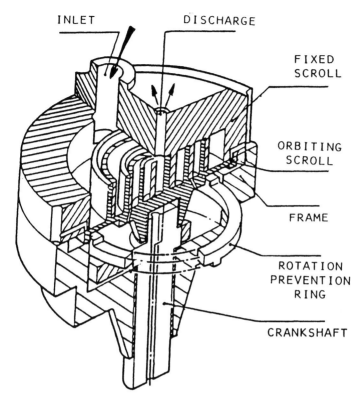

Figure 5.49 A schematic cut-away view of a scroll compressor.

signer's preference, which can produce more pumping speed or more compression. It is also possible, for high-vacuum use, to have double-sided orbiting scrolls (and two stationary), and have different configurations on the two sides in series—one made to give a higher pumping speed and the other to give a higher compression.

Two different versions of scroll pumps have been used in high-vacuum industry. One (developed by Normetex, France) employs a scroll device as the first stage followed by diaphragm-type backing pump. This pump was originally developed for pumping hazardous gases, such as tritium, in which case the pumped gas must be isolated from outside environment and any leakages prevented. Therefore, nutating bellows are used to connect the pump to the motor with a crank mechanism and, in addition, the oscillating spiral with its plate is attached to a large bellows with the other end attached to the stationary base

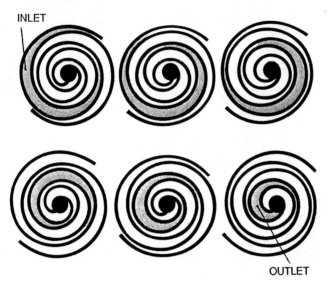

Figure 5.50 Progress of a captured gas pocket (shaded area) during a compression cycle for a dual entry orbiting scroll mechanism.

(see Figure 5.51). Using metallic seals for internal and external joints and metal bellows, such pumps can be made to expose only a metallic surface to the pumped gas. Because the backing pump tends to have a lower pumping speed, the speed of such pumps is low at atmospheric pressure, has a peak near 3 torr inlet pressure, and falls off sharply on both sides of the peak (Figure 5.52).

The other version (Iwata), is made to discharge directly to atmosphere as an ordinary rough vacuum pump (Figure 5.53). It may have compression ratios of the order of 10^5, as long as the seals remain in good condition. (They may need to be replaced once a year or more frequently.) This is impressive for a dry pump but the compression ratio is 10 to 100 times lower compared to oil sealed pumps. This difference may be of no significance for an ordinary roughing pump but should be considered when partial pressures of certain gases are involved (for example, in leak detection working with helium). Figure 5.54 shows the usual pumping speed performance and contains no surprises. Figure 5.55 shows an example of an evacuation progress for a 100-liter volume with a nominal 310 L/m scroll pump.

Scroll pumps have a few intrinsic advantages: gas is compressed at a steady rate without pulsations; the pressure difference between adjacent pumping pock-

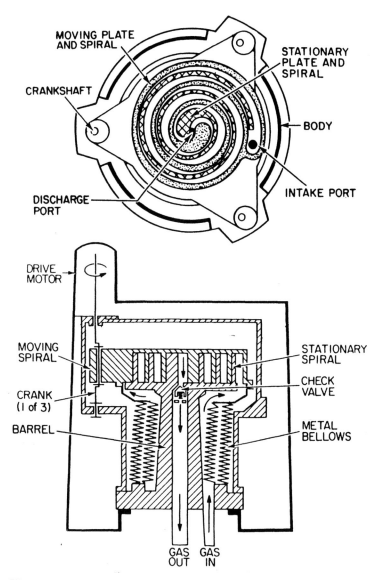

Figure 5.51 Scroll pump design for isolation of the pumped gas from outside environment, (Normetex).

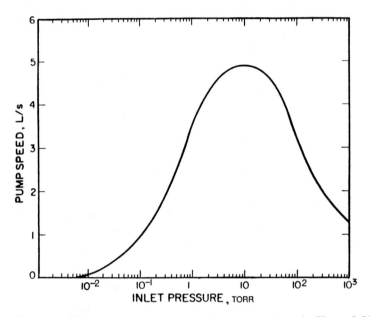

Figure 5.52 Pumping performance of the pump shown in Figure 5.51.

ets is small and there is a long path between inlet and discharge, which reduces back leakage; there are no discharge valves, which reduces noise and back pressure; and wear is reduced because in an orbiting motion the relative speed between moving and stationary surfaces is low, which results in lower temperature and longer wear without lubrication. The pump is totally oil free, it has few parts, is compact, and can be disassembled (for example, for seal change), in a few minutes. The disadvantages are that the mechanism is less practical for larger sizes and that the special spiral-shaped seal "gaskets" may have to be replaced in a few months, depending on particle content in the gas stream.

5.9.7 Diaphragm Pumps

The major advantage of diaphragm pumps is the complete separation of the driving mechanism from the pumping spaces where gas flow occurs. A thin flexible diaphragm fixed to both the cylinder and the piston is used for this purpose. If the valves and the diaphragm are made of compatible materials, the pump is free of lubricants and can be considered to be clean. The diaphragm

Figure 5.53 A view of a scroll-type vacuum pump.

pump is essentially a positive displacement piston pump. The basic design is shown in Figure 5.56. The rocking motion produced by the driver in the figure is not a necessary requirement but is used for practical engineering purposes. The inlet valve is on the right and the exhaust on the left. The leaf (reed) valves are actuated by pumped gas pressure differences and the compression is limited by the residual volume between the piston and the valve (dead space) at the end of the piston motion. Also, when the gas pressures are so low that the pressure forces are too small for actuating the valves, the pumping action ceases. The pumping cycle is demonstrated in Figure 5.57 for a two-stage pump, starting in the top figure and ending in the bottom. In this case the two

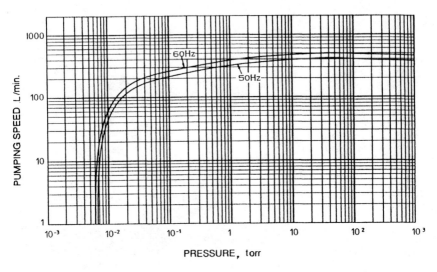

Figure 5.54 Pumping speed versus inlet pressure of a 150 L/m scroll pump.

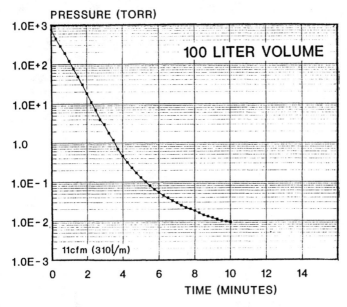

Figure 5.55 Evacuation progress for a 100 liter chamber using a 310 L/m scroll pump.

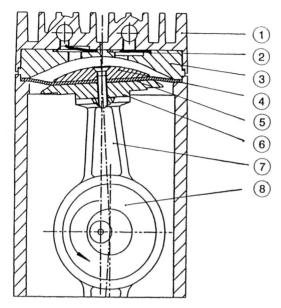

Figure 5.56 Schematic sectional view of a diaphragm pump unit; 1) body, 2) valves, 3) cylinder head, 4) diaphragm clamping disk, 5) diaphragm, 6) diaphragm support disk, 7) connecting rod, and 8) eccentric cam (Vacuubrand Co.).

pistons are out of phase, for mechanical reasons, but this is not an intrinsic requirement.

A study of Figure 5.56 may also indicate the major disadvantages of diaphragm pumps. The extent of the piston travel is limited by the flexibility and the elasticity of the diaphragm, the cycle frequency is limited by the dynamics of the leaf valves, which also must be flexible. Therefore, from an engineering point of view, it is difficult to make pumps of high pumping speed. Typical commercial pumps have pumping speeds below 200 L/m. Typical compression ratios of single-stage diaphragm pumps are between 10 and 15. For high-vacuum work, the usual arrangement is to use four pistons, in a single pump body, driven by the same motor, two inlet pistons connected in parallel and the third and fourth in series with the first two. Such three-stage pumps can achieve ultimate pressures of a few torr (see Figure 5.58, curve 3). If interconnections between cylinders are made of metal tubing (instead of the usual plastic), if all the connections are adequately leak free, and if four stages are connected in series, inlet pressures below 1 torr can be obtained, as seen in

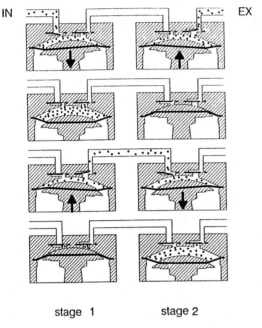

stage 1   stage 2

Figure 5.57 Progress of gas (dots, from top to bottom, pistons phased 180°) and movement of valves during a pumping cycle for a two-stage diaphragm pump.

Figure 5.58. As many as eight cylinders can be incorporated in a single pump, in which case the usual connection is:

first stage, four cylinders in parallel,
second stage, two in parallel, and
third and fourth stages, one each.

Limitation of size can be alleviated by attaching the diaphragm pump to a Roots-type or dry rotary vane pump to serve as the first stage, as shown in Figure 5.59. In this way, a variety of pumping trains can be arranged, depending on the desired pumping performance and the degree of cleanliness. The usual limitation of durability of the flexible diaphragm and leaf valves has been alleviated by better mechanical design (for example, longer connecting rods to reduce the rocking angle), and by the use of carefully selected materials, regarding their mechanical qualities and chemical inertness. Service life of diaphragms has been improved in modern pumps to more than 5000 hours, which

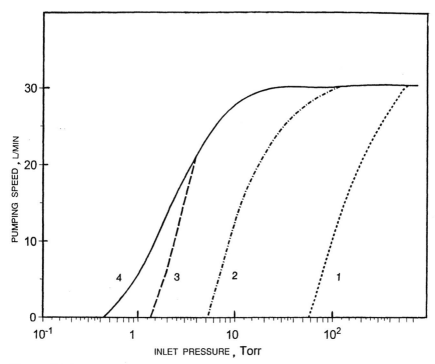

Figure 5.58 Pumping speed versus inlet pressure for a 33 L/m diaphragm pump; the number of stages connected in series are shown at each curve (Vacuubrand).

in some applications may represent approximately one year of use. If the pump is used only for initial evacuation of vacuum chambers (which remain in vacuum for a long time) and then shut off, the diaphragm will last many years. If used as a backing pump for continuously operated high-vacuum pump, a service schedule of six month periods is advisable. As in oil-sealed pumps, water vapor may condense in the last (exhaust) stage and accumulation of particles can occur, which may interfere with the proper operation of valves. In such cases, gas purge (gas ballast) can be arranged to lengthen the time period between service but leaf valves may need to be changed more frequently.

During the last 5 to 10 years, the use of diaphragm pumps has increased in high-vacuum industry partly because of the demand for oil-free pumping technologies and partly because the new hybrid (compound) turbomolecular pumps have reached allowable discharge pressures to permit the use of diaphragm

Figure 5.59 View of three-stage (four-piston) diaphragm pump connected with a Roots-type pump (Vacuubrand).

backing pumps (see Figure 5.60). Higher discharge pressures of high-vacuum pumps permit not only higher ultimate pressures for backing pumps, but also permit the use of smaller backing pumps as well as smaller roughing pumps. This is because the transfer from roughing to high-vacuum pumps can be effected at a higher pressure (and therefore sooner during evacuation), and because the gas is compressed to a higher density (and therefore smaller volume) before arriving into the backing pump.

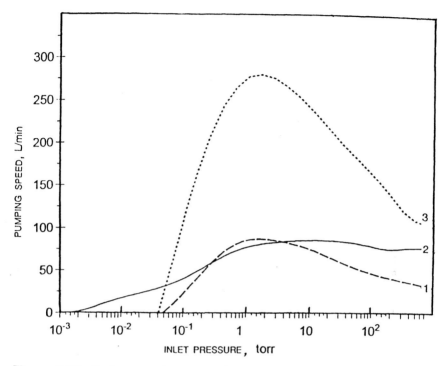

Figure 5.60 Performance of pump combinations in series: 1) a 170 L/m Roots pump with a 33 L/m diaphragm pump; 2) a combination of 100 L/m dry rotary-vane and diaphragm pumps; 3) a 500 L/m Roots pump with a 100 L/m diaphragm pump (Vacuubrand Co.).

REFERENCES

1. H. Wycliffe, *Vacuum,* 37(8/9), 603 (1987).
2. H. Wycliffe, *J. Vac. Sci. Technol.,* A5(4), 2546 (July/Aug. 1987).
3. Z. C. Dobrowolski, in *Vacuum Physics and Technology,* G. L. Weissler and R. W. Carlson, eds., Academic Press, New York, 1979.
4. M. H. Hablanian, *J. Vac. Sci. Technol.,* 19(2), 250 (July/Aug. 1981).
5. P. Bachman and H. P. Berges, *Vak.-Tech.,* 36(2), 41 (Mar. 1982).

6. *Hazard's Manual*, American Vacuum Society, New York, 1979.
7. D. B. Frazer, ed., *Pumping Hazardous Gases*, American Vacuum Society, New York, 1980.
8. P. Duval, *Vide,* 43(240) 19 (Jan./Feb. 1988).
9. P. Duval, *High Vacuum Production in the Microelectronics Industry*, Elsevier, Amsterdam, 1988.
10. M. Hablanian, *J. Vac. Sci & Technol., A4*(3), 286 (May/June 1986).
11. P. Duval, *J. Vac. Sci. & Technol., A7*(3), p. 2369, 1989.
12. M. Hablanian, *J. Vac. Sci. & Technol., A6*(3), 1177 (May/June 1988).
13. A. Troup and D. Turrell, *J. Vac. Sci. Technol., A7*(3), 2381 (1989).
14. M. Hablanian, E. Bez, and J. L. Farrant, *J. Vac. Sci. & Technol., A5*(4), 2612 (1987).
15. D. Coffin, *J. Vac. Sci. Technol.,* 20(4), 1126, (Apr. 1982).
16. P. Duval, A. Raynaud, and C. Saulgest, *J. Vac. Sci. Technol., A6*(3), 1187, (1988).
17. L. Laurenson and D. Turrell, *Vacuum, 38,* (8/10), 665.
18. G. Lewin, *J. Vac. Sci. Technol., A3*(6), 2212 (Nov.Dec. 1985).
19. D. M. Hoffman, *J. Vac. Sci. Technol., 16,* 1, (Jan.Feb. 1979).
20. N. S. Harris, *Vacuum, 28,* 6/7, 261, (1978).
21. R. Sherman and J. Vossen, *J. Vac. Sci. Technol., A8*(4), 3241 (Jul.Aug. 1990).
22. D. J. Santeler et al., *Vacuum Technology and Space Simulation*, NASA, 1966, (reprinted by AIP, 1993).
23. M. H. Hablanian, *Vacuum, 41,* 7-9, 1814, (1990).
24. M. H. Hablanian and E. Bez, *Vuoto, 20,* 2, 535, (Apr./Jun. 1990).
25. R. P. Donovan, *Particle Control for Semiconductor Manufacturing,* Marcel Dekker, Inc., New York, 1990.
26. H-D. Bürger et al., *Vacuum, 41,* (7-9), 1822 (1990).
27. H-A. Berges and M. Kuhn, *Vacuum, 41,* (7-9), 1828, (1990).
28. D. Chen and S. Hackwood, *J. Vac. Sci. Technol., A8,* 2, 933, (Mar./Apr. 1990).
29. J. Zhao et al., *Solid State Tech.,* 85, (Sept. 1990).
30. "Contamination in Vacuum" *Amer. Vac. Soc. Workshop,* Atlanta, *35th Nat. Symp.,* (Oct. 1988).
31. J. F. O'Hanlon, *J. Vac. Sci. Technol., A5,* (4), (1987).
32. J. F. O'Hanlon, *35th Nat. Symp. Am. Vac. Soc.,* (Oct. 1988).
33. C. M. Van Atta, *Vacuum Science and Engineering*, McGraw-Hill, New York, 1965.
34. H. E. Hesketh, *Fine Particles in Gaseous Media*, Ann Arbor Science, Ann Arbor, Mich., 1977.
35. R. J. Miller et al., *J. Vac. Sci. Technol., A6*(3), 2097, (May/Jun. 1988).
36. C. Hayashi, *J. Vac. Sci. Technol., A5*(4), 1375, (Jul./Aug. 1987).
37. H. Wycliffe, *J. Vac. Sci. & Technol., A5,* 2608, (1987).
38. S. D. Cheung, *Semiconductor Int.,* (Oct. 1988).
39. P. G. Borden, *Microcontamination*, (Oct. 1987).

40. R. Milgate, R. Simonton, *Nuclear Instruments and Methods in Physics Research*, *B1*, 381, North-Holland, 1987.
41. P. Rossi, *Microcontamination*, 65, (Nov. 1988).
42. M. Hablanian, *Res. & Dev.*, 81, (Apr. 1989).
43. G. Strasser et al., *J. Vac. Sci. Technol.*, *A8*(6), 4092, (Nov./Dec. 1990).

6
Vapor Jet (Diffusion) Pumps

6.1 INTRODUCTION

"Diffusion" pumps are vapor jet pumps or vapor ejector pumps designed for pumping rarefied gases in the high-vacuum range of pressures. The only justification for calling them diffusion pumps is due to the observation that the molecules of the pumped gas penetrate some distance into the vapor jet in a manner resembling diffusion of one gas into another. There exist theoretical studies of this initial gas capture process (notably by G. Toth, M. Wutz, A. Rebrov, and V. Skobelkin). The theory is complex because the density and velocity distributions in the pumping fluid and the pumped gas cannot be assumed to be isotropic.

First designs of diffusion pumps were by W. Gaede in Germany (1915). The pumping action was assumed to be due to diffusion of the pumped gas into the vapor stream and subsequent removal by the mechanical pump in a section where the vapor was condensed. The first designs did not resemble modern configurations and the device was inefficient. The basic design was evolved approximately 10 years later. Still, the essential mechanical pumping action of the high-velocity vapor jet was not clearly understood, as evidenced by another lingering name for the pump—"condensation" pump (I. Langmuir, 1916). The assumption was made that the pumped gas is condensed together with the vapor, resulting in the pumping action.

The original pumping fluid was mercury. It has the advantage of not breaking down at elevated temperature but the disadvantage of toxicity and a rather high vapor pressure at room temperature (near 1.5×10^{-3} torr). Oil-like substances were introduced in 1928. At least 99% of vapor jet pumps used today have oil as the pumping fluid and the subsequent discussion pertains primarily to oil pumps.

High-vacuum pumping systems based on vapor jet pumps include at least one vapor jet pump and one mechanical pump connected in series. Mechanical pumps remove about 99.99% of the air from the vacuum chamber. The remaining air and water vapor, down to any pressure from 10^{-3} to 10^{-9} torr, is removed by the vapor jet pump discharging into the mechanical pump (less than 10^{-10} torr can be obtained with the aid of liquid-nitrogen traps).

Vapor jet pumps are used when constant high speeds for all gases are desired for long periods of time without attention. Such pumps cannot discharge directly into the atmosphere. A mechanical pump is required to reduce the pressure in the vacuum system to the correct operating range. The mechanical pump is then used to maintain proper discharge pressure conditions for the vapor jet pump.

Vapor jet pumps were developed in response to the need of achieving higher vacuum after the capabilities of oil-sealed mechanical pumps were exhausted (below approximately 0.1 mtorr). Mechanical pumps, for practical reasons, have a rather low pumping speed, and even if, in principle, they were capable of achieving higher vacuum, their pumping speed is insufficient to overcome the gas evolution from the walls of a sizable vessel to maintain a pressure below about 10^{-4} torr. Vapor jet pumps have much higher pumping speed but at the low pressure level. When used in series together, they produce the high pumping speed and high pressure required to exhaust the gas to atmosphere.

At somewhat higher inlet pressure—in other words, higher gas loads "booster diffusion" pumps are used. These pumps generally resemble other vapor jet pumps. They are designed to displace the performance range to inlet pressures roughly 10 times higher than is common for vapor jet pumps. They use pumping fluids with higher vapor pressure so that their boiler pressure and forepressure tolerance are correspondingly higher. They have peak pumping speeds near 10^{-2} torr inlet pressure. In the past they were also used for backing common vapor jet pumps. During the last 20 years, the use of oil booster pumps has been reduced due to the development of more powerful vapor jet pumps and more common use of mechanical booster pumps (Roots blowers).

6.2 PUMPING MECHANISM

The basic fluid dynamic cycle involved in a vapor jet pump is shown schematically in Figure 6.1. The vapor is produced in the boiler and is expanded through

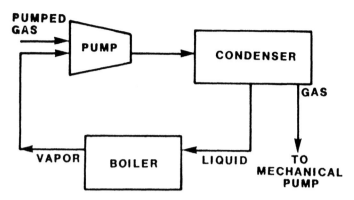

Figure 6.1 Pumping vapor cycle and gas flow in a vapor jet pump.

a nozzle achieving a high velocity (typically, near 300 m/s). The high-velocity jet entrains gas molecules that happen to enter it, due to their usual thermal motion. The gas and vapor are separated in the condenser, the gas is pumped away by the mechanical pump, and the condensate returns to the boiler.

A schematic representation of the capture mechanism is shown in Figure 6.2. The arrows represent pumping fluid molecules; the black dots, the pumped gas molecules. Because the vapor jet is directed down, it will be more difficult for

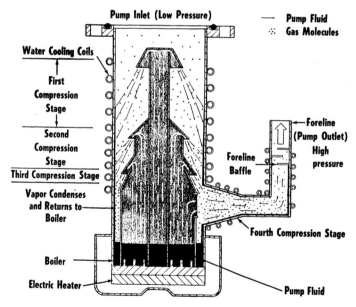

Figure 6.2 Pumping mechanism of a vapor jet pump.

molecules in region P_2 to cross the vapor jet in the upward direction, against the vapor stream, than in the downward direction. Once a molecule has entered the vapor stream from above, the collisions with vapor molecules will, on average, impart a downward motion. The gas molecules entering the vapor stream from below will be also deflected downward, after colliding with much heavier vapor molecules.

The actual picture is more complex. The density of the vapor jet is not uniform. Immediately downstream of the nozzle there exists a core of the jet which is still accelerating and expanding. This core is surrounded by vapor of lesser density where the entrainment of gas molecules occurs.

The boiler pressure upstream of the nozzle is typically 1 to 2 torr; the vapor density between the nozzle and the wall at the center of the vapor stream may be 0.1 torr. The pumped gas density is normally below 10^{-3} torr. The vapor stream is not in the molecular flow condition but is rarefied enough to permit gas molecules to enter. The capture efficiency of the vapor jet depends on its density, velocity, and the molecular weight. If the jet is too dense, the gas penetration into the jet will not be as effective, and if the jet is too rare, it will not be able to maintain a sufficient pressure difference.

As noted earlier, in regard to their basic pumping action, high-vacuum vapor pumps are related to oil or stream ejectors. The potential energy of elevated pressure inside the jet assembly (boiler pressure) is converted to the kinetic energy of the high-velocity vapor stream or jet after it passes through a nozzle. The gas is pumped by the jet by momentum transfer in the direction of pumping. In another way, the pumping action can be related to the molecular pumps. Instead of a solid surface moving at high speed, a jet of vapor imparts the necessary collisions. Because the pumping fluids used in vapor jet pumps are easily condensable at room temperature, multistage nozzle-condenser system can be fitted in a compact space.

The interaction of pumping vapor and the pumped gas can be illustrated experimentally by exploring the density distribution of both species in the pumping region. The gas density distribution obtained by traversing the pumping region with a vacuum gauge probe is shown in Figure 6.3, and the distribution of vapor arriving at the pump wall, in Figure 6.4. The pattern of diffusion of pumped air into the vapor jet, its relative absence in the core of the jet near the nozzle exit, and the gradual compression of the gas can be deduced from Figure 6.3. Compare, for example, 0.18 near the nozzle to 0.42 near the wall. Also, note the gradual increase of gas pressure at the wall from 0.4 at 2 cm from the nozzle exit level to 2.4 at a distance of 13 cm.

The theoretical analysis of the basic pumping mechanism of a vapor jet pump has been considerably refined. The basic performance aspects relating to speed, throughput, and pressure ratio are well understood. The nature of vapor flow from the nozzles and the gas capture mechanism have been analyzed with three-

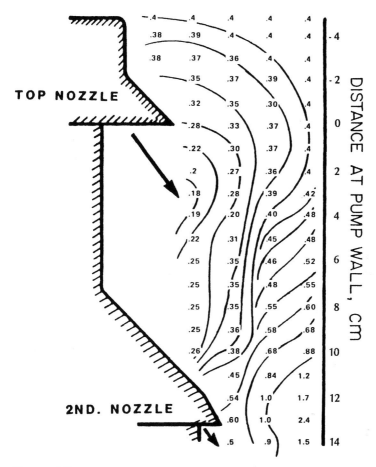

Figure 6.3 Pressure distribution of the pumped gas in the vicinity of the nozzle. Inlet pressure, 0.4 mtorr.

dimensional mathematical treatment. From the series of papers on vapor jet pump theory the following observations can be made: (1) further increases in the specific pumping speed are not likely; (2) the first stage must be as rare as possible but strong enough for a sufficient compression ratio to match the second stage; (3) a single stage cannot produce both high speeds and high compression ratios; (4) from the gas dynamic considerations the possibility of the increase of vapor velocities is limited; (5) in modern pumps the speed is independent of heat input, although the latitude is narrower for hydrogen than for air, and (6) the factor of 3 higher speed for hydrogen derived from molecular flow arrival rate is not expected in practice.

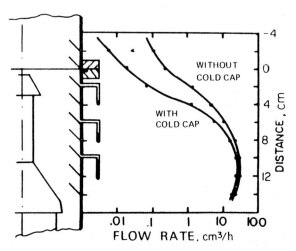

Figure 6.4 Distribution of pumping fluid condensed on the wall (see Section 6.8.1 for cold cap).

The pumping action in a vapor jet pump results from collisions between vapor and gas molecules. As noted earlier, it is more difficult for gas molecules to cross the vapor stream in the counterflow direction. Thus a pressure (or molecular density) difference is created across the vapor stream. The maximum compression ratio created by the vapor stream can be approximately expressed by

$$\frac{P_2}{P_1} = \exp\left(\frac{\rho V L}{D}\right) \quad (6.1)$$

where ρ is the vapor density, V its velocity, and L the width of the stream. D is related to the diffusion coefficient and depends on molecular weights of the vapor and gas and their molecular diameters,

$$D = \frac{3}{8(2\pi)^{1/2}} \left(RT \frac{M_1 + M_2}{M_1 M_2}\right)^{0.5} \left(\frac{\sigma_1 + \sigma_2}{2}\right)^{-2} \quad (6.2)$$

where subscript 1 refers to the pumped gas and 2 to the pumping fluid.

It is more likely that a smaller molecule can penetrate the vapor stream in the upstream direction without colliding. It must be appreciated that the compression ratio is much lower for lighter gases. The pumping speed of a vapor jet pump depends on the inlet area, the molecular weights, and the velocity and density of the vapor jet. There is an optimum for the vapor density that produces the greatest capture efficiency of molecules entering the pump.

6.3 BASIC DESIGN

Most modern pumps have three stages (i.e., three pumping nozzles in series). Larger pumps can have as many as six stages. A typical design is shown in Figure 6.5. The inlet to the pump is at the top. The discharge is at the side near the bottom section. This outlet is always connected to another pump to provide the necessary preliminary vacuum to make it possible for the vapor jet pump to function. Usually, this is a single- or double-stage oil-sealed mechanical pump. When the vapor jet pump is operating, the mechanical pump must be operating, too. Most common designs consist of a vertical, cylindrical body with a flange at the inlet for attachment to the pumping duct and the vacuum system to be evacuated.

At the bottom of the cylinder is the boiler where the pumping fluid is placed. There is an electrical heater at the bottom of the pump with sufficient power to evaporate the pumping fluid and maintain desired pressure in the boiler. The upper part of the body is cooled by cooling coils brazed or soldered to the body. Usually, water is used for cooling. Small pumps are often air-cooled with a fan

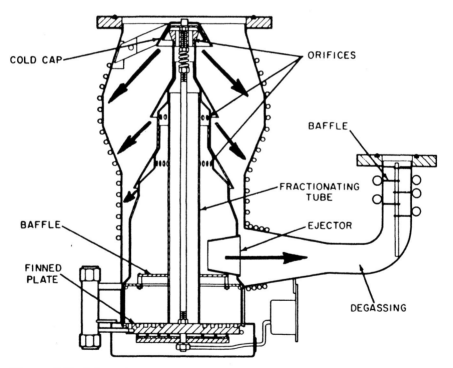

Figure 6.5 Schematic cross section of a four-stage pump (Varian VHS type).

or blower and the cooling coils are substituted by fins. The cooled wall is the condenser.

The internal structure, called the jet assembly, consists of cylindrical and conical pieces shaped to produce a set of nozzles directed downward and toward the cooled wall. The lower cylinder of the jet assembly is immersed in the pumping fluid in close proximity to the outer walls of the pump. This piece has small slots or other means of liquid passage underneath, making it possible for the condensed oil to return to the boiler.

When the pumping fluid is evaporated, the entire jet assembly is filled with vapor. The various orifices and nozzle throats are small enough to permit the buildup of pressure inside the jet assembly. This pressure is rather low, only 1 to 2 torr (1 or 2 mmHg absolute). But this is sufficient to produce high-velocity vapor jets. The nozzles are usually annular in shape so that the vapor jet has the form of truncated conical skirts emerging from the nozzles. Upon reaching the wall, the vapor condenses and the liquid flows down by gravity to reenter the boiler. There are no moving mechanical parts, except the pumping fluid.

Vapor jet pumps normally are operated for long periods of time, an entire day at the minimum, and sometimes continuously for a few years. So the fluid cycle is a continuous process. Power is almost entirely consumed to reevaporate the condensed fluid. Condensation is a required step for separation of the pumping fluid and the pumped gas. Thus the cooling water or air must remove almost all of the heat supplied to the boiler.

There are no principal differences between air cooling and water cooling. The major concern is the temperature at the pump inlet which determines the rate of reevaporation of the condensed pumping fluid. Ideally, the inlet region of the pump or the baffle should be the coldest part of the vacuum system that does not have a cold trap. A system designer should pay attention to inlet duct temperature distribution, especially with air cooling when the vacuum system is placed in a cabinet. Cool air should be drawn from outside for pump cooling. Additional fans may sometimes be required to cool the inlet duct near the pump, keeping in mind the tendency of condensable vapors to migrate toward the coldest areas. Only small pumps are usually air cooled (maximum diameter 10 cm). Air cooling of larger pumps is inconvenient (especially in summer time) because of heat exhausted into the room.

The number of pumping stages or vapor nozzles depends on particular performance specifications. Normally, the first stage at the inlet has high pumping speed and low compression ratio and the last (discharge) stage vice versa. The initial stages have annular nozzles; the discharge stage sometimes has a circular nozzle and is called an ejector. There are no principal differences in such variations of geometry, although certain advantages are gained by one or the other choice.

Pumping speed is proportional to the area where pumping action occurs. Because at any moment the pumped gas mass flow is constant, the second stage requires less pumping speed after compression in the first stage. Thus the distance from the last nozzle and the pump wall can be small compared to the upper stages. With less expansion, the vapor stream in the lower stages is denser. It is more difficult for gas molecules to penetrate the denser jet in the upstream direction, which results in a higher compression ratio.

Figure 6.5 is an example of a modern vapor jet pump. There are a number of notable features. Appreciation of these details may be helpful in maintenance and operation of the pump. The boiler plate is finned to provide a larger heat transfer area and to reduce the temperature difference between the upper surface of the boiler plate and the temperature of the bulk liquid (typically, 30°C). The rate of thermal decomposition is highly dependent on temperature, and unnecessary overheating of the liquid must be avoided.

The internal baffle above the boiling liquid serves as a liquid-vapor separator. During vigorous boiling, droplets of liquid are ejected upward and can interfere with nozzle performance if permitted to reach the nozzle. The baffle forces the fluid into a sharp change of direction, so that the much heavier liquid droplets are diverted from the main vapor stream.

The horizontal nozzle facing the foreline produces an efficient degassing stage and isolates the pump boiler from the mechanical pump environment. The oil used in mechanical pumps has a 100- to 1000-fold higher vapor pressure than that of the pumping fluid in the vapor jet pump and should not be permitted to enter the boiler. The ejector section acts as a vapor stripper of lighter fluid fractions because the condensed pumping fluid has to reenter the boiler by flowing against the jet stream from the ejector nozzle. The midsection of the foreline is kept at high enough temperature to degas the condensate before it flows into the boiler. The degassing products cannot flow back against the vapor jet. The foreline baffle ensures complete condensation of the pumping fluid.

Modern pumps often have a drain and fill arrangement to permit convenient replacement of the fluid without removing the pump from the vacuum system and thermal switch protection against overheating. The heaters should have some elastic clamping provision so that even if the relaxation of clamping forces occurs due to high temperature, no gap develops between the heaters and the pump.

The tie rod that connects and holds down the jet assembly should be made of the same material as the jet assembly, to avoid differences in thermal expansion coefficients. Otherwise, there should be an elastic arrangement to maintain the clamping force, such as a spring. It is generally desirable to tie the jet assembly to the boiler plate. If this is not done, the entire jet assembly can be propelled upward if the pump, when under vacuum, is accidentally air-released from the foreline side.

6.4 BASIC PERFORMANCE AND OPERATION

The most important general performance aspects of vapor jet pumps are pumping speed, maximum permissible gas load (maximum throughput), maximum permissible discharge pressure (tolerable forepressure), and the best achievable vacuum (ultimate pressure). The conventional method of displaying the pumping speed is shown in Figure 6.6 in the form of a graph showing pumping speed at various inlet pressures. In high-vacuum technical literature the pressure is usually plotted on the horizontal axis and the flow rate on the vertical. This deviation from the general mechanical engineering practice is a historical tradition and has no particular significance or advantage.

If we multiply speed and pressure at any point of the curve in Figure 6.6, we obtain the corresponding throughput, $Q = SP$. Note that the throughput at any given pressure in the horizontal (constant pumping speed) section of the curve is not a new characteristic of the pump. It is simply another way of reflecting the pumping speed. A throughput curve at different pressures can simply be reconstructed from Figure 6.6 without any additional measurements.

A significant property of the pump is the "maximum throughput," point MQ in the graph. This is the maximum gas mass flow the pump is able to handle without an overload due to its power limitations. Details of this limit will be discussed later. To the right of point MQ the throughput (or the mass flow rate)

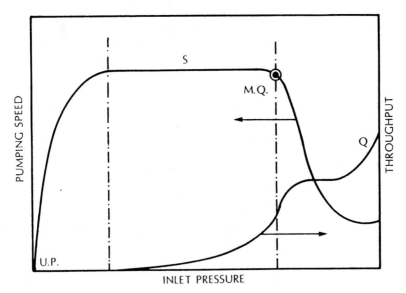

Figure 6.6 Usual representation of volumetric capacity (pumping speed).

is almost constant. In other words, to the right of point MQ the same amount of gas molecules is pumped by the vapor jet pump regardless of pressure.

A vapor jet pump is designed for high-vacuum application. Since its boiler pressure usually is 1 to 1.5 torr, this implies that the maximum pressure the pump can be expected to produce is 1.5 torr. In addition, the working fluids cannot be boiled at high pressures because the resulting high temperatures will decompose the fluid at an unacceptable rate.

Maximum discharge pressure of modern oil vapor jet pumps ranges from 0.5 to 1 torr. This means that such pumps cannot function at all unless the discharge pressure is maintained below that value by an auxiliary vacuum pump, or temporarily, by a preevacuated reservoir. The maximum permissible discharge pressure (tolerable forepressure) is lower when the gas load is high. Near the maximum throughput conditions the tolerable forepressure is typically 0.5 torr.

The ultimate pressure (UP on the graph) is the lowest pressure obtained by the pump. This is the point where the net flow is zero because the internal back leakage equals the forward pumping rate. However, in vapor jet pumps, this concept cannot be applied in a simple fashion because the the ultimate pressure usually depends not only on pump behavior but also on the existence of "parasitic" gas loads in the measurement system. Also, the ultimate total pressure cannot be simply defined without regard to the partial pressures of gases which are commonly present at high vacuum.

Generally, because of gas evolution from walls and fluids that may be present, high-vacuum pumps need a period of conditioning. Once started, they are usually operated for a longer period of time. Vapor jet pumps need 20 to 30 min of warm-up time before the full pumping action of the jet is established. During that time, the vacuum system can be evacuated by mechanical pumps. The pressure must be below 0.5 torr before the pumping fluid reaches boiling temperature, which is typically near 200°C. For proper function, the entire internal jet assembly must reach that temperature.

It will be understood that a period of time is also required to cool the pump, usually 30 to 60 min, depending on pump size, unless auxiliary means of cooling are introduced. It may take several hours to reach the ultimate pressure. Usually, in common vacuum systems, pressures in the 10^{-6} torr range are reached quickly, but to reach 10^{-8} torr, both the pump and the system need a period of degassing. Preevacuation or rough pumping of the entire system can be done through the vapor jet pump before it is heated, but this is impractical if the vacuum system must periodically be opened to atmosphere.

To permit atmospheric exposure of the chamber, valves are used to isolate the vapor jet pump to keep it under vacuum. If the period is short (less than a few minutes), the vapor jet pump can be isolated from both sides until the chamber is reevacuated. Otherwise, a separate auxiliary small mechanical pump called a holding pump is used to maintain the necessary vacuum level at all times. A separate rough vacuum line is often used (in parallel to the vapor jet pump) to evacuate the chamber with a rough vacuum pump, often called the roughing pump. By means of valve arrangements, the roughing pump can also be used to back the vapor jet pump during high-gas-load pumping, in which case it is called a backing pump. A typical valve arrangement is shown in Figure 6.7.

6.5 PUMPING FLUIDS

A variety of "organic" liquids have been used as motive fluids in vapor jet pumps. Criteria for the selection of the fluid are low vapor pressure at room temperature, good thermal stability, chemical inertness, nontoxicity, high surface tension to minimize creep, high flash and firepoints, reasonable viscosity at ambient temperature, low heat of vaporization, and of course, low cost. A list of the presently popular fluids and their properties is given in Table 6.1. The selection of the fluid should be made giving due consideration to its operational stability in the pump boiler. Many of the fluids presently used have been developed within the last 25 years. Until 1960 most fluids used had vapor pressures (at 20°C) of 10^{-7} to 10^{-8} torr and the ultimate pressure without

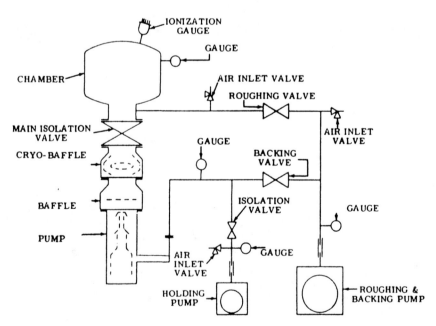

Figure 6.7 Typical arrangement of valves for a diffusion pump system.

Table 6.1 Properties of Vapor Jet Pump Fluids

Trade name	Chemical name	Molecular weight	Vapor pressure at 20°C (torr)	Flash point (°C)	Viscosity at 20°C (cS)	Surface tension at 20°C (dyn/cm)
Octoil	Diethyl hexyl phthalate	391	10^{-7}	196	75	<30
DC-704	Tetraphenyl tetramethyl trisiloxane	484	10^{-8}	216	47	30.5
Apiezon C	Paraffinic hydrocarbon	574	4×10^{-9}	265	295	30.5
DC-705	Pentaphenyl trimethyl trisiloxane	546	5×10^{-10} (25°C)	243	170 (25°C)	>30.5
Santovac-5	Mixed five-ring polyphenyl ether	447	1.3×10^{-9} (25°C)	288	2500 (25°C)	49.9
Neovac SY	Alkyldiphenyl ether	405	$<1 \times 10^{-8}$	230	250 (25°C)	<30
Fomblin 25/9	Perfluorinated polyether	~3300	2×10^{-9}	None	270	20 (25°C)
Krytox 1625	Perfluoro polyether	~4600	2×10^{-9}	None	250	19

the use of cryogenic trapping was limited to this range. A breakthrough in the selection of fluids was made when K. C. Hickman used five-ring polyphenyl ethers consisting of chains of phenyl groups interbonded by oxygen. This fluid offered exceptional thermal and chemical stability and enabled reaching ultimate pressures of 10^{-9} torr (approaching its vapor pressure at ambient temperatures) with only water-cooled baffles. Operational characteristics of another low-vapor-pressure silicone fluid (DC-705) are also excellent. Ultimate pressures of 10^{-9} torr with water-cooled baffles and 10^{-10} torr with baffles at $-20°C$ are possible with such fluids.

Whatever fluid is used, its vapor may pervade the pumped system, depending on its vapor pressure, the pump design, and the type of trapping used. The vapor can be broken down by the presence of hot filaments and bombardment by charged particles. The polymerization of silicone fluids resulting from bombardment by charged particles may cause an insulating film to be produced on electrode surfaces, changing the characteristics of the electronic instrumentation. Octoil or polyphenyl ethers are usually recommended to eliminate this problem in applications where mass spectrometers and other electron optical devices are used. Recently, fluorinated oils have been developed for use in vapor jet pumps. They have the added advantages of compatibility with corrosive gases used in some processes.

The use of mercury as a pumping fluid permits a greater latitude in boiler pressure and inlet pressure range. Small mercury pumps exist with a discharge pressure as high as 50 torr. In principle, even atmospheric pressure can be achieved. The use of mercury pumps may be of advantage, for example, when the pumped device is to be filled with mercury vapor, or when occasionally high gas loads must be handled by the pump and hydrocarbon contamination cannot be risked.

Mercury vapor pressure at room temperature is near 1.5×10^{-3} torr. Thus backstreaming and trapping of mercury vapor must be given more careful consideration. Usually, mercury pumps are used with baffles and liquid-nitrogen traps. Rapid accumulation on the trap surfaces compared to oil pumps cannot be prevented.

Mercury vapor does not condense on water-cooled surfaces as easily as oil vapor and previous hydrocarbon contamination (e.g., from forevacuum pumps) may interfere with condensation and result in reduced pumping speed. Mercury vapor is toxic. Handling of mercury liquid must be done carefully to avoid spillage (especially on rough concrete floors). Also, the venting of mechanical pump exhausts and vacuum chambers after defrosting of traps must be made with special precautions. The detailed operation and performance of mercury pumps is discussed in the technical literature.

6.6 PERFORMANCE CHARACTERISTICS

The pumping performance of a vapor jet pump is displayed in the form of a plot of pumping speed versus inlet pressure in Figure 6.8. The graph consists of four distinct sections. To the left, the speed is seen to decrease near the limit of obtainable vacuum. The constant speed section results from the constant gas arrival rate at molecular flow conditions and a constant capture efficiency of the vapor jets.

As noted before, at molecular flow, the gas molecules arrive into the pump due to their normal molecular velocities, which depend on temperature and the molecular weight. The rate of arrival also depends on the conductance of the inlet ducts and the geometry of the pump entrance. A certain percentage of molecules reaching the vapor jets will be captured. The capture rate is usually constant until the vapor jets become overloaded. The part marked "overload" is a constant-throughput section which indicates that the maximum mass flow capacity of the pump has been reached. In the last section, at the right, the performance is highly influenced by the size of the mechanical backing pump.

The performance of vapor jet pumps is fundamentally similar to any other pump, compressor, blower, ejector, and similar devices. The essential elements of flow and pressure relationships are analogous.

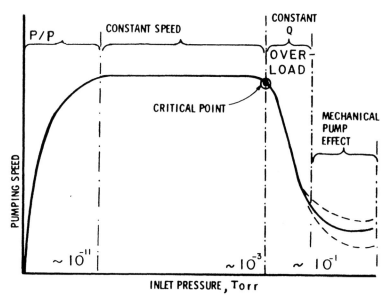

Figure 6.8 Pumping speed versus inlet pressure in four different performance regions.

6.6.1 Pumping Speed

The pumping speed of vapor jet pumps is nearly proportional to the inlet area, but the larger pumps are somewhat more efficient, as shown in Figure 6.9. The entrance geometries of small and large pumps are not strictly similar. Thus it is possible that the largest pumps can be constructed with a speed efficiency of 50% referred to the inlet plane (rather than the conical surface where pumping action occurs). This can become an important consideration in systems where the desired pumping speed is so high that there is simply not enough wall space available for attaching additional pumps.

The pumping speed must be considered in relation to the partial pressure of each gas species. When pumping speed is measured, the values obtained near the ultimate pressure of the measurement system become meaningless. The total pressure values cannot be used for obtaining speed due to the uncertainties of gas composition and the condition of the gauge. The composite picture should look as shown in Figure 6.10. Each gas has its separate speed, and what is more important, separate ultimate pressure. The normally measured "blank-off" is due to pump fluid vapor, cracked fractions, and perhaps water vapor remaining in the system. Connecting such "ultimate" pressure points by a smooth curve to the horizontal section of the air curve does not serve a functional purpose and can be misleading.

Although the arrival rate, expressed in liters per second, should be inversely proportional to the square root of molecular weight, the lighter gases are not pumped with the same efficiency as air (or nitrogen) and pumping speed val-

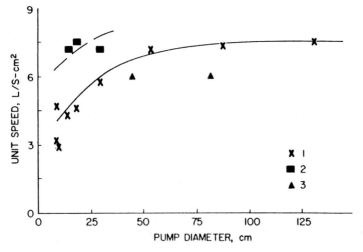

Figure 6.9 Pumping speed per inlet area: 1, conventional pump; 2, bulged-body pumps; 3, high-throughput pumps.

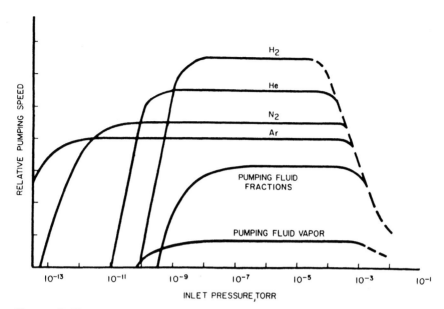

Figure 6.10 Typical performance with various gases present in the vacuum system.

ues for different gases do not peak at the same power setting. Despite expectations, helium and hydrogen speeds can be much lower than air for poorly designed pumps. Usually, the gases present in vacuum systems such as hydrogen, helium, water vapor, carbon monoxide and dioxide, nitrogen, and argon are pumped at approximately the same speed. It is common to see helium speed about 20% higher than air and hydrogen speed about 30% higher. The impedance of baffles and traps is lower for lighter gases compared to air. Thus the net system speed values for lighter gases are relatively higher than those obtained for the same baffle with air. Figure 6.11 shows a typical set of results for helium and hydrogen in a large pump.

6.6.2 Throughput

It may be observed that maximum throughput is often the important aspect of vapor jet pump performance, rather than pumping speed. The value of the maximum throughput determines the amount of power required to operate a given pump. Dimensionally, throughput and power are equivalent. For pumps of current designs using modern pumping fluids, approximately 1 kW of power is required to obtain a maximum throughput of 1.2 to 1.5 torr · L/s.

For systems that remain under high vacuum for long periods of time, the maximum throughput is of little value. In such cases provisions can be made to reduce the power after the initial evacuation or to use lower-power heaters.

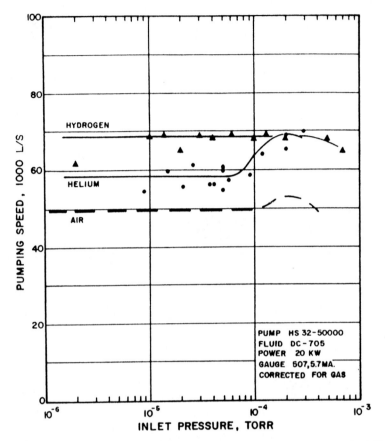

Figure 6.11 Typical results of speed tests with various gases (nominal 35-in. pump, HS-35).

Operation of the pump at half power is possible without changes in pump design Figure 6.12. Special designs can be made for low-power (low-throughput) operation, with appropriate attention to a corresponding reduction of forepressure tolerance.

For rapid frequent evacuation and for high-gas-load application, the value of maximum throughput determines the choice of pump size. Knowing this value and the inlet pressure at which it is reached (the point where the speed suddenly decreases), it is possible to reconstruct both the speed versus pressure and throughput versus pressure characteristics. The speed is constant below, and the throughput above, this prominent pressure point.

It is much simpler to see these relationships by looking at a throughput versus inlet pressure graph (Figure 6.13). Any combination of throughput and pres-

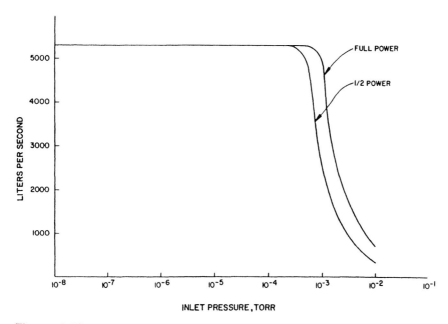

Figure 6.12 Performance at half power with only Q_{max} decreased (VHS-10 pump).

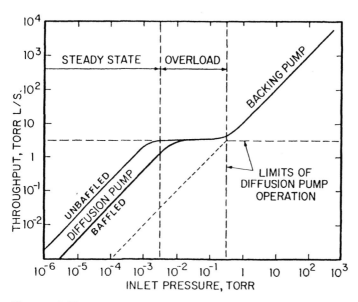

Figure 6.13 Plot of throughput versus inlet pressure, illustrating the overload condition (nominal 6-in. pump, VHS-6).

sure in the ABCD region (Fig. 6.14) can be chosen for operation, provided that the pumped gas admittance is restricted (throttled) at pump inlet pressures above approximately 1×10^{-3} torr. In this particular example, whenever the gas load is higher than 3 torr · L/s, the vapor jet pump becomes useless and the pumping reverts to the mechanical pump.

6.6.3 Gas Load

If the graph in Figure 6.13 is replotted interchanging the axes, a more customary arrangement of dependent and independent variables is obtained (Figure 6.14). Then it will be simpler to view the maximum throughput and the pressure limits and the concept of overload when those limits are exceeded. The more convenient way of looking at this is to keep in mind that for a given system gas load presented to the pump, there is a resulting inlet pressure. This will help in selecting the required pump size and in distinguishing between the

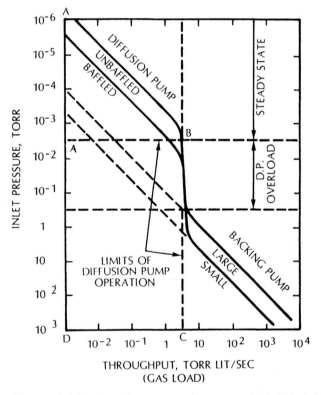

Figure 6.14 Plot of inlet pressure versus gas load. This delineates clearly the transition between the vapor jet pump and the backing pump.

VAPOR JET PUMPS

requirements of evacuating an empty chamber and maintaining a desired operation pressure at a given process gas load.

A common example of trade-off between pumping speed and pump inlet pressure range is given by the net pumping speed curve when the pump is used with baffles and traps. The constant-speed region is extended to higher pump inlet pressures at a cost of some reduction of net speed. These pressure and speed trade-offs can be made easily as long as the maximum throughput is not exceeded.

6.6.4 Tolerable Forepressure

Tolerable forepressure of a pump is the maximum permissible pressure at the foreline. Let it be repeated that the expected pumping action of the pump collapses when the tolerable forepressure is exceeded. Essentially, the vapor of the discharge stage of the pump does not have sufficient energy and density to provide a barrier for the air in the foreline when its pressure exceeds about 0.5 torr. Then this air will flow across the pump in the wrong direction carrying with it the pumping fluid vapor. As noted before, modern vapor jet pumps have a boiler pressure of about 1.5 torr. Approximately half of this initial pressure is recovered in the form of tolerable forepressure. The term "recovery" is referred to the discussion in Section 3.4, and the pressure profile shown in Figure 3.5. When the tolerable forepressure is measured, a conventional non-ultrahigh-vacuum system usually does not reveal the dependence between the discharge and the inlet pressures. This is because the maximum pressure ratio for air is usually not exceeded unless the experiment is conducted under ultrahigh-vacuum inlet pressures and the pump is underpowered.

The most important rule of vapor jet pump operation is: Do not exceed tolerable forepressure! In an operating pump, the maximum permissible discharge pressure should not be exceeded under any circumstances. Observance of this most basic requirement will eliminate most of the difficulties encountered with vapor jet pumps, especially problems with noticeable backstreaming of the pumping vapor into the vacuum system. High-vacuum systems should be designed with interlocks, fail-safe valve arrangement, or clearly marked instructions to preclude the possibility of exceeding the tolerable forepressure. As in most engineering considerations, a safety factor should be included in establishing the maximum permissible discharge pressure. A factor of 2 is a good general recommendation. As much as a 25% reduction of tolerable forepressure can be expected near maximum throughput operation (full load), and some reduction can be expected from low heater power. The selection of the mechanical backing pump must be made with these points in mind. Also, it should be noted that the desired discharge pressure must be obtained near the vapor jet pump to avoid errors caused by limited conductance of forelines and the reduction of mechanical pump speed at lower pressures.

The tolerable forepressure for various gases is approximately the same as far as the complete collapse of vapor jet (last stage) is concerned. However, in regard to the appearance of lighter gases at the inlet of the pump, when they are introduced in the foreline, the maximum pressure ratio effect becomes noticeable.

The results of pressure ratio measurements are shown in Figure 6.15. The data were obtained by introducing helium into the foreline of the pump and ob-

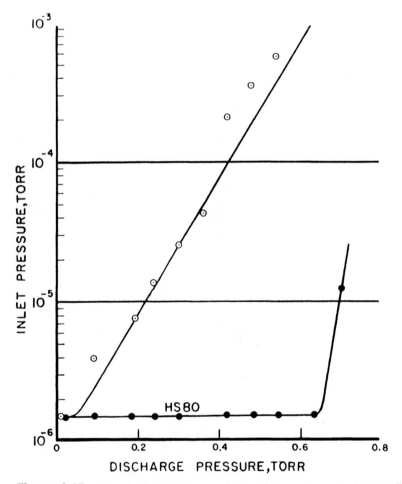

Figure 6.15 Dependence of inlet pressure on discharge pressure for two different pumps. The gas used is helium.

serving the resulting effect on the inlet pressure. The upper curve represents an older 5-cm pump with a 400-W heater. The comparison between the two pumps indicates a thousandfold improvement in maximum pressure ratio for helium which can be sustained across the pump. This should permit the use of smaller mechanical backing pumps and the elimination of small oil-booster pumps sometimes used in instruments that depend on the helium compression ratio for their performance.

For the older pump in Figure 6.15, it is difficult to establish a concept of forepressure tolerance for helium, at inlet pressures below 10^{-6} torr. The newer pump can be said to have a forepressure tolerance of 0.6 torr for helium. This pump had a forepressure tolerance for air of about the same value.

6.6.5 Backing Pump Selection

To select an appropriate backing pump for a given vapor jet pump, several questions must be considered. First, what is the size of the initial roughing pump, and is it to be used for both roughing and backing? Second, is the backing pump expected to perform at the maximum throughput of the vapor jet pump? Third, what is the tolerable forepressure of the pump? Also, what is the volume of the foreline ducts (or a special reservoir that could be used in the line)? The nominal pumping speed of the required backing pump for the full-load condition is obtained as follows:

$$S = \frac{Q_{max}}{TFP}$$

where Q_{max} is the maximum throughput of the vapor jet pump and TFP stands for its tolerable forepressure. Aside from consideration of safety factors and conductances of forelines, it should be noted that mechanical pumps often have less than nominal speed when their inlet pressure is equal to the tolerable forepressure of the jet pump. A safety factor of 2 is required to make sure that the tolerable forepressure value is never exceeded, even if the mechanical pump and jet pump are not operating at their best. A numerical example follows. Assume a pump with a maximum throughput of 4 torr · L/s and tolerable forepressure of 0.5 torr at full load (at maximum throughput). The required backing pump speed then becomes

$$S = \frac{4 \text{ torr} \cdot L}{\text{seconds } (0.5 \text{ torr})} = 8 \text{ L/s} = 17 \text{ ft}^3/\text{min}$$

A good choice for the backing pump would be a nominal 30-ft^3/min pump, provided that the conductance between the two pumps is not severely limited.

6.6.6 Maximum Compression Ratio

Tolerable forepressure and the maximum pressure ratio achievable with a given pump should not be confused. Tolerable forepressure is directly related to the boiler pressure, while the pressure ratio has a logarithmic dependence on vapor density in the jets. This distinction is indicated in Figure 6.16. The various performance aspects are shown in relation to heat input of the pump. Pumping speeds show an optimum level at certain power, but the peak speeds for different gases do not coincide. Also, the power needed for the highest tolerable forepressure does not coincide with the power for the best ultimate pressure, and so on. Therefore, the normal power level is chosen by practical

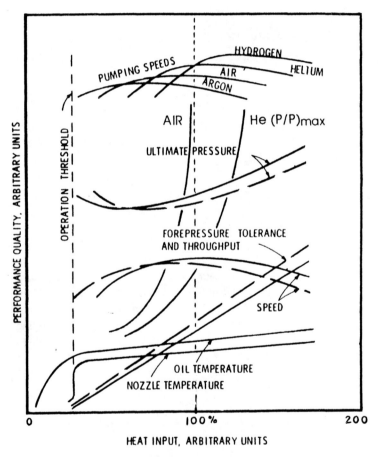

Figure 6.16 Dependence of various performance qualities on power (normal power is 100%; dashed lines for higher power).

considerations and compromises. For air, the maximum pressure ratio is usually so high that it cannot be measured under normal circumstances. Only when inlet pressure is near 1×10^{-10} torr may a dependence between discharge and inlet pressure be observed. However, for hydrogen and helium the dependence can be observed in the high-vacuum range.

The effects of boiler pressure variation on pump performance are summarized, qualitatively, in Figure 6.16. Due to the strong dependency, clearly, small changes in vapor jet density may cause large variations in the maximum pressure ratio. In applications where stable pressure is required with lighter gases, this may require attention. Ordinary pumps often display inlet pressure variations exceeding 5% for helium. This is unacceptable for some applications, such as highly sensitive mass spectrometer leak detectors and analytical mass spectrometers where helium is used either as a tracer or a carrier gas. Thus special pump designs having very stable boiler pressure and stable pumping action may be required.

As noted, the pressure ratio can be sufficiently small for the light gases to reveal the dependence of inlet pressure on the discharge pressure. Measurements of pressure ratios for various gases have been reported as follows: hydrogen, 3×10^2 to 2×10^6; helium, 10^3 to 2×10^6; neon, 1 or 2×10^8; CO and argon, 10^7; oxygen and krypton, 3 to 5×10^7; and hydrocarbons (n-C_2H_3), 7×10^8. In modern pumps the helium pressure ratio is closer to 10^7 and it can be increased as high as 10^{10} by doubling the heat input. In practice, even an ion gauge operated in the foreline can produce sufficient hydrogen to cause an increase in the inlet pressure. Exactly the same occurrence has been observed in a turbo-molecular pump system. The supply of hydrogen comes from the mechanical backing pump oil and other sources.

In regard to pumping speed, it may be observed that the curves for helium and hydrogen may have a rather steep slope near the nominal heat input value (100%). Therefore, if variations of vapor density occur, they may result in noticeable pumping-speed variations for these gases.

As far as ultimate pressure is concerned, hydrogen can be a substantial part of the residual gas composition due to its presence in metals, in the pumping fluid, and water vapor. This can be an important consideration for ultrahigh-vacuum work, where some vapor jet pumps may need a second pump in series.

The same consideration may apply to helium in leak detectors, mass spectrometers, molecular beam experiments, and so on. The fully developed relationship between discharge and inlet pressure in the presence of some gas flow is shown in Figure 6.17. To obtain the data shown on this graph, first a certain gas flow is arranged at the inlet to the pump using, for example, a standard speed measuring dome. Then the discharge pressure is slowly increased using a leak valve in the foreline until an effect on inlet pressure is noticed.

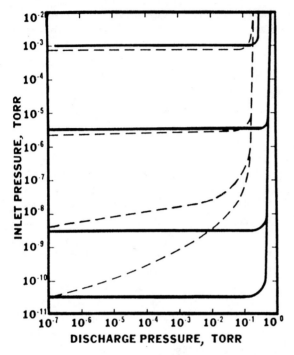

Figure 6.17 General dependence of inlet pressure on discharge pressure (see Figures 6.16 and 6.16). Solid lines, higher power; dashed lines, lower power; (helium gas).

The envelope of the family of curves obtained at various gas loads is related to the basic pressure-flow characteristic. On one end it shows the maximum pressure difference. This maximum pressure difference which the pump is capable of sustaining is the tolerable forepressure.

If the power of the pump is changed, the entire family of curves will shift. The same will happen if the pumping fluid of different vapor pressure is used. Also, when the pumped gas has a molecular weight different from air, the maximum compression ratio will shift, but the tolerable forepressure will remain the same. This is indicated by dashed lines in Figure 6.17.

6.6.7 Ultimate Pressure

The inlet pressure range for the use of vapor jet pumps is between 10^{-10} and 10^{-1} torr. Without the assistance of cryogenic pumping and without baking, the lowest inlet pressures conveniently achieved are near 10^{-8} torr. The steady-state pressure (at the pump inlet) normally should not exceed about 1×10^{-3} torr.

With the aid of cryogenic pumping, for example liquid-nitrogen-cooled traps, inlet pressures below the 1×10^{-10} torr range can be obtained. The significance of the ultrahigh-vacuum capabilities of vapor jet pumps is in the reduction of contamination from the pump to a low enough level so that other sources of contamination become predominant. However, for many ultrahigh-vacuum applications, ion pumps are often preferable and vapor jet pumps are usually chosen for their higher mass flow or high-gas-load capacity. Baking and subsequent cooling of pumped chambers and inlet ducts can also extend the pressure range because it can reduce outgassing as well as produce sorption pumping effects.

Two observations can be made regarding the relationship between the pressure of a pump and of a system. The ultimate pressure may be considered to be a gas load limit or a pressure ratio limit. Both are of significance in practice, the latter usually only with light gases. The pumping action of the vapor jet does not cease at any pressure, however low. The ultimate pressure of the pump depends on the ratio of pumped versus backdiffused molecules, plus the ratio of gas load to the pumping speed. In addition, the pump itself can contribute a gas load either through backstreaming of pump fluid vapor and its cracked fractions or the outgassing from its parts. Thus, in practice, the observed total ultimate pressure is a composite consisting of several elements.

The commonly observed first limit is due to the pumping fluid, although with the best fluids a clean gauge connection and some system degassing (baking) may be necessary to observe this limit when it is below 1×10^{-8} torr. When liquid-nitrogen traps are employed, the limiting pressure is usually given by various gas loads in the system even if the system consists only of the measuring dome and a pressure gauge. Below 5×10^{-9} torr, thorough baking is usually required. For helium and hydrogen, a true pressure ratio limitation can be observed in most pumps, although special designs can be made to improve the limit beyond observable level.

The ultimate pressure in a vacuum system consists of the partial pressures of various gases. Components arising from gas evolution from sources other than the pump (including the inlet gasket) are not discussed here. From the components arising in the pump, the major constituent is the vapor of the motive fluid, the products of its decomposition in the boiler, the contaminant vapors from other system parts, and backdiffused gases. In a clean system without a cryogenic trap, the ultimate pressure due to the vapor pressure of the fluid is the minimum achievable. The other components of the total pressure can be reduced by optimizing the pump design.

Degassing of the condensed fluid can be achieved by keeping the temperature of the pump wall near the boiler high enough to permit fluid degassing before it enters the boiler. A significant improvement in ultimate pressure can be obtained in this way. This is particularly important in pumps that do not have separated boiler compartments producing vapor for inner stages (fractionation). In pumps that have a horizontal ejector stage in the foreline, in addition to increasing the tolerance forepressure the ejector stage maintains low pressures in the region under the lower annular stage. With this arrangement, the volatile components of the returning condensate are more easily removed to the forevacuum region. Pumps with a strong side ejector stage are very little influenced by the addition of a booster diffusion pump for the improvement of ultimate pressures even in the 10^{-10} torr range.

With the use of cryogenic traps and modern fluids, such as DC-705 or Santovac-5, ultimate pressures of 10^{-10} torr and below can be achieved in vapor jet pump systems where contaminants from sources other than the vapor jet pump are controlled.

An increase in heat input will usually increase the pressure ratio obtainable with a pump. However, depending on pump design, higher power can either improve or worsen the ultimate pressure. This depends on the ratio between the fluid thermal breakdown rate and the purification ability of the particular pump. Small pumps present, generally, a more challenging problem in this respect because all distances are smaller and the control of fluid circulation and the temperature distribution is more difficult than in larger pumps.

6.6.8 How to Measure Ultimate Pressure

Sometimes it is necessary to demonstrate the ultimate pressure capability of the pump to establish its performance divorced from the condition of the pumping system. This is not a simple, straightforward matter even if the requirements are limited to total pressure measurement.

A standard dome used for measuring the pumping speed usually cannot be utilized unless it is designed with ultrahigh vacuum in mind. This means that gaskets and valves must be metallic and the dome made of stainless steel with a provision for at least mild baking using an auxiliary pumping system prior to the vapor jet pump operation.

A compromise solution, in the absence of a standard procedure, is suggested in Figure 6.18. A plate is used instead of a dome to minimize the surface area and reduce the size of the equipment. The plate should be made of stainless steel and have attached to it one or two turns of water cooling coils to keep the temperature near normal room ambient. The vacuum gauge connection should have a 90° bend in it to prevent a direct trajectory for pumping fluid molecules. The aim is not to measure the vapor pressure of the fluid but rather, the performance of the pump. Generally, if oil molecules are permitted to enter the gauge freely, the measurement will be severely distorted because the interaction of the gauge with oil vapor is not a simple matter.

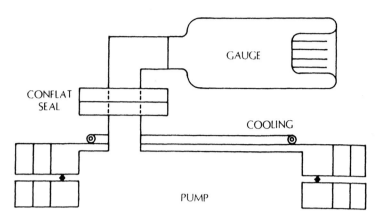

Figure 6.18 A simple method for measuring the total ultimate pressure.

It should be understood that this setup essentially measures the ultimate pressure of a baffled pump, the 90° bend in the gauge connection acting as a baffle. The seal connecting the test plate to the pump should preferably be metallic, for example, a continuous soft copper gasket with a diamond cross section. This is not always easy because such gaskets are not readily available, and the sealing surfaces are not always smooth enough for a tight seal with a metal gasket. However, an elastomer gasket can be used with reasonably satisfactory results. It should be of low-permeability, low-outgassing-rate material, such as Viton. It is more important to use a metal gasket at the gauge connection. The conventional compression seal with an O-ring cannot be used because of its outgassing products, which usually limit the measurement to the low-10^{-8}-torr region. This is insufficient for modern jet pumps using pumping fluids of lowest vapor pressure. When making ultimate pressure measurements it is usually necessary to condition the pumping fluid and outgas the pump surfaces by continuous operation for several hours. In this way, ultimate pressures below 5×10^{-9} torr can be demonstrated for well-designed pumps using fluids such as Santovac-5 or DC-705 without the aid of liquid-nitrogen traps.

6.6.9 Backstreaming

Any transportation of the pumping fluid into the vacuum system may be called backstreaming. The possibility of this backflow is perhaps the most undesirable characteristic of vapor jet pumps. It is in this area that most misunderstanding and misinformation exists. Pump manufacturers usually report the backstreaming rate at the inlet plane of an unbaffled pump. System designers are concerned about the steady-state backstreaming above liquid-nitrogen traps. The users are often troubled by inadvertent high-pressure air inrush into the discharge end of the pump and the breakup of the frozen liquid films on liquid-nitrogen traps. The quantities of backstreaming fluid involved in the foregoing cases can range over many orders of magnitude.

As far as the pump itself is concerned, there can be several sources of backstreaming: (1) the overdivergent flow of vapor from the rim of the upper nozzle; (2) poorly sealed penetrations at the top nozzle cap; (3) intercollision of vapor molecules in the upper layer of the vapor stream from the top nozzle; (4) collisions between gas and vapor molecules, particularly at the high gas loads (10^{-3} to 10^{-4} torr region); (5) boiling of the returning condensate just before the entry into the boiler (between the jet assembly and the pump wall), which sends fluid droplets upward through the vapor jet; and (6) evaporation of condensed fluid from the pump wall. Most of these items can be corrected by good design.

All the items listed above which can be stopped or intercepted by a room-temperature baffle may be called primary backstreaming. The reevaporation of the pumping fluid from the baffle and the passage through the baffle may be called secondary backstreaming. Primary backstreaming can be controlled effectively by the use of cold caps surrounding the top nozzle.

In systems with liquid-nitrogen traps (barring accidents and high-gas-load operation) the backstreaming level can be controlled at such a low level that contaminants from sources other than the vapor jet pump will predominate. Properly operated and protected vapor jet pump systems can be considered to be free of contamination from the pumping fluid for many applications.

The backstreaming rates measured by the American Vacuum Society Standard and given in pump specifications refer to the inlet plane of the pump. It is important to understand that this rate quickly diminishes at some distance from the pump inlet. At a distance from inlet equivalent to two pump diameters, the rate is reduced typically 50 times (Figures 6.19 and 6.20). A 90° bend in the inlet duct will act as a baffle and reduce backstreaming to the level of the natural evaporation rate of the fluid at the ambient temperature. Without cryogenic traps or similar devices, this rate cannot be reduced much further.

Figure 6.21 shows the qualitative relationship between conflicting requirements of minimum backstreaming and maximum retained pumping speed with baffles at ambient temperature. If a series of elements that reduce backstreaming are introduced above the pump, the backstreaming rate approaches the evaporation rate of the fluid. It may be observed that in many applications, complete opaque baffles are redundant. With efficient cold caps and low-vapor-pressure pumping fluids a system can be operated for a year or more with continuous liquid-nitrogen cooling of the trap. The fluid will be returned to the pump and the buildup on the trap will not be detrimental.

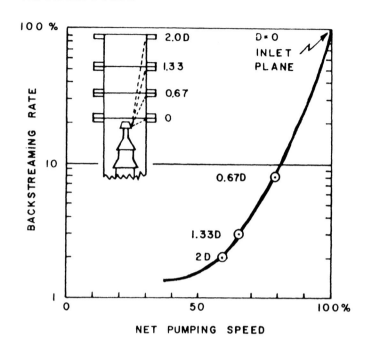

Figure 6.19 Plot showing relative backstreaming rates at various distances from the pump.

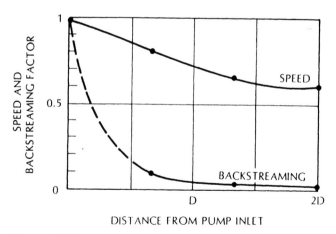

Figure 6.20 Reduction of pumping speed and backstreaming at various distances from the pump inlet (D is diameter).

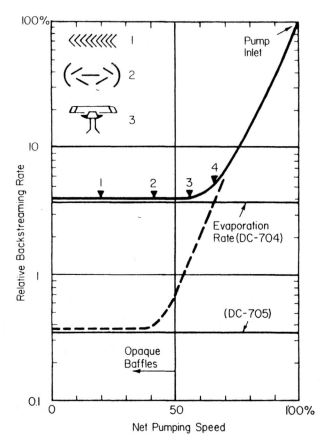

Figure 6.21 Typical relative values for backstreaming and pumping speed for a variety of baffle shapes.

In normal practice, an optimum design for a vapor jet pump and trap combination can have nearly 40% net pumping speed and reduce backstreaming to less than 1×10^{-10} g/cm^2 · min (at the inlet plane of the trap). Values of this magnitude have been measured.

It is important to remember that malfunction and misoperation can destroy the intentions of most intelligent designs. The most common causes of gross backstreaming are: "accidental" exposure of the discharge side of the pump to pressure higher than the tolerable forepressure, high inlet pressure exceeding maximum throughput capacity for long periods of time, incorrect startup, and incorrect bake-out procedures.

Providing proper interlocks and protection in system design is an important as training of operating personnel. The worst possibility is to air-release the system through the foreline when the pump is operating. Fast-acting fail-safe valves in the foreline should be considered whenever the equipment is left unattended. The signal for closure and power cutoff can be obtained from a gauge with a fast time response.

From lowest pressures to about 1×10^{-4} torr, the backstreaming rate appears to be independent of pressure, indicating that oil-gas collisions are not significant in this region. Between 10^{-4} and 10^{-3} torr, a slight increase may be noticed. Above the critical pressure point when the maximum throughput is reached (pumping speed begins to decline), the backstreaming rate may rise markedly.

Most measurements of backstreaming rate are conducted at the blank-off operation of the pump. One of the important questions is what happens during operation at higher inlet pressures and particularly during startup of the pump. Measurements of the backstreaming rate at various inlet pressures made under such undesirable conditions are plotted in Figure 6.22, and it can be seen that the rate does not change significantly as long as the inlet pressure is below about 10^{-3} torr. This is the point where most pumps have an abrupt reduction in speed, indicating that the top jet essentially stops pumping. The sudden increase of backstreaming at excessive throughputs indicates that the pump should not be operated in this range unless the condition lasts a very short time.

It is common practice in vacuum system operation to open the high-vacuum valve after the chamber has been rough pumped to about 10^{-1} torr, exposing the vapor jet pump to this pressure. In well-proportioned systems this condition lasts only a few seconds. Extended operation at high pressure can direct unacceptably high amounts of oil into the vacuum system.

The tests shown in Figure 6.22 were obtained with a collecting surface in the immediate vicinity of the top nozzle. If the measurements were made at the inlet plane as specified by the American Vacuum Society standard, the rate would have been 10 or 100 times lower. The higher-pressure conditions exist at the inlet to the pump briefly after the high-vacuum valve is opened. The associated backstreaming peak normally lasts only a few seconds unless the pump is severely overloaded.

The same condition will exist when the pump is initially heated. In this case the duration is a few minutes and the amplitude a few times higher than steady-state backstreaming at lower pressures. The generalized picture is shown in Figure 6.23. At pressures higher than about 0.3 torr, depending on diameter and length of ducts, the pumping fluid vapor may be swept back into the pump. This can be utilized for reduction of initial backstreaming and prevention of mechanical pump backstreaming by arranging a flow of air (or other gas) through a leak valve. However, such complications are not necessary for normal operation.

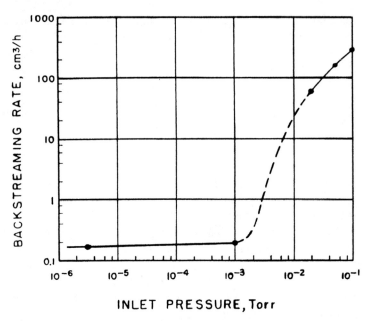

Figure 6.22 Backstreaming due to overloading a 35-inch pump, measured directly at the upper nozzle.

Another possibility for hydrocarbon contamination arises from the roughing pump. The pump should not be left connected to the system for long periods of time. After initial evacuation, the roughing valve should be closed and pumping switched to the jet pump.

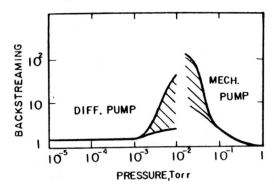

Figure 6.23 A qualitative representation of the danger zone for excessive backstreaming.

In systems where only a mechanical pump is used for maintaining vacuum conditions in the range 10^{-2} to 10^{-1} torr, sometimes it is advisable to pump periodically rather than continuously. A valve and a gauge are needed to start and stop pumping whenever necessary. This prevents continuous backdiffusion of lubricating oil and, in addition, keeps the oil cold reducing its vapor pressure.

Primary backstreaming can be measured by relatively simple means for pumps without baffles or traps. A recommended test dome is shown in Figure 6.24. The condensed pumping fluid collects in the trough around the periphery of the dome and drains into the measuring glass. Usually, it takes several days to collect sufficient fluid for satisfactory measurements (see the AVS standard). Typical results are shown in Figure 6.25, depending on whether or not the test chamber had been previously wetted with oil.

Backstreaming above the baffle cannot be measured with the standard test apparatus. The rates are so low that the collecting surfaces must be refrigerated to prevent reevaporation. An experimental system for measuring backstreaming rates from baffles and traps is shown in Figure 6.26. In this case,

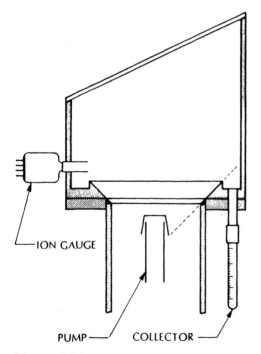

Figure 6.24 A method for backstreaming measurement.

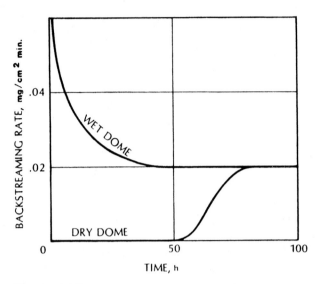

Figure 6.25 Typical results of backstreaming measurements for an unbaffled 6-inch (15 cm) pump.

the collector is made of aluminum foil in good thermal contact with a liquid-nitrogen trap. The foil is chilled to prevent reevaporation of the condensed backstreaming fluid. The collected amount can be assessed by weighing the foil before and after the test. It may be necessary to wait a few days or a few weeks to accumulate a sufficient quantity for unambiguous measurements. Figure 6.27 shows results made with the apparatus shown in Figure 6.26. The two black dots represent measurements made without (liquid nitrogen) cooling. Apparently, the rate of reevaporation at ambient temperature equals the rate of arrival and no accumulation is obtained. In general use of this system, the internal wall of the chamber above the cross section A-A (Figure 6.26) is free of visible fluid deposits.

An analysis of residual gases left in a vacuum system above a trapped vapor jet pump cannot be reliably performed with mass spectrometers having poorer detectability than 10^{-9} torr. When the spectrometer tube and other parts of the system are baked, the results can be very misleading. It may take weeks before equilibrium conditions are established.

The errors and uncertainties associated with tubulated ionization type vacuum gauges are perhaps even more serious with the mass spectrometer residual gas analyzers. This is particularly true in regard to condensable vapors of high molecular weight, so that it is extremely difficult to correlate the ion currents indicated by the mass spectrometer with the rate of backstreaming through the baffle.

VAPOR JET PUMPS

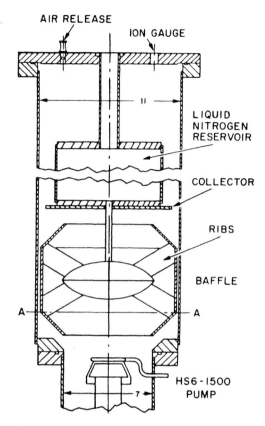

Figure 6.26 A method of backstreaming measurements above a cryotrap.

Qualitative measurement can be made, however, with no more difficulty than ultrahigh-pressure measurements made with total pressure gauges such as an ionization gauge. A typical mass spectrum from a large baffled pump system, unbaked, operating in the ultrahigh-vacuum region with DC-705 motive fluid is shown in Figure 6.28. The mass numbers 16, 19, and 35 are unusually high, due to the characteristic properties of the spectrometer tube, and the hydrogen peak is not shown. Peaks 50, 51, 52, 77, and 78 are characteristic for the pumping fluid. Another spectrum for a small pump is shown in Fig. 6.29.

6.7 OTHER PERFORMANCE ASPECTS

6.7.1 Design Features

Manufacturers of vapor jet pumps traditionally report in their specifications only pumping speed, ultimate pressure, forepressure tolerance (measured at the

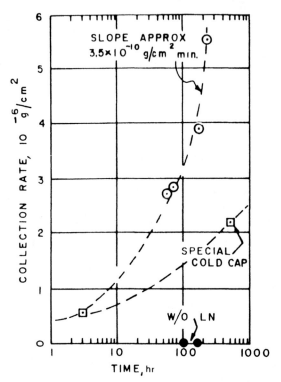

Figure 6.27 Backstreaming measurements made with the system shown in Fig. 6.26. (Black dots-uncooled collector).

foreline), throughput, and sometimes the backstreaming rate (referred to the inlet plane of the pump). With the help of the American Vacuum Society standards, most of these properties can be checked. However, there are other performance aspects listed below which can be used as guides in the selection of pumps for particular applications. Most of these characteristics can be improved in case of special need, but not without sacrificing others.

Pumping Speed Per Unit Inlet Area. This can be called speed efficiency, and it is not likely to exceed 50% compared to a hypothetical pump with 100% capture probability for molecules that cross the inlet plane into the pump.

Power Required to Obtain a Given Maximum Throughput Without Overloading and Without Oversized Forepumps. Nearly 1 kW per 1.2 torr L/s is used in common designs.

Ratio of Maximum Throughput and Forepressure Tolerance. This determines the minimum required forepump speed which must be provided at the discharge of the pump (at its full load).

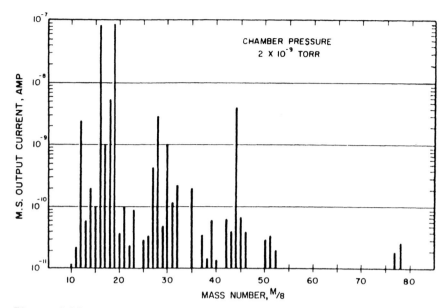

Figure 6.28 Residual gas analysis for a test done above a 35-in. pump.

Maximum Pressure Ratio for Light Gases. Of particular interest are helium and hydrogen. This pressure ratio can be sufficiently small to reveal the dependence of inlet pressure on the discharge pressure. Pumps vary widely in this regard; reported figures range from 300 to 2,000,000 for hydrogen and 1000 to 10,000,000 for helium.

Pumping Speed for Helium and Hydrogen. Compared to the speed for air, the ratio should be near 1.2. Note that the baffle conductances are greater for lighter gases. Hence, the net speed of a given pumping system for helium and hydrogen will have a greater relative value compared to air.

Ratio of Evaporation Rate of Pumping Fluid at Ambient Temperature and the Actual Pump Backstreaming Without Traps. This ratio is approaching unity in some modern pumps which have efficient vapor-condensing cold caps surrounding the top nozzle.

Ratio of Vapor Pressure of Working Fluid at Ambient Temperature and Ultimate Pressure Obtained by Pump Without Cryogenic Traps. The target for this ratio should also be near unity. To achieve this, the pump must have a low pumping fluid breakdown level and a high degree of fluid purification.

Ratio of Forepressure Tolerance and Boiler Pressure. This is usually about 0.5. The significance of this figure is in keeping the fluid temperature as low

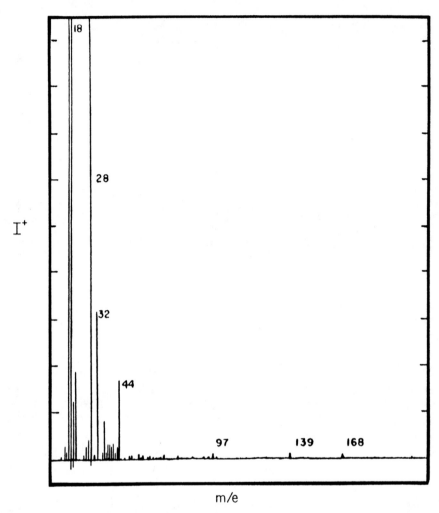

Figure 6.29 Residual gas spectrum for a 2-in. pump using Santorac-5 fluid. Total pressure, 5×10^{-7} torr.

as possible to reduce thermal breakdown while keeping the forepressure tolerance as high as possible.

Ratio of Pump Diameter and Height. The height is normally minimized for the sake of compactness, but some performance improvements could be realized if pumps were allowed to be taller.

Pressure Stability in Constant-Speed Region. This can be expressed as a percent variation referred to an average value. Pressure instability is more common in smaller pumps and with lighter gases. It is probably associated with fluctuations of boiler pressure, which can occur in a boiler that has relatively small volume for a given heat transfer area and power.

Sensitivity to Heat Input Fluctuations. In addition to variations of heater power, cooling water flow rate and temperature may have some significance. Modern pumps are not seriously affected by heat input variations.

Sensitivity to Pumping Fluid Level. See the discussion below.

6.7.2 Loss of Pumping Fluid

The pumping fluid level in a well-designed pump need not be controlled precisely. Generally, 30% above and below normal level should be tolerable without noticeable effects. When the level is too low, the boiling process may pass from nucleate boiling to partial film boiling, which leads to overheating of the boiler heating surface. If this condition is continued for an extended period of time, particularly for large pumps, it can cause distortions of the boiler plate. This, in turn, may expose the center of the boiler plate above the liquid level and lead to further overheating. The resulting poor contact between heaters and the boiler plate may also overheat the heating elements and cause their failure. If the liquid level is too high, the boiling process may foam the fluid and raise its level as high as the foreline opening.

Excluding normal backstreaming, the fluid may be lost out of the pump in several ways: prolonged operation above or near maximum throughput; accidental high-pressure, high-velocity airflow through the pump in either direction; and evaporation of higher-vapor-pressure fluids due to incorrect temperature distribution.

With relatively low gas loads, modern pump fluids, and correct systems design and operation, pumps can be operated for many years without adding or changing pumping fluid. Operation exceeding 10 years has been reported in the case of a particle accelerator. Large pumps usually have means of monitoring the fluid level. To reduce fluid loss, some pumps have built-in foreline baffles.

6.7.3 Size Effects

Vapor jet pumps are made from 2- to 48-in. inlet flange sizes. An obvious difference between the smaller and larger pumps is the distance the pumping fluid (oil vapor) must travel from the nozzle to the condensing surface or pump wall. It is apparent that the density of oil vapor by the time it arrives at the pump wall is lower in a large pump than in a small one. This is why the pumping speed plateau in the case of a 5-cm pump may be extended up to a pressure of 3×10^{-3} torr, while for a 48-in. pump it may be only 3×10^{-4} torr. This is a significant difference of an order of magnitude in the steady state or stable operation region and must be considered in system design. Special measures are necessary to improve the higher-pressure operation of large pumps (relatively higher heat input, higher number of stages).

It may be noted that the small and large pumps are not geometrically similar. The boiler pressure in small and in large pumps is approximately the same because we are limited in maximum temperature of the oil to avoid thermal breakdown. Therefore, the vapor density at the nozzle exit is nearly the same for all pumps. However, the vapor expands in both axial and radial directions, and we may assume that the density is inversely proportional to the square of the distance from the nozzle. Thus, near the pump wall the jet is rare enough to be less effective in pumping gas molecules at higher pressures.

6.7.4 Boiler Design

Vapor jet pumps of conventional design usually require a maximum throughput and a high tolerable forepressure. This requirement, together with the necessity for prevention of pumping fluid breakdown, focuses attention on heat transfer conditions in the boiler. The low degree of fluid breakdown and self-purification qualities of the pump can be judged by the degree with which the ultimate pressure of the pump follows the expected vapor pressure of the pumping fluid at room temperature. Figure 6.30 illustrates the performance of various fluids in a pump which must have a given discharge pressure (e.g., 0.5 torr). The pressure of the oil vapor inside the jet assembly may be 1 to 1.5 torr. From Figure 6.30 it can be seen that above the temperature range of 250°C, the performance of the pump deteriorates as far as ultimate pressure is concerned. Generally, it is useless to employ fluids of very low vapor pressure because of resulting excessive boiler temperatures. However, one of the requirements for a pump fluid is low vapor pressure at the ambient temperature of the vacuum system, so that ultimate pressure without traps and the migration of oil is sufficiently low. Satisfactory design must reconcile these conflicting requirements.

When using extremely low vapor pressure fluids (e.g., Santovac-5) it is sometimes necessary to adjust the water cooling rate or power to obtain optimum performance. Some pumps may have to operate with more effective heat input, some with less, depending on whether they are below or above the valley of the curve in Figure 6.30.

The boiler is often divided into two or three compartments to provide separate sections for supplying vapor to the various nozzles. This is called fractionation, borrowing the term from the chemical engineering field. The separation of the boiler assures that only the purest fluid reaches the inner section, which supplies vapor to the top jet. Therefore, the top vapor jet does not contain absorbed gases and products of thermal decomposition of the pumping fluid.

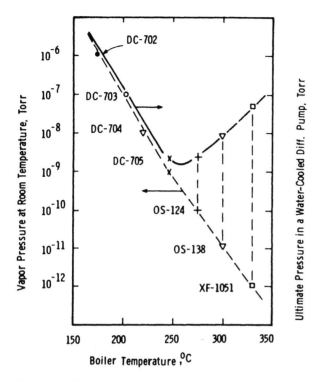

Figure 6.30 Dependence of boiler temperature at a given boiler pressure for fluids of various vapor pressures and the actual achieved ultimate pressure.

6.7.5 Pressure Stability

Pressure fluctuations occasionally seen in vapor jet pump systems can originate in the pump as well as from sources outside the pump. The sources outside the pump are (1) gas bubbles from the elastomer seals; (2) liquid dripping from the baffle; (3) high foreline pressure, light gases; (4) throughput overload; (5) trap defrosting; and (6) explosive breakup of a frozen layer in the trap. The sources inside the pump are (1) eruptive or unstable boiling; (2) boiling outside the jet assembly; (3) low pressure ratio, light gases; (4) liquid droplets in the nozzles; (5) cold top nozzle; and (6) leaks near the boiler.

Unstable boiling is one of the early recognized causes. This produces a fluctuating density of the vapor ejected from the nozzle causing variations in the pumping speed as well as in the pressure ratio for light gases. Vigorous boiling in a vacuum causes droplets to splash upward and be blown out through the nozzle opening, causing momentary blockages as well as liquid issuing from the nozzle. A baffle located inside the boiler is used to eliminate this effect.

A low pressure ratio for light gases can cause pressure instabilities. An increase in the heater power as well as an increase of the number of stages will be beneficial. Random pressure instabilities can arise from overcooling of the top jet cap, leading to discontinuities in the vapor stream.

Outside the pump, gas bubbles from the elastomer seals are the major and frequent source of pressure bursts. Gas evolved from and permeating through the elastomer seals is sometimes trapped by oil films. As the pressure builds up in the bubbles, they break the oil film and release the gas into the system. The inlet flange of the pump is the most vulnerable location. Careful design of the O-ring groove can minimize this instability. Fluid dripping from the baffles on the hot surface of the top jet cap can evaporate, producing vapor, and temporarily affect the pumping speeds. Pumps having pressure fluctuations for light gases of $\pm 1\%$ at the pump inlet have been reported. A typical pressure record with pressure bursts is shown in Figure 6.31.

6.8 BAFFLES AND TRAPS

Room-temperature baffles placed above the pump do not affect the ultimate pressure except by shielding the gauge from a direct entry of pumping fluid vapor. Water-cooled baffles suppress the rate of reevaporation of condensed or intercepted fluid, thereby reducing the density of vapor in the space between the baffle and the trap. For substances such as pump fluids, each 20°C temperature difference near room temperature will account for about an order-of-magnitude change in vapor pressure and hence the rate of evaporation. The lowered vapor density will reduce the possibility of intercollisions and consequent bypass through the trap without touching a refrigerated surface.

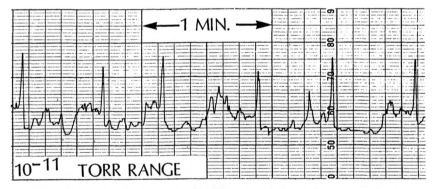

Figure 6.31 Pressure-time record showing a few periodic pressure bursts with different intensities.

Cryogenic or refrigerated traps have basically two effects. They act as barriers for the flow of condensible vapors from pump to system, but they also act as cryopumps for condensible vapors emanating from the system. The latter may be the primary effect on the ultimate pressure in many cases. In the high-vacuum region and for unbaked systems using modern low-vapor-pressure pumping fluids the reduction in pressure (when traps are cooled) is due primarily to water vapor pumping. In unbaked systems after the initial evacuation, water may constitute 90% of the remaining gases, and cooling of the trap simply increases the pumping speed for water vapor (usually a factor of 2 or 3). In rapid-cycle leak detection apparatus, for example, the trap serves to protect the mass spectrometer tube from contamination from test samples rather than from pump backstreaming.

6.8.1 Cold Caps

Cold caps surrounding the top nozzle of the pump reduce backstreaming rates 50 or more times. In medium-sized pumps, they are usually made of copper and cooled by radiation or conduction through supports contacting the water-cooled pump wall. This allows the fluid drops from the cold cap to fall into the pump, preventing evaporation into the regions outside the pump. Figure 6.32 shows the evolution of cold cap design.

Figure 6.33 shows the dependence of backstreaming on the temperature of the cold cap. The difference between the two curves is due to the degree of interception of the overdivergent portion of the vapor jet emanating from the top nozzle. It can be seen that temperatures below approximately 80°C are adequate for efficient condensation on the cold cap, while above 105°C the cold cap function is impaired.

To ensure proper operation of a cold cap, it must be thermally isolated from the hot top nozzle parts (jet assembly). In addition, it should be concentric with the nozzle and have adequate distance from it to prevent accumulation of pumping fluid of high viscosity between the hot and cold parts.

Cold caps do not significantly reduce pumping speed. Figure 6.34 shows the pumping speed effect, depending on pump size. In small pumps the reduction in pumping speed is due to the reduction of entrance area. In large pumps, the increase is probably due to the elimination of upward-directed portions of the vapor jet. This enhances the capture probability of entering gas molecules by the jet.

6.8.2 Baffles

The backstreaming rate of pumps equipped with cold caps can be measured with the standard test recommended by AVS. The test duration can be rather long,

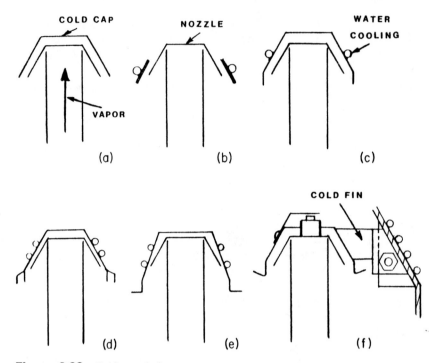

Figure 6.32 Cold cap designs.

particularly for a cold cap with an extended lip (Figure 6.32). An alternative method is analogous to a pinhole camera arranged above the pump which can trace the trajectories of backstreaming molecules. Such an arrangement is shown in Figure 6.35. The evidence suggests that the major contribution to backstreaming is associated with the vicinity of the cold cap, although the distribution profile is not as sharp as in similar tests made without a cold cap.

With this knowledge of the geometrical distribution of backstreaming patterns, partial baffles can be designed which almost eliminate the primary backstreaming. Such a baffle is indicated in Figure 6.21, item 3. Some commercially available versions of such baffles are called "halo" baffles. They retain approximately 60% of the pumping speed.

Most common baffles are so-called "optically tight" baffles, which implies that their internal geometry is such that light cannot pass directly through them. This ensures that a molecule will collide at least once with a surface regardless of the incoming direction. The pumped gas passes through the baffles with some resistance but without adsorption. The pumping fluid molecules condense inside and the condensed fluid eventually returns to the pump by gravity.

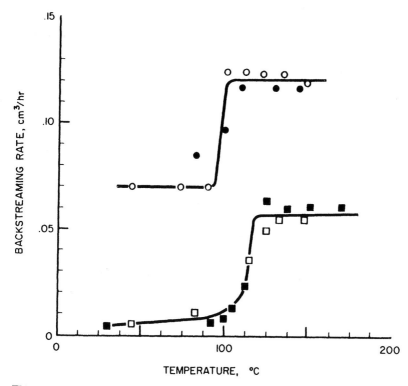

Figure 6.33 Temperature effect on cold cap performance.

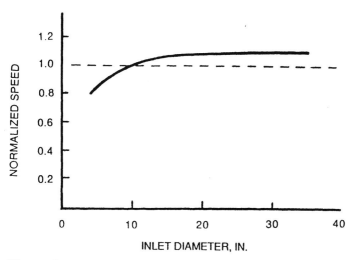

Figure 6.34 Effect on pumping speed due to the introduction of a cold cap depending on pump size.

253

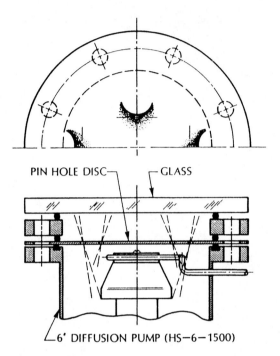

Figure 6.35 Pinhole camera method for observing backstreaming sources.

Baffle design can take many forms, but the essential geometry is passage "around a corner." The important design aspects are size, conductance, temperature, and cost. Some baffles are kept at ambient temperature, some are water cooled. Water cooling has the advantage of reducing the reevaporation rate from the baffle surfaces. Near room temperature, the vapor pressure of most substances is reduced by about a factor of 10 for each 20°C reduction in temperature. Because of radiation from the pump, an "ambient" baffle can acquire temperatures of 30 to 40°C, while cooling water is often 20° lower. The heat load at the baffle is small, so minimal amounts of cooling water are required to keep the baffle cool. It may be convenient to connect the baffle water cooling in series with the pump, the cold water going first into the baffle and then into the top of the pump.

Baffles can also be refrigerated using mechanical Freon refrigerators. The advantage would be further suppression of reevaporation; the disadvantage is the increase of viscosity of the pumping fluid so that it may not return to the pump. At about –40°C most pumping fluids will freeze.

VAPOR JET PUMPS

Conductances of baffle geometries are well understood. They can be derived theoretically and also measured in the form of small models by comparing against a known orifice conductance. The conductances or transmission probabilities of common baffle shapes are given in textbooks (see, e.g., Figure 6.36).

To have a high conductance, a baffle must have a certain height-to-diameter ratio. Chevron baffles have the advantage of small height but they have comparatively poor transmission probability. The practical design target for baffle conductance is a numerical value equal to the speed of the pump. This implies a net pumping speed of 50%.

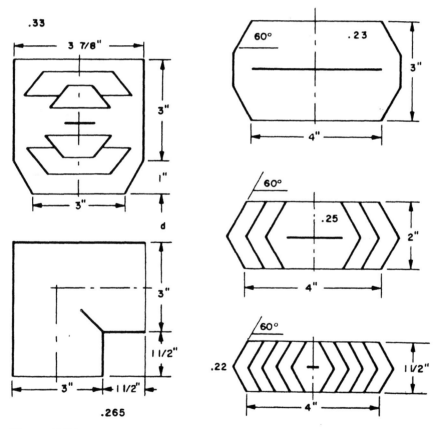

Figure 6.36 Transmission probabilities of typical optically opaque baffles (dimensions are given only to clarify the geometric shape).

6.8.3 Liquid-Nitrogen Traps

Liquid-nitrogen traps were used at very early stages of development of the vapor jet pump technology. Their use was associated with mercury, which was the initial pumping fluid. Mercury vapor pressure at room temperature is approximately 1.5×10^{-3} torr. A well-designed mercury vapor jet pump will produce ultimate pressure in the 10^{-5} torr region because the pumping jets have some pumping speed for their own vapor.

Liquid-nitrogen traps were used to reduce the pressure of mercury vapor in the vacuum chamber. With traps, pressures in the 10^{-7} torr range were obtained. It is not really necessary to cool the trap to such a low temperature but liquid nitrogen (or liquid air) happened to be a convenient cryogenic fluid available in the laboratory.

In the early days, because of the high vapor pressure of mercury, the pumping effects of water vapor were often overlooked. When oils were used instead of mercury, liquid-nitrogen traps would reduce the ultimate pressure of typical vacuum chambers by about one order of magnitude. This was due to the combination of oil vapor and water vapor pumping.

In recent years, after the introduction of very low vapor pressure pumping fluids, the reduction of pressure, after the trap is cooled, is typically only by a factor of 3. This is due primarily to pumping of water vapor. Liquid-nitrogen traps placed between the vapor jet pump and the vacuum chamber typically have a cryopumping speed for water vapor two to three times higher than the water vapor pumping speed of the pump. The pressure reduction is inversely proportional to this increase of pumping speed because in an unbaked system, 90% of residual gas is water vapor.

Liquid-nitrogen traps are used in two distinct varieties. One is a trap that is placed anywhere in the vacuum chamber. This can take a form of a cryopanel, a spherical or a cylindrical bottle, or a tubular arrangement. The other form is an optically opaque baffle that is cooled to cryogenic temperatures. The distinction between the two types of traps can be seen in Figure 6.37, where the right column shows an optically opaque geometry that ensures at least one collision with the cold surface.

Traps placed in the chamber some distance from the walls have the advantage of a higher pumping speed. If access to the cryogenic surface is not conductance limited, each square centimeter of the surface can be assumed to have a pumping speed of about 11 L/s for water vapor. The disadvantage of such traps is the inconvenience of exposure to atmospheric pressure. When the chamber is released to air, the trap must be either warmed up or isolated by a special valve to prevent excessive condensation of moisture in the air.

Cryobaffles are intended to act as barriers to prevent transfer of pumping fluid into the vacuum chamber. It should be understood that cryobaffles act as

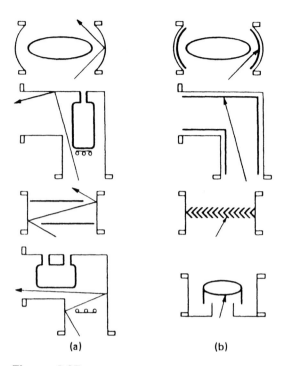

Figure 6.37 Examples of traps (a) and cryobaffles (b).

cryopumps in both directions. They pump condensable vapors, including hydrocarbon contamination, from any source in the vacuum chamber. Properly designed and continuously cooled, liquid-nitrogen cryobaffles produce backstreaming rates of about 1×10^{-7} mg/cm² · min when measured directly above the baffle by a device which itself is refrigerated to prevent reevaporation. However, this amount should not be regarded as a continuous accumulation because whatever bypasses the trap from the pump into the chamber can be repumped by the trap. The potential pumping speed of the trap is higher than the rate of backstreaming. The residual backstreaming in optically opaque continuously cooled liquid-nitrogen traps is due primarily to the small (but not zero) chance of collision and redirection of oil molecules such that they pass through the trap without coming in contact with a cold surface. The condensation coefficient of oil molecules striking the liquid-nitrogen-cooled surface can be regarded to be 1.0. When hydrocarbon partial pressures are monitored for periods of 2 to 3 months in well-trapped vacuum systems, no significant increases are noted.

Liquid-nitrogen traps can be used directly above the pump as long as the primary backstreaming is controlled by means of a cold cap or a partial baffle. This will permit continuous operation for about 1 year before a sufficient amount of the pumping fluid is transferred to the trap to cause a significant depletion of the pumping fluid level or too heavy an accumulation in the trap. For better protection and longer operation, an opaque baffle should be placed between the pump and the trap. Because of the additional conductance limitation, the net pumping speed for noncondensable gases will be reduced to about one-third of the pump speed. The aim should be maximum protection while retaining as much pumping speed as possible. The actual choice depends on application; Figure 6.38 shows a typical arrangement.

When liquid-nitrogen traps are cooled for the first time, the pressure at which this is done should be considered. For best protection against any initial transfer of pumping fluid into the vacuum chamber, the trap should be cooled as soon as high-vacuum pressure is obtained (approximately 10^{-5} torr). This, however, will produce excessive cryosorption of partially condensable gases and result in a higher ultimate pressure. To obtain the lowest possible ultimate pressure it is better to wait until the pressure is 10^{-7} torr or less before cooling the trap.

An unusual mechanism of gross backstreaming has been observed in vapor jet pump systems with liquid-nitrogen traps. It occurs sometimes during or following the charging of the trap with liquid nitrogen and can be recognized by the appearance of small droplets upstream from the trap and by accompanying pressure fluctuations. The amount of pumping fluid that may reach the base plate of a typical bell-jar vacuum system due to this mechanism may be few orders of magnitude greater than the normally expected steady-state backstreaming rate. The phenomenon is produced by fracture of the frozen pumping fluid film due to the unequal temperature expansion coefficients between it and the metal surfaces on which it was previously deposited. The elastic energy stored in the film is sufficient to impart high velocity to the fragments resulting from the fracture. This phenomenon has been observed in many traps of conventional design. However, once it is recognized, the design can be improved by avoiding certain geometric configurations which tend to accumulate heavy fluid films and produce highly stressed films which are likely to fracture. Also, the qualities of the trapping surfaces can be modified to minimize the tendency of the film fracture.

With the combination of the design and procedural remedies mentioned, the difficulties associated with the phenomenon described can be avoided. Figure 6.39 shows a record of typical pressure variation due to this effect.

Liquid-nitrogen traps generally should not be placed too close to unprotected pumps. Ambient or water-cooled baffles, partial baffles, or efficient cold caps which essentially remove primary backstreaming should be used whenever

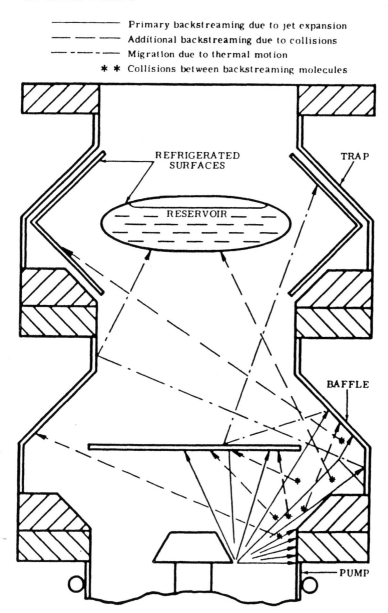

Figure 6.38 Common arrangement of a trap and a baffle.

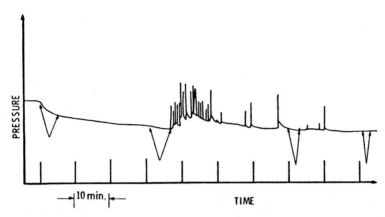

Figure 6.39 Common pattern of pressure bursts due to occasional cracking of the frozen liquid layer on the trap surfaces. Arrows indicate liquid-nitrogen filling periods.

possible. This is to reduce the accumulation of the pumping fluid on cryogenic surfaces. Traps that have removable internal structures should be cleaned periodically. The frequency will depend on the vapor pressure of the pumping fluid and possibly the degree of accumulation of condensables such as water vapor, assuming that the trap is continuously refrigerated.

A thin film (less than 10^{-3} mm in thickness) is not likely to fracture. In this respect, traps that have a long liquid-nitrogen holding time are particularly recommended.

If the trap has to be desorbed occasionally, such as during weekends, the restarting procedure needs some attention. It is a good practice to keep the high-vacuum valve closed during and after the trap filling periods. In this manner, the fractured frozen film particles and products of desorption can be prevented from reaching the vacuum chamber. The valve should be opened a certain time after recooling the trap to allow time for repumping of condensable matter. Figure 6.40 shows a trap of modern design that maintains stable temperatures for nearly 24 hours. Temperature measurements after a single filling are shown in Figure 6.41.

6.8.4 Surface Migration

Some pumping fluids may have a tendency to spread on metal surface as a thin film. This spreading is discussed in the literature dealing with lubrication of small instruments. The spreading may sometimes be noticed as wetness on rough (sandblasted) surfaces. Modern low-vapor-pressure pumping fluids having surface tensions above 30 dyn/cm do not spread on ordinary metal surfaces.

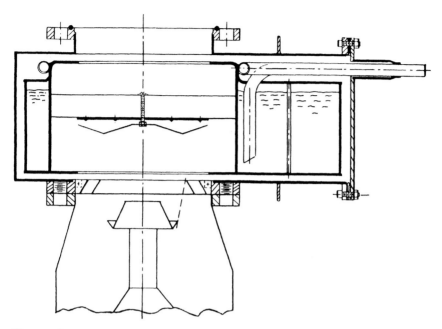

Figure 6.40 Long-life liquid-nitrogen trap with removable cold parts.

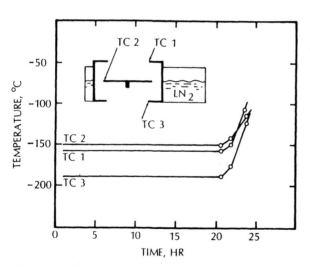

Figure 6.41 Temperature history of the liquid-nitrogen trap shown in Figure 6.40.

They are so-called "autophobic" liquids because they do not spread on their own monolayers covering a metal surface. There remains, however, the possibility of molecular surface diffusion, sometimes referred to as "creep," in conjunction with backstreaming. The possibility of surface diffusion as a mechanism of backstreaming may be estimated as follows.

The residence time associated with surface diffusion steps is shorter than the period between evaporative events. The values of heat of adsorption for surface diffusion range from 0.55 to 1.0 of the values for evaporation. A sojourn-time ratio between evaporation and surface diffusion can be used to estimate the importance of surface migration. For a 1 cm gap surface diffusion is assumed to be significant when the sojourn-time ratio is above 10^{15}. For a pumping fluid, such as DEC-705, the heat of adsorption values may be taken to be 28 and 16 kcal/mol, respectively. This gives a sojourn-time ratio of only 10^5. The residence time corresponding to 16 kcal/mol is approximately 0.1 s at room temperature. The distance between a pump (20 cm in diameter) and the chamber is approximately 50 cm. Assuming a single-step distance to be about 10 Å, a molecule has to make about 10^9 steps traveling in the same direction to cross a baffle or a trap wall. Multiplying 0.1 s per step times 10^9 steps, we obtain 10^8 s or 3 years. It may be concluded that surface diffusion is not likely to be a significant backstreaming mechanism on surfaces at room temperature (or below) for pumping fluids with vapor pressures in the 10^{-9} torr range and with a high surface tension.

The practical significance of creep barries is not in arresting surface diffusion (and spreading) but in stopping the bypass, possibly in the narrow space between cold outer parts of the trap and the warm wall. The geometry of a well-designed trap should provide opportunity for reevaporated molecules to go toward the cold surface. When this is done and when autophobic fluids are used, the major remaining steady-state backstreaming mechanism is due to collisions and possibly thermal breakdown, resulting in lighter fragments that are not fully adsorbed on the cold trap surfaces.

6.9 MAINTENANCE

The vapor jet pump is basically a relatively trouble-free device. It has no moving parts and requires only proper water flow, proper power input to the heaters, and the correct charge of clean pumping fluid. Extensive damage to the internal jets or incorrect assembly of the parts can have an effect on pumping speed and backstreaming. A routine check should be made whenever a problem is encountered to be sure that the heaters are producing the rated power. The outlet water temperature should be usually 40 to 75°C. Water flow should be adjusted to keep the outlet water temperature in this range. If a durable pump fluid is used such as the Dow Corning 700 series or Monsanto Santovac-5, a check of the oil level is sufficient. If, however, one of the lower-cost fluids is used, it may have been decomposed and require cleaning of the pump and a fresh change of fluid. The life of the fluid dependends on the exposure to relatively high temperature and pressure. Excessive loss of pump fluid suggests misoperation of the valves, that is, opening either the foreline valve or the high-vacuum valve at pressures in excess of the tolerable pressures specified by the pump manufacturer.

A slow pump-down is normally associated with high gas loads. This is evident by high foreline pressures if the pump is operating correctly. The vapor jet pump will be pumping a relatively large mass of gas when pumping at an inlet pressue of 100 to 1 mtorr. It is normal for the foreline pressure to rise to 200 or 250 mtorr during this part of the pumpdown, and the pressure will decrease rapidly as the chamber pressure is reduced to below 5×10^{-4} torr. Previous records should indicate "normal" operation and the foreline pressure can be a key to diagnosis. If the foreline pressure is higher than normal, high gas loads in the chamber or leaks in any part of the system may be suspected.

High gas loads in the system can come from two sources, leakage and outgassing. Outgassing is any gas that is adsorbed in the chamber, work fixtures, substrate, or component surfaces. The gas evolved is usually predominantly water vapor but may also be from volatile materials, such as plastics or lubricants. As both leaks and outgassing have the same effect on pump-down or on the minimum pressure achieved, a check should be made to determine the nature of the problem. Checks should be made both with an empty chamber and with fixture and the product inserted.

Vapor jet pumps generally require little attention when operated correctly. However, it is advisable to perform periodic checks to ensure continued trouble-free operation. By simple preventive maintenance, costly downtime and cleaning procedures can be avoided. A day-to-day log of pump and system performance will indicate the condtion of the pump, and marked variations will show the need for corrective action.

The frequency of inspection will depend on the type of system, its operation, and its utilization. The maximum interval between inspections is established on the basis of experience. Complete cleaning of the pump may be required periodically because of gradual deterioration of some pump fluids. Removal of the pump from the system is then necessary. Some large pumps have special cleanout ports for this purpose (Figure 6.42).

6.9.1 Power Requirements and Cooling

As has been noted elsewhere, there is no ideal power at which a given pump must be operated. A look at Figure 6.16 should be sufficient to see that various performance qualities have conflicting requirements in regard to the power

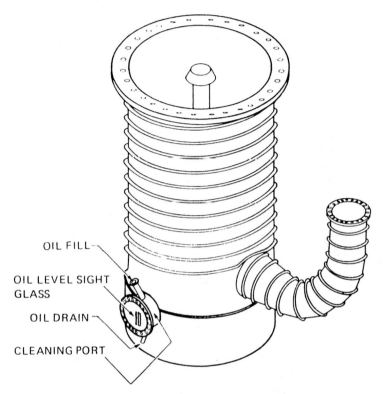

Figure 6.42 Pump with a special cleaning port at the boiler section.

level at which the specific quality is optimized. The information contained in Figures 6.16 and 6.17 should be reviewed before considering performance changes due to variations of line voltage or substitutions of pumping fluids.

The voltage rating of a heater is only a nominal value in terms of the expected power and temperature it will produce. For a given pump, the heat transfer rate will vary depending on how tightly the heaters are attached to the boiler plate, the degree of cooling of condensed liquid, the emissivity of outer surfaces of the jet assembly, and other factors.

When different pumping fluids are used, the fluid with the lower vapor pressure will have the same effect as the lowering of power. Depending on pump design, the pumping speed may increase or decrease, but the variation between common pumping fluids is near 10%, hardly a significant value for pumping system performance. The most definite changes when power is varied are associated with the maximum gas-handling capability (maximum throughput). These will change with power. Thus a performance associated with a 220-V line voltage will change approximately 14% when the line voltage is changed to 208 or 240 V. Considering that any system should have a safety factor for at least 20%, the variation is not detrimental except in cases when the system is operated close to overload conditions.

Fluids mechanics and heat transfer devices such as vapor jet pumps cannot be designed and manufactured with high precision. Some variation in performance is to be expected from pump to pump and for the same pump under varying conditions. The major performance parameters of well-designed pumps are not strongly affected by small changes in heat input, pumping fluid level, cooling water flow rate, and temperature. For small changes in heater voltage, the various effects can be considered to be linearly dependent on power. The only exception is the maximum pressure ratio that the pump can sustain, which is highly dependent on the density of the vapor jets (see Figure 6.16). Generally, $\pm 10\%$ variations of major specifications are possible. High-vacuum systems should be designed with some safety factors to avoid marginal performance. The performance variation may also be affected by changes in experimental conditions and instrumentation. In some pumps the sharp increase in helium pressure ratio may occur within the $\pm 10\%$ band shown in Figure 6.16 around the nominal power value (100%), which may become noticeable in some applications.

Normally, for best results pumps should be operated continuously. This will produce the lowest outgassing environment in the pump or in the vacuum system and will eliminate starting and stopping chores. It will also eliminate a brief period of increased backstreaming rate during the period when the vapor jets are newly formed.

However, it is desirable to reduce the power to the pump to save the cost of energy. When the pump is idling or, in other words, when the gas load is low, it can be operated at a lower power. This can be done by reducing the voltage to the heaters or by switching off some of the heaters if more than one are used.

The condensing wall temperature in oil vapor jet pumps is not critical. The pumping fluid condensation coefficient is so close to 100% that the only requirement is to remove the heat energy arriving at the wall from the vapor jets. About 80% of pump power is directly transferred to the cooling water. Most of the rest is lost to the surrounding air. Very little power is used for the actual work of compressing the gas because the mass flows are very small even at the maximum-throughput condition.

Normally, it is recommended that the cooling water be circulated from top to bottom. To achieve lower ultimate pressure, it is better to have the top of the pump the coldest. It is also better to have the bottom sections of the pump warmer in order to degass the fluid returning to the boiler. The exhaust water temperature can be as high as 55°C or 130°F (near the tolerance limit of touching with one's bare hands), without detrimental effects.

The requirement of removing about 80% of pump power must be kept in mind when recirculation cooling systems are utilized. The same is true for small fan-cooled pumps. If an air-cooled pump is placed inside a closed cabinet, there should be provision for a supply of cool room air at the entrance to the fan.

6.9.2 Special Recommendations

The following recommendations should generally be followed unless compelling circumstances require deliberate departures. The maximum permissible discharge pressure (the tolerable forepressure) should under no circumstances be exceeded when the pump is operating. Strict observance of this most basic rule is mandatory and should be explained clearly to all equipment operators. This will eliminate most of the gross backstreaming difficulties encountered in vapor jet pump systems.

The maximum gas load capacity of the pump (maximum throughput) should not be exceeded in steady-state operation. The transient overload period immediately after the high-vacuum valve is opened should be kept as short as possible. If this period is longer than a few seconds, the roughing should be continued for a longer period to a low pressure before returning to pumping with the vapor jet pump.

An operating pump should normally not be air-released to atmospheric pressure, to prevent exposure to oxygen of the hot pumping fluid. In systems without valves, air release should not be undertaken from the discharge side. Otherwise, even if the high pressure suppresses boiling, vapor in transit between the nozzle and the pump wall will be carried into the chamber.

High-vacuum valves upstream of vapor jet pumps should be opened slowly to reduce the degree of transient overload. If the system or process tends to operate near or sometimes above the overload point, the conductance of inlet ducts (baffles, traps, valves, etc.) should be reduced. This can be done by choosing baffles of lower conductance and better baffling properties. In sputtering work, throttle valves are introduced to allow chamber pressures in the range 10^{-2} to 10^{-1} torr without exceeding the maximum throughput of the pump.

Baffles and traps or, in their absence, the pump itself, should be the coldest part of the vacuum system. Otherwise, the pumping fluid vapor will tend to migrate to the colder area in the chamber.

Liquid-nitrogen traps should be defrosted periodically to avoid heavy pumping fluid deposit, which may break up when the trap is refilled. Radiation of thermal energy directly into the cryogenic surfaces of the trap should be avoided.

6.9.3 Safety, Hazards, and Protection

Vapor jet pumps perform very well within their normal operating range. They have been widely used for many years as a simple and efficient means for attaining high vacuum. Knowledge of their limitations—particularly at the higher-pressure end of their operating range—can be extremely helpful in achieving best performance.

All pump fluids, with the exception of mercury, are heat sensitive to some extent. In general, if the pressure in a vapor jet pump rises above a few torr near the operating temperature, the power should be turned off. When hot, hydrocarbon fluids may be damaged even by a brief exposure to air. Silicone fluids are much more resistant to oxidation, but even they can be damaged when the pressure and temperature are significantly higher than normal for a period of time. It is possible under certain conditions of temperature and pressure for most pumping fluids to ignite and cause an explosion. Such explosions are usually not strong enough to burst the vacuum chamber, but damage to high-vacuum valves and sudden opening of doors can occur. This has been reported, in most cases, when an overheated pump was exposed to air. Vapor jet pumps may be operated for brief periods of time at inlet pressures as high as 0.5 torr without damage to the fluid. But a steady gas flow at pressures above the normal operating range of the pump may result in a rapid depletion of pump fluid. This may result in overheating. Thermal switch protection, fail-safe valve arrangements, and means of checking fluid level, particularly for large pumps, should be provided.

REFERENCES

1. W. G. Brombacher, *Bibliography of Index on Vacuum and Low Pressure Measurements*, Monograph 35, National Bureau of Standards, Washington, D.C., 1961 and Supplement 1966.
2. M. H. Hablanian and J. C. Maliakal, *J. Vac. Sci. Technol.*, 10(1), 58 (1973).
3. M. H. Hablanian, *Proc. 6th Int. Vac. Cong.*, Kyoto, Japan, 1974, IUVSTA.
4. G. Toth, *Vakuumtechnik 16*, 41 (1960).
5. M. H. Hablanian, *J. Vac. Sci. Technol.*, A12, 4, 897, (1994).
6. A. K. Rebrov, *Vacuum*, 44, 5–7, 741–752 (1993).

7
Turbomolecular Pumps

7.1 INTRODUCTION

Turbomolecular pumps are essentially axial-flow compressors designed for pumping rarefied gases. Original designers adapted more or less traditional axial-flow compressor stage arrangements using mathematical modeling that was based on studies of molecular trajectories inside alternating rotating and stationary blade rows. Recent design trends lean toward hybrid stage arrangements, which incorporate turbomolecular and turbodrag stages within the same body and mounted on the same shaft. The new pumps achieve much higher compression ratios (10 to 100 times), permitting higher discharge pressures and allowing the use of oil-free backing pumps.

When engineers first encounter high-vacuum technology, they sometimes have difficulty in understanding molecular flow concepts but, after a few years, they seem to have more difficulty in associating the rarefied gas flow with the more common higher pressure viscous flow. There are similarities, however, between the two and the appreciation of such similarities can lead to better designs. High-vacuum pump technology was developed mostly by experimental physicists, electrical engineers, and some chemists. As a result, unique design concepts developed and the descriptive terminology is not well-related to mechanical engineering and, specifically, to fluid mechanics. Even worse, because different vacuum pumps have been developed by different persons in different times, the terminology used to describe the performance of various pumps is different. For example, even though turbomolecular and vapor jet pumps are very similar in their basic performance, the compression ratio values are firmly associated with turbomolecular pumps but almost never mentioned in relation to vapor jet pumps.

In high-vacuum technology, we speak of forepressure tolerance instead of maximum permissible discharge pressure or, even more simply instead of the pressure achieved versus the mass flow rate, we speak of pumping speed at various inlet pressures, thus associating the concept with the rate of evacuation of a chamber rather than with bulk velocity (or volumetric flow) at the inlet of the pump. For the last 30 years, the technical committees whom write standards, recommended practices, and glossaries, have struggled to define concepts such as tolerable forepressure, ultimate pressure, intrinsic speed, or maximum mass flow. We plot speed versus pressure instead of pressure versus speed, etc. The different images, compared with the usual fluid-mechanic terminology, often produce cause-and-effect confusions that interfere with the complete understanding of the pumping mechanism and cloud the vision for improvement and achieving an optimum design.

In this respect, we are somewhat lucky with turbopumps because the initial description of the pumping mechanism of turbomolecular pumps was done in a technical university (Massachusetts Institute of Technology), but even in this case the common initial period of confusion and fallacies was not entirely avoided. For example, it is not fully understood why pumping speed rapidly decreases at pressures above a certain level (near 10^{-3} mbar), and why stators are placed between each rotor. The arrangement of sequential stages is often made without consideration of the pumping speed requirements of each stage, according to the rule of preservation of mass or according to the reduction of bulk volume flow in each compression stage.

In general, turbine-type pumps are very flexible with regard to the possibilities for different designs. Established during the last 25 to 30 years as the prime method of high-vacuum pumping, they appear to be entering into a second period of creative design from which a few typical varieties will emerge in the near future.

At the present time, there are four types of turbine pumps: multistaged axial-flow turbomolecular pumps; molecular drag pumps (usually of the Holweck type); hybrid pumps with modified downstream stages (but the same number of rotors); and compound pumps that combine the axial stages with additional drag pumps in the same body. Turbine-type high-vacuum pumps, unlike vapor jet pumps and cryopumps, permit relatively simple engineering solutions for increasing their pressure regime to an almost arbitrarily high level, including the possibility of discharging directly to atmosphere. This provides opportunities for further improvements.

7.2 MOLECULAR DRAG PUMPS

Molecular pumps are pumps in which the pumping action is produced by momentum transfer from a fast moving surface directly to gas molecules. In diffusion pumps, the gas molecules are "pushed" from inlet to discharge section of the pump by a fast, directed stream of vapor. In molecular pumps, collisions occur directly with a moving solid surface. First mention of such pumps is attributed to W. Gaede in 1912.

The pumping action can be visualized by considering the rotor in Figure 5.5 without the vanes and without the slots for the vanes. When the pressure is high (near atmospheric) and the rotor velocity is low, the pumping action will be extremely inefficient. If any pressure difference is established due to drag at the surface, it will be lost immediately because of backflow some distance from the rotor. But under molecular flow conditions and the peripheral rotor velocity approaching the average velocity of gas molecules, a significant pressure difference can be maintained. Every collision with the rotor will send the molecule back to the discharge area.

Early versions of molecular pumps are shown in Figure 7.1. The basic pumping action is the same in all three versions. The disk type has the disadvantage of lower tangential velocity in the spiral grooves near the center, but it is easier to balance and has a single rotor. Figure 7.2 is an example of an old Siegbahn-type pump.

The pumping speed of early pumps was low. The art of making high-speed rotating machinery had not yet been developed. It took nearly 50 years before a molecular pump was made as an industrial device of adequate reliability.

The simple schematic designs of the three original pumps (known as Gaede, Holweck, and Siegbahn) are shown in Figure 7.1. Despite occasional claims to the contrary, they have no principal differences. All have a surface rotating at velocities comparable to the ordinary thermal velocities of gases (usually at room temperature) and a channel in the stationary part surrounding the rotor. Molecules enter the channel and are driven to the exit as a result of imparted additional velocity when they collide with the rotor. This process works very well under molecular flow conditions producing rather large compression ratios. In the absence of real pressure gradients, as far as individual molecules are concerned, it is easy to achieve high compression ratios compared to viscous flow conditions. Figure 7.3 demonstrates the difference in behavior of the two flow conditions. In viscous flow, if the gap between the rotor and stator is sufficiently large, reverse flow will develop due to the pressure gradient to which the molecules are subjected. In general, for obvious reasons, molecular flow is distinguished by very high pressure ratios and very minimal pressure differences, while in viscous flow the reverse is true. It makes more sense to

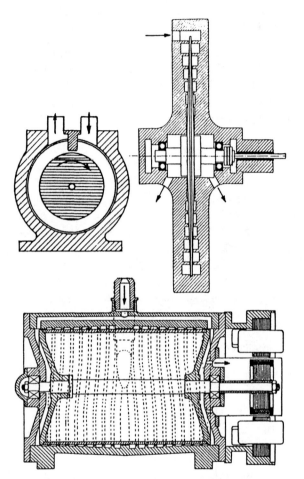

Figure 7.1 Three early versions of molecular drag pumps.

characterize the molecular range performance using compression ratios, but in viscous flow the pressure difference is a more meaningful parameter.

Note that the Holweck pump can be easily converted into a multistage Gaede pump by simply making concentric grooves with axial connectors instead of helical grooves, as shown in Figure 7.4. In the same way, a Siegbahn pump can be converted into a Gaede pump of varying diameters with radial connectors. The difference is in the geometrical arrangements that may be deemed more convenient for a particular mechanical design.

It is somewhat difficult to clearly define a number of stages for drag pumps. In viscous flow, the accumulated pressure difference is proportional to the

Figure 7.2 An example of old Siegbahn-type pump (Beaudouin Co., Paris, 1957) which had a pumping speed of 60 L/s and a compression ratio for air over 10^5.

viscosity of the gas, the velocity of the rotor, the length of the channel, and inversely proportional to the square of the channel height. It is then better to talk of pressure rise per unit length at a given velocity rather than a pressure ratio per "stage." In molecular flow, the compression ratio is proportional exponentially to the molecular weight, rotor velocity, channel length, and inversely proportional to channel height. In practice, none of these relationships is preserved because of lateral leakages along the channel. For direct comparison of pumps, a list of channel lengths and the peripheral velocities of the rotor at the channels (assuming a constant rotational speed), is necessary.

7.2.1 Performance

Figure 7.5 shows a cross section of a modern molecular drag pump. The rotor is a cylindrical cup open at the bottom. It rotates at tangential velocity comparable to that of air molecules at room temperature. There are two cylindrical stator surfaces containing helical grooves, one outside and one inside the rotor. The pumped gas molecules are propelled inside the stator grooves, first

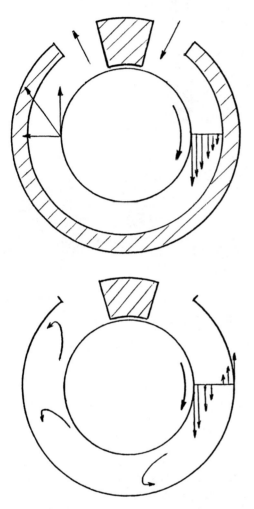

Figure 7.3 Basic gas velocity patterns in a Gaede-type pump: top left, molecular flow; top right, viscous flow; bottom, reversed (viscous) flow when the rotor-stator spacing is too great or pressure is too high.

downward at the outside of the rotor and then upward in the internal grooves. The long path provides multiple "stages of compression," the actual value of which depends on the rotor-to-stator distance. This distance should be as small as possible (within the practicality of the mechanical design) to prevent excessive leakage from one groove to the adjacent groove, which is at a lower pressure.

TURBOMOLECULAR PUMPS

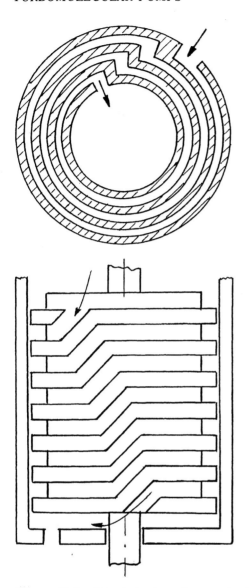

Figure 7.4 Concentric channels approximating Siegbahn-type (top) and Holweck-type patterns (bottom); with strippers in each loop and interconnecting ducts in the stationary wall this is like a Gaede pump.

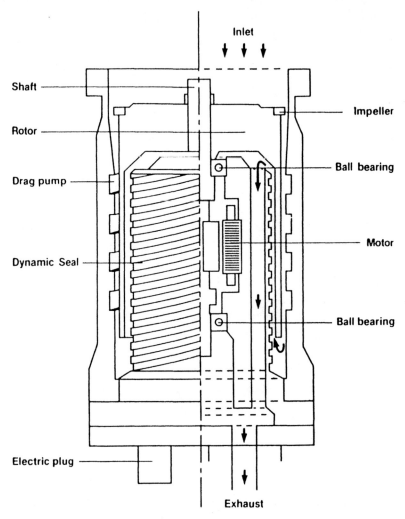

Figure 7.5 Cross section of a molecular drag pump (Alcatel).

Pumps of this type can maintain a compression ratio of 10^9 (for nitrogen) while discharging into an environment that is at 10 to 40 torr. Very special designs have been made that could discharge directly to the atmosphere, but this requires extremely close air-bearing fits which are not only difficult to manufacture but are subject to damage in the presence of small particles.

There are advantages to having high tolerable forepressure in molecular drag pumps. One advantage is that the backing pumps can be small. Also, the possibility of backstreaming from the lubricating oil or grease is reduced because at 10 torr both the evaporation and the mobility of oil molecules are suppressed. In addition, high tolerable forepressure permits the use of oil-free backing pumps.

The pumping speed of a typical molecular drag pump is shown in Figure 7.6. A typical compression ratio for helium is over 10,000, and for hydrogen, 1000. Molecular drag pumps can be used for many applications where pumping speed and low-pressure requirements are modest and the functional inlet pressure range extends from 10^{-6} to 10^{-1} torr. They can be useful, however, even for large chambers, as shown in Figure 7.7, which illustrates the evacuation of a 650-L mild steel chamber with a molecular drag pump of nominal speed 27 L/s. The crossover pressure is 1 torr, although for continuous operation the inlet pressure must be kept below 0.1 torr. Switching sooner from the roughing pump (which in this case happens to be a 450 L/min oil-free pump) will only slow down the evacuation process because the drag pump, at a pressure above 1 torr, acts as an impedance.

7.2.2 Pumping Mechanism

The mathematical expression for the pumping mechanism in channels or ducts adjacent to moving surfaces is rather complex and is certainly beyond the scope of this book. The recent basic treatment can be found in articles by Helmer,

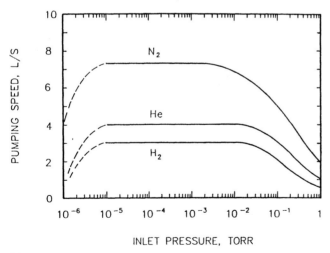

Figure 7.6 Pumping performance of a drag pump for nitrogen, helium, and hydrogen.

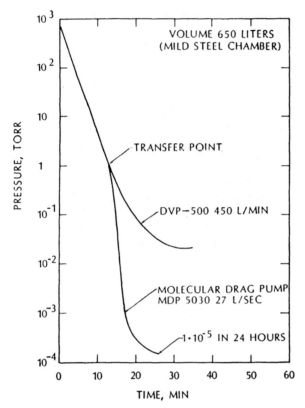

Figure 7.7 Evacuation of a relatively large chamber (650 L) using a 450-L/min oil-free pump and a 27-L/s drag pump.

Levi, and De Simon for pumps with geometry resembling Gaede disks, and by Sawada for pumps of the Holweck type.

Molecular Flow

The maximum possible pumping speed is obviously limited by the cross sectional area of the pumping channel and its conductance, and by the velocity of the moving surface or the rotor. In most textbooks, the pumped gas leakage at the obstacle (stripper) in the pumping channel, which separates the inlet and discharge, is neglected. This leads to values for the maximum compression ratios too high (at zero flow) by a few orders of magnitude. A typical derivation for pumping speed and the maximum compression ratio is included here, not for design purposes but to demonstrate the pumping mechanism and the trend effects of various parameters.

For simplicity, consider a pumping channel profile as shown in Figure 7.8. It has a relatively great perimeter of the moving surface. An extreme example of this type of profile would be a hollow tube, rotating around its center in its plane (as a bicycle tire), and the stationary stripper may be considered held in place by external magnets, but this would not make a practical pump. We examine a small differential element of channel length (dx) at some point of the ring. (The basic derivation follows the discussion by Rozanov). The differential equation for gas flow in steady-state can be written as the difference between forward flow (due to the velocity of the moving surface) and the backward flow (due to the developing pressure difference):

$$Q = S_0 p - C_x dp \qquad (7.1)$$

where Q is the throughput, S_0 is the maximum pumping speed, p is the pressure, C_x is the conductance, and dp the pressure of a differential element. In our case, the maximum pumping speed is given approximately by the cross sectional area of the pumping channel multiplied by the velocity of the moving surface

$$S_0 = \text{Area} \times V.$$

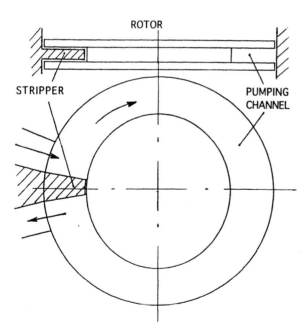

Figure 7.8 A basic schematic drawing of a molecular drag pump.

In practical cases, where the moving surface comprises only a part of the cross sectional periphery, a weighing function can be used to express the maximum pumping speed, for example, the ratio of the moving part to the total. To reflect the pressure change along the channel, we can multiply the top and bottom of the second term of Eq. 7.1 by dx and notice that for a channel of constant shape and area, $C_x dx$ can be replaced by CL, where C is the conductance of the entire channel and L is its length. Then, the equation can be written

$$Q = S_0 p - CL dp/dx \qquad (7.2)$$

This equation is in the form $dp/dx + Ap - B = 0$, where $A = -S_0/CL$ and $B = -Q/CL$. Considering the boundary condition $p = p_1$ when $x = 0$, the solution obtained is in the form

$$p = (B/A)(1 - e^{-Ax}) + p_1 e^{-Ax} \qquad (7.3)$$

Then, at the end of the channel, where $x = L$ the pressure p_2 will be

$$p_2 = (Q/S_0)(1 - e^{S_0/C}) + p_1 e^{S_0/C} \qquad (7.4)$$

Since $Q = p_1 S_1$, from Eq. 7.4

$$S_1 = S_0 (p_2/p_1 - e^{S_0/C})/(1 - e^{S_0/C}) \qquad (7.5)$$

When $p_1 = p_2$, the pumping speed is greatest and $S = S_0$. When the net pumping speed is zero (for example, when the exit valve is closed), the pump reaches its maximum compression ratio. Setting the nominator of Eq. 7.5 to zero, we obtain

$$K_{max} = e^{S_0/C} \qquad (7.6)$$

The dependency of compression ratio on conductance demonstrates that the compression ratio will be higher for gases of higher molecular weight and lower gas temperatures.

The compression ratio obtained using Eq. 7.6 is an exaggerated value. When leakage through the gaps between moving and stationary surfaces is taken into account, the result obtained for the maximum compression ratio is not zero flow but rather equal flow through the main channel and through the leak. It must be noted that the leakage around the stripper in Gaede-type pumps, or between adjacent grooves in Holweck-type pumps, is not due to the existing pressure difference but actually is driven across by the motion of the rotor surface. Thus, the situation can be viewed as having two pumps in series with each other. Therefore, the maximum compression ratio achieved will be mainly due to the ratio of the cross sectional areas of the pumping channel and the cross sectional areas of various leakage paths. For small and medium-sized pumps (rotor diameters 5 to 10 cm), the compression ratio is likely to be 50 per circular loop instead of millions as predicted by Eq. 7.6.

In addition to the effect of unavoidable leakages, other departures from the simple theory exist. Molecular drag pumping channels usually have very low pumping speeds, typically a few liters per second. Therefore, even if the inlet is closed, the outgassing rate of the channel walls (after few hours of pumping) can limit the inlet pressure to about 10^{-6} torr. Also, the pumping action may depend on the ratio of the molecular velocity of the pumped gas to the velocity of the moving surface. When molecular velocities of the gas are lower than the velocity of the rotor surface, the discontinuities may appear in gradual progression of density (built-up along the channel length), producing a rarefaction region downstream of the stripper (obstacle) and a densification region ahead of the stripper.

Viscous (Laminar) Flow

A similar derivation can be made for viscous flow. Here, the conductance depends on pressure and is proportional to pressure and diameter to the fourth power (for a circular cross section) and inversely proportional to the viscosity of the gas and the channel length. For a rectangular cross section of the pumping channel, analogous to Eq. 7.1, the pumping speed can be expressed as

$$S = whV - wh^3 dp/\mu dx \qquad (7.7)$$

where w is the width, h the height, μ the viscosity, and V the rotor velocity. When net flow is zero,

$$whV = wh^3 dp/\mu dx \quad \text{or} \quad dp = \mu V dx/h^2 \qquad (7.8)$$

The Eqs. 7.7 and 7.8 need a coefficient relating to the geometry of the channel and the used unit quantities. For equal moving and stationary cross sectional perimeter sections, with linear dimensions in centimeters, velocity in cm/s, and viscosity in poise, the coefficient is 0.5 and

$$\Delta P_{max} = 6 \; \mu VL/h^2 \qquad (7.9)$$

This solution is much closer to real values obtained for small pumps with a pressure difference of 1.5 to 3 torr per circular loop. Generally, from an engineering point of view, it may be more convenient to speak of the maximum pressure difference for the viscous flow range and of the compression ratio in molecular flow range (see Section 3.8).

7.2.3 Experimental Examples

Figures 7.8 to 7.12 (after G.Levi), provide experimental results for the performance of molecular drag channels in transition flow regime. The following pump parameters were used:

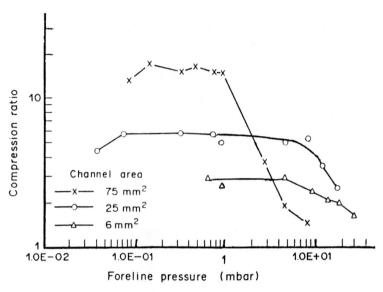

Figure 7.9 Compression ratio versus discharge pressure for a single loop molecular drag pump (from G. Levi).

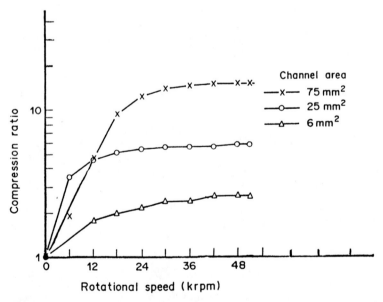

Figure 7.10 Compression ratio versus peripheral velocity of the rotor for a single loop molecular drag pump (from G. Levi).

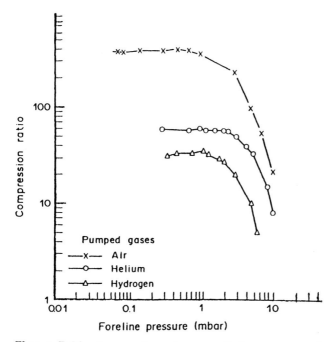

Figure 7.11 Compression ratio versus discharge pressure for a three-loop molecular drag pump with different pumped gases (from G. Levi).

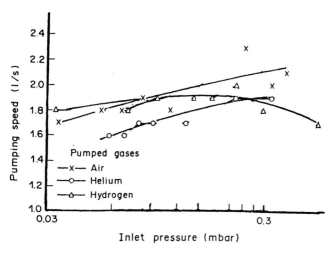

Figure 7.12 Pumping speed versus inlet pressure for a three-loop molecular drag pump with different gases (from G. Levi).

Diameter of the rotor (disk) 9.4 cm
Length of the pumping channel 22 cm
Peripheral velocity the of rotor 250 m/s
Active moving surface (rotor) height 0.5 cm
Gap between rotating and stationary parts 0.2 mm

The results demonstrate the basic trends. The highest compression ratio of 16 per loop was obtained with a channel cross section of 0.75 cm^2, indicating the importance of the pumping channel size relative to the leakage areas, but smaller channels maintained a high compression ratio at the higher pressure levels (see Figure 7.9). This was expected because the smaller channels are less likely to produce reverse flow (see Figure 7.3).

Figure 7.10 demonstrates the typical leveling of the compression ratio with the increase of peripheral velocity (an important engineering aspect), which is implied in Eq. 7.6 because conductance increases in transition and in the viscous flow region. Figure 7.11 shows compression ratios for different gases. The values are lower for lighter gases mainly because the conductance of the pumping channel is higher and thus the relative pumping speed is lower. Figure 7.12 indicates that the actual pumping speed for different gases is not very different, presumably because the ratio of pumping speed to conductance is similar.

The effect of leakage and the sudden reduction of compression ratios at high discharge pressure can be appreciated from Figures 7.13 and 7.14 (from Helmer

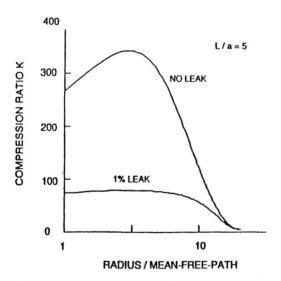

Figure 7.13 Compression ratios computed for a tubular channel with and without leakage L/a, length to radius ratio (from Helmer and Levi).

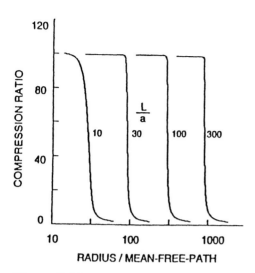

Figure 7.14 Calculated compression ratios for a circular channel, depending on the ratio of length to radius under various conditions of transitional flow (from Helmer and Levi).

and Levi). The first (Figure 7.13) shows a theoretical estimate with a model similar to Figure 7.8 and rotor velocity equal to the average molecular velocity of the pumped gas. Figure 7.14 shows theoretical estimates of the sudden reduction of compression ratios, depending on flow regime for a pumping channel of idealized circular cross section and a flat disk rotor protruding into the channel to 3/4 of its diameter. These and similar experiments can be used to arrange a sequence of turbine-type and drag-type pumping stages to obtain a set of desired performance aspects for a turbomolecular pump (i.e., pumping speed, compression ratio, maximum sustainable pressure difference, maximum gas flow without overload, power requirements, etc.).

Normally, measurements of compression ratios with different gases are done by introducing a more or less pure gas at the discharge and observing the effect on the inlet pressure. However, when mixtures of gases are present at the discharge part of the pump, where flow conditions are in the viscous range, the compression ratio may be affected by the presence of other gases and will no longer follow the relationship established at lower pressures.

The sudden reduction of the compression ratio at higher pressures in molecular drag pumps (and turbopumps) can be associated with the general properties of compressors as discussed in Section 3.8 and Figures 3.13, 3.14, and 3.15; the approach of the limit of the maximum sustainable pressure difference (Figures 10.10 and 10.11); and power issues reflected in Eq. 10.8 and Figure

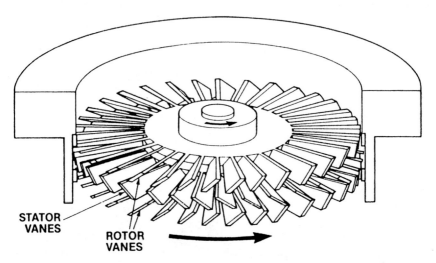

Figure 7.15 View of an axial-flow pump. Two rotors and a stator are shown.

3.15. It is related to the similar effects in vapor jet pumps discussed in Section 6.6.6 and shown in Figure 6.17.

7.3 TURBOMOLECULAR PUMPS

Modern turbomolecular pumps of an open, thin-bladed, axial-flow type appeared in the early sixties following the publication of the article on the performance of an axial flow compressor under rarefied gas conditions (1st International Vacuum Congress, Ref. 2). Prof. A. Shapiro of MIT and his associates provided the theoretical model for the pump and shortly thereafter the first engineering designs were produced by a few commercial vacuum companies.

Despite the existence of molecular flow at the entrance of turbomolecular high-vacuum pumps, they can be viewed as a special adaptation of an axial-flow compressor. Molecular dynamics must be considered for detailed mathematical modeling but the general conceptual pictures of bulk flow patterns are rather similar. In some analogy with vapor jet pumps, the major difference is the longer distance from which the molecules enter the pump parts, which impart energy to the flow. Also, due to the very high compression ratio in high-vacuum pumps, the flow conditions are not always molecular at the discharge end.

In general, compared to other compressors, the axial-flow turbine has the highest volumetric efficiency for a given size or diameter. The rotor in this type of device is concentric with the stator. External forces on the rotor from the flowing fluid are entirely symmetric around the periphery. Balanced forces permit high-speed rotation. With the peripheral velocity at the tip of the blades approaching the speed of sound in air, a very high entrance speed for the pumped gas can be obtained. Because of the high rotational speed for a given diameter, the turbocompressors are small and light, which is the major reason they are used in aircraft jet engines. At ordinary pressures, multistaged axial-flow compressors produce a pressure ratio of 3 to 5. Typically, about 10 stages of axial rotors are needed to produce the same compression obtained by a single centrifugal-flow rotor.

An axial-flow compressor consists of a set of alternating bladed rotors and stators, as shown in Figure 7.15. When such compressors are tested in a high-vacuum environment, they produce a compression ratio per stage that is nearly 10 times higher than that of ordinary pressures. This produces an adequate compression ratio for, say, 10 stages to make a useful high-vacuum pump. Of course, at an extremely rarefied gas density, the motor necessary to drive the device is much smaller, the forces on the blades are much less, and because of molecular flow, there is no need to use blades with air-foil shapes.

Turbomachinery compressors have one more significant advantage over other types of vacuum pumps, including molecular drag pumps and vapor jet pumps. Due to the symmetrical position of the rotor and the gas flow pattern around the periphery, there is no surface that is alternately exposed to high- and low-pressure areas. Under steady-state conditions each area of the rotor (and stator) has an established constant pressure. This is of some significance at pressures approaching ultrahigh vacuum, due to the avoidance of sorption and desorption effects resulting from alternate exposure to high- and low-pressure regions. In vapor jet pumps, for example, the pumping fluid is alternately exposed to the foreline environment and must be degassed before entering the upper nozzle.

Turbomolecular pumps first appeared in the high-vacuum industry in 1958. At first, they were derived from molecular drag pumps and had thick blades and thin slots for the gas flow passage. About 10 years later thin-bladed rotors were developed, which resemble compressors used in aircraft. Hybrid and compound pumps appeared around 1985.

7.3.1 Pumping Mechanism

There are a few ways to explain the pumping mechanism of an axial flow turbopump in high-vacuum environment. It can be considered as a momentum transfer device where moving inclined blades collide with individual pumped gas molecules and accelerate them in the downstream direction. To be effective, the velocity of the moving surface must be of the same order as the thermal velocity of the gas. Another way is to consider the action of the rotating blade row as a volumetric displacement, not positive as in a piston pump but a partial displacement. The momentum transfer model can be visualized as a moving semi-chevron baffle, which (as a result of its velocity relative to the "stationary" gas), has a higher conductance in one direction and lower in the opposite. Then the pumping action can be obtained as a net result of molecular movements in forward and backward directions. The mathematical methods can be the same as used for determining conductances of baffles.

For the displacement model, we can (for a moment) visualize the rotating blade row to be like a propeller in an aircraft moving ("flying") into the stationary gas. Then, the displacement will be

$$S_d = \text{(velocity in pumping direction)} \times \text{(projected pumping area)}$$

where pumping velocity is rotational velocity multiplied by the cosine of the blade angle, and the projected pumping area is the annular area of the bladed section multiplied by the sine of the blade angle. Then, the pumping speed is

$$S = k(\pi D f \sin \alpha)(\cos \alpha \times \pi D \, l) \tag{7.10}$$

where k is the coefficient of volumetric efficiency (which may be called the capture coefficient), D is the average blade diameter (at the midpoint of the blade), f is the frequency of rotation (cycles per second), l is the length of the blade, and α is the blade angle. To account for the blade thickness, subtract the diameter associated with the pumping area of the thickness of the blades (zt, number of blades times their thickness). Thus,

$$S = k \sin \alpha \cos \alpha \, f \, \pi \, D \, (\pi D - zt) \, l \tag{7.11}$$

The important design parameter here is the coefficient k, which depends on the ratio of rotor velocity to the molecular velocity of the gas, the ratio of blade height to the distance between adjacent blades, back leakage around the blade tips, etc. For pumping speed estimates of typical turbomolecular pumps with first row blade angles of 45°, k is near 0.6 for air or nitrogen and the corresponding compression ratios are typically between 5 and 10 per rotor-stator pair.

7.3.2 Design Features

Turbomolecular pumps are employed in the same pressure region as that of vapor jet pumps. Throughout most of a turbomolecular pump, molecular flow conditions prevail, but at the point of discharge the flow is usually in the transition region and may even approach the low end of the viscous flow region.

The basic pumping mechanism of turbomolecular pumps differs from that of the axial-flow compressor only in degree, not in kind. However, the analytical techniques used in analyzing and optimizing the design for molecular flow conditions may be different. The early analytical approach for the design of blades (width and height), blade angles, and distances between blades was developed at MIT under the direction of A. H. Shapiro. Figure 7.16 indicates the existence of net flow from left to right for a row of blades moving downward. A molecule colliding with the moving blade at the entrance to the blade now can pass to the right side (angle 3) or return to the entrance (angle 1). A molecule at the discharge side has a much greater chance of remaining at the point of discharge (angle 1). Thus the inclination of the blades produces a net flow toward the right.

To obtain a maximum pumping speed, the blades should have less inclination; that is, they should have a more open semi-chevron baffle shape, which has a higher conductance. But to produce the highest possible compression ratio, the blade inclination should be higher. This leads to the obvious conclusion that at the entrance to the pump, the blade angles should be small, and they should increase at the later stages. The most efficient design would require changes in all blade design parameters for each rotor and stator, but for practical rea-

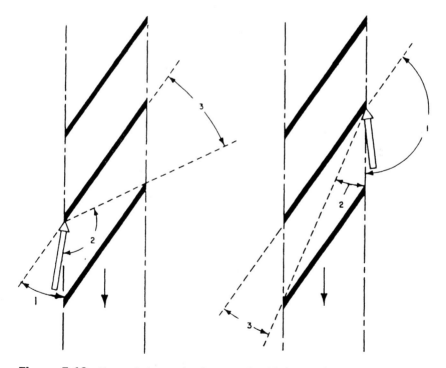

Figure 7.16 Transmission angles for a moving blade row. Pumped gas flows from left to right.

sons, blade heights and blade angles are usually changed only once or twice. The number of stages depends on the desired performance (the overall compression ratio). Typically, the number of pumps in commercial production is between 9 and 13. A common arrangement of the entire train of rotors and stators is shown in Figure 7.17.

We may ask: What is the role of the stator? In viscous flow conditions, the stator serves the same function as a diffusor in case of the nozzle flow (Section 3.4). An illustration of stator function is shown in Figure 7.18.

The role of the stator in turbomolecular pumps is, in principle, not really different from the role of the stator in axial flow compressors (such as in front

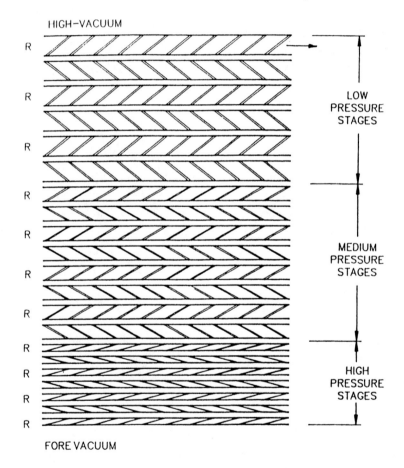

Figure 7.17 Typical assembly of rotors and stators for conventional pumps.

TURBOMOLECULAR PUMPS

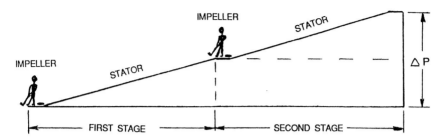

Figure 7.18 Deceleration of flow (stator function) between two rotors.

of an airplane jet engine). If the stator is omitted and the second rotor, is placed close to the first rotor, it cannot impart additional velocity to an already accelerated molecule. A crude analogy of this situation is shown in Figure 7.18, where the problem is to kick a hockey puck to the top of a high, ice-covered hill. It is obvious that a second kicker must be placed where the puck velocity decreases (by gravity and friction), so that additional velocity can be imparted. The role of the stator is to decelerate the flow in the shortest possible distance but to retain sufficient conductance for flow into the next stage. The rotor accelerates the flow and the stator decelerates the flow. This does not change under molecular flow conditions. In principle, the stators could be entirely omitted if we placed randomizing chambers between each rotor, but that would not make a practical pump.

The pumping mechanism of a turbomolecular pump can be viewed in a quasi-continuum way if, instead of considering the individual molecules and their trajectories, we take a distant view and consider molecular concentrations or densities in the rotors and stators, as indicated in Figure 7.19. Thus, it should be clear that rotors are impellers and stators are baffles, and a rotor and a stator together should be considered as a single stage as is normally done in axial flow compressors.

In Figure 7.18, the inclined plane between the impellers serves the function of the stator (i.e., to decelerate the flow and increase the pressure). If the following rotor were placed too close to the former, it would essentially be idle because its velocity will be nearly equal to the bulk velocity of the pumped gas.

When the turbomolecular pumping stage is analyzed by considering the actions of individual molecules, the flow patterns in the stator can be considered by imagining the stator rotating relative to the rotor. This has prompted many technical writers to assign equal pumping roles to both the rotor and stator, leading to occasional confusion in counting the number of stages in the pump. It seems preferable to consider a rotor and stator together (as a pair) as one stage. Relative to the body of the pump, relative to the entrance and exit

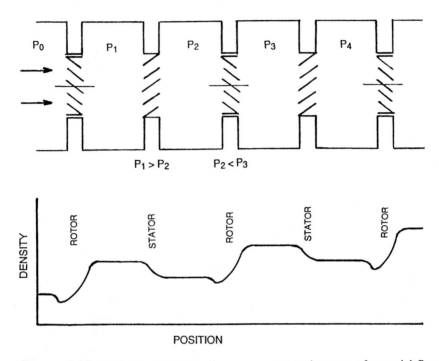

Figure 7.19 Bulk density distribution near rotors and stators of an axial-flow turbopump.

chambers, the stator is a passive element (relative to the bulk flow of the pumped gas) and should not be considered as another rotor. Rather, the stator should be thought of as a baffle that redirects, straightens, or readjusts the gas flow to make the next rotor more effective. The stators should be entirely omitted if the distance between rotors were perhaps 10 times greater than the diameter. For practical reasons, the stator geometry should be such as to accomplish its function in a shortest possible distance while providing adequate conductance to flow. At the very last stage, the stator can be and often is omitted. It would serve no useful function because any surface in the discharge area is adequate to decelerate the flow.

The actual number of stages in a turbomolecular pump depends almost entirely on the design requirements for pumping hydrogen. The thermal velocity of the hydrogen molecules is nearly 3.85 times higher than that of air molecules. The blade tip speed is low compared to the molecular velocity, and therefore hydrogen is not pumped as effectively as air. In principle, this is also true in the case of vapor jet pumps, although for vapor jet pumps the pumping speed and compression ratio for hydrogen are often higher, because it is easier to increase the width of the vapor jet as indicated in Eq. 6.1.

7.3.3 Performance

A cross section of a typical turbopump is shown in Figure 7.20. Pumps of this general type are manufactured with pumping speeds from 50 to 9000 L/s and, for special purposes, up to 20,000 L/s. There are 13 rotors assembled on a common shaft and 12 stators with spacers fitted at the outer wall. The rotor assembly is supported by two bearings and driven by a high-frequency motor.

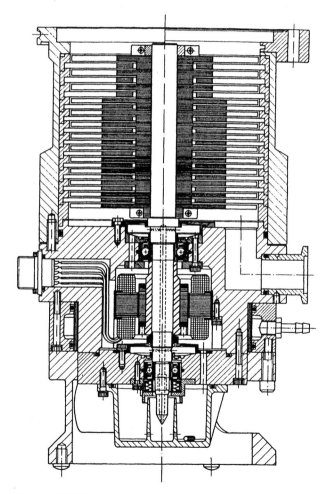

Figure 7.20 Cross section of a typical, conventional turbopump.

The bearings are lubricated by oil, which is supplied to the bearings by the centrifugal action of the conical section at the bottom end of the shaft. The detail of oil pickup is shown in Figure 7.21. The excess oil drips back into the oil container. The returning oil helps in cooling the motor, which is placed in the foreline pressure environment to avoid dynamic seals and friction losses at

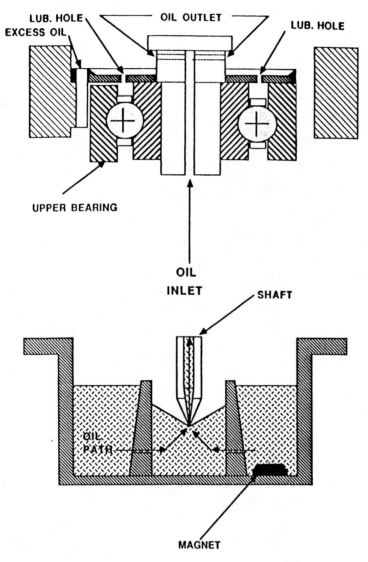

Figure 7.21 Detail of oil flow in an oil-lubricated turbopump.

atmospheric pressure. The bottom section of the pump is cooled by water, although air cooling is also available. For air cooling, the water channel is replaced by fins and a fan mounted so as to blow air across the finned section.

The basic performance of such pumps is very similar to the typical pumping speed performance of a vapor jet pump, but there are important differences. A typical traditional pumping speed performance is shown in Figure 7.22. The ultimate pressure, the range of the constant-speed section, and the dependence of the speed at the higher pressures on the size of backing pump are all very familiar. If the information contained in Figure 7.22 is replotted in the form of achieved pressure versus the flow rate (i.e., the gas load), the graph again looks familiar (Figure 7.23). This graph shows clearly the gas load at which the turbopump can be considered to be overloaded and above which the pumping action is due to the backing pump.

Overloading a turbopump for a short period of time is not as detrimental to its performance as is the case with vapor jet pumps, specifically in regard to the possibility of backstreaming. The stored energy in the rotor is high enough to prevent sudden deceleration. For example, during the evacuation of a relatively large chamber the crossover to the turbopump can be accomplished at a higher pressure, as shown in Figure 7.24. After evacuating a 650-L chamber

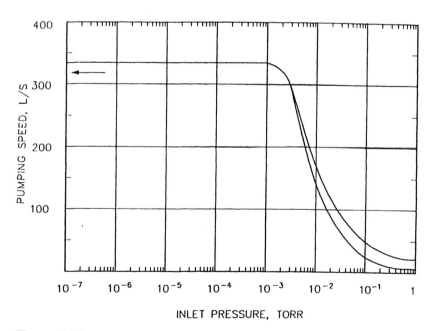

Figure 7.22 Pumping speed of a turbopump with (alternative) backing pumps of different size.

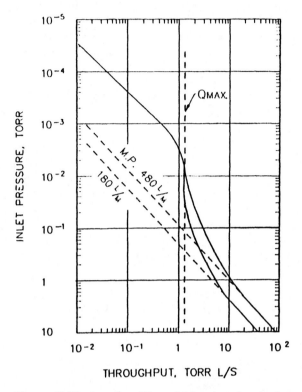

Figure 7.23 Data from Figure 7.22 replotted as the degree of achieved vacuum (pressure) versus flow rate (throughput or gas load).

with a 450 L/min oil-free pump, the high-vacuum valve was opened at a crossover pressure of 1 torr. The turbopump speed is 260 L/s. The evacuation time response $V/S = 650/260 = 2.5$ s, which means that a new pressure, after the valve is opened, should be reached in 10 s. If the crossover is attempted at a pressure higher than 1 torr, nothing will be gained because at such high pressure the turbopump will be idling and will represent an impedance to the backing pump.

Turbopumps and the associated backing pumps may transmit vibrations to the vacuum system. Vibration peaks at 60 and 120 Hz (50 or 100 Hz with 50-Hz line frequency) usually originate from mechanical backing pumps. When necessary, these can be reduced by using flexible bellows to isolate the turbopump from the backing pump. In very sensitive applications, part of the foreline can be encased in a block of concrete to dampen the transmission of vibrations.

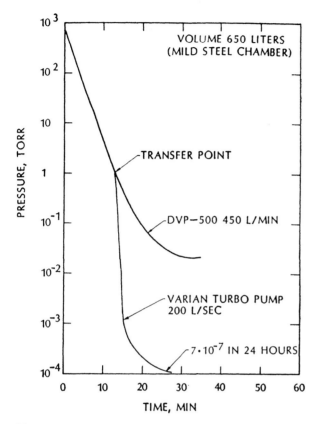

Figure 7.24 Example of evacuation of a large chamber (650 L) using a 260-L turbopump and a 450-L/min oil-free backing pump.

The primary source of vibration in an operating turbopump is the residual imbalance of the rotor. The magnitude of lateral (radial) displacement for small and medium-sized pumps should be less than 0.02 μm (2×10^{-6} cm, i.e., of the order of 50 atomic dimensions). In the axial direction the magnitude is typically 10 times lower, and in this case, it is a function of bearing tolerance and the degree of perpendicularity between the bearing and the shaft.

When extremely low vibration amplitudes are required, special vibration isolators can be used to attach the inlet flange of the pump to the system. These can reduce vibration amplitudes typically by a factor of 30.

When a pump is mounted on an actual system the vibration levels at the inlet flange will be usually less than 0.02 μm. The exception to this is the case when the system components have a natural frequency (including harmonics) near the operating frequency of the pump. Oil-lubricated pumps must be mounted in vertical position (inlet at the top). The grease-lubricated pump can be mounted in any position, including upside down.

7.3.4 Speed for Different Gases

The pumping speed of a turbomolecular pump depends essentially on four basic parameters:

1. The diameter of the rotor and the blade height, which determines the entrance area
2. The peripheral rotational velocity
3. The blade angle of the first rotor
4. The blade spacing ratio [i.e., the distance between blades divided by the width (not thickness)]

It should be obvious that the blade height for a given diameter must be limited due to a consideration of the peripheral speed at the root of the blade. Once the geometry is fixed, the pumping speed will depend on the rotational velocity. In other words, the pumping speed depends on capture efficiency and the entrance area to the pumps. The capture efficiency is higher when the ratio of blade speed to the speed of molecular motion is high. Because the peripheral speed is limited by mechanical considerations, the capture efficiency for light gases is poor compared to air. Thus the pumping speed for hydrogen and helium in typical turbopumps is 25 to 30% lower than for air (Figure 7.25).

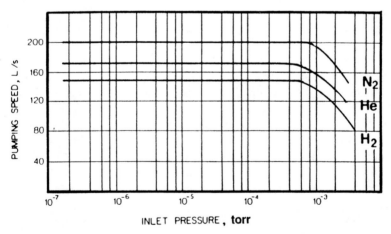

Figure 7.25 Pumping speeds of a conventional turbopump for nitrogen, helium, and hydrogen.

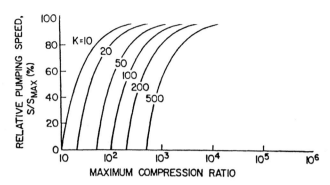

Figure 7.26 Influence of pressure ratio K on pumping speed.

Generally, hydrogen pumping problems may produce a significant limitation and system design, and operation should be conducted with some attention to the presence of hydrogen in the vacuum chamber as well as in the discharge side of the turbopump. For example, the pumping speed for hydrogen will be affected by the size of the backing pump. If the hydrogen pressure ratio (discharge to inlet) approaches the maximum compression ratio (at zero flow), the pumping speed will be diminished. The operation will be near the left side of the pumping speed curve where the speed decreases (Figure 7.26). These effects are also present in vapor jet pumps, although they may not be as noticeable except in cases where the pump has a relatively low heat input. For small changes in practical design parameters, the effect on pumping speed may be considered to be linear. For example, reduction of the peripheral velocity should produce a corresponding reduction in pumping speed.

A somewhat neglected parameter is the ratio of blade height to the distance between the blade tip and the pump body. In small turbopumps this gap is approximately 0.3 mm. Its effect on pumping speed is nearly linear. In practice, this means that small changes in the width of the gap will not have a significant effect on pumping speed. However, the effect on the compression ratio tends toward an exponential relationship, so that the gap must be kept as small as practicable.

7.3.5 Compression Ratio for Different Gases

The dependence of the achievable maximum compression ratio on the basic design parameters tends to be exponential. Again, this is very similar to the behavior found in vapor jet pumps. For a turbopump, approximately,

$$K_{\max} = \exp(\alpha V \sqrt{M}) \tag{7.12}$$

where α is a constant that depends on the blade angle and the blade aspect ratio (s/b, distance between blades/blade width), V the peripheral velocity, and M the molecular weight of the pumped gas. Figure 7.27 shows typical values of the maximum compression ratio for different gases. The dots represent hydrogen, helium, and nitrogen. The lower curve represents a standby operation at about two-thirds of the normal speed.

Typical maximum compression ratios of industrial pumps are approximately 1000 for hydrogen, 10,000 for helium, and 10^8 to 10^9 for nitrogen. Some pumps may have a hydrogen compression ratio as low as 300. When very high or ultrahigh vacuum is desired, such pumps will be very sensitive to the amount of hydrogen in the foreline. Mechanical backing pumps may be a source of hydrogen because of decomposition of the lubrication oil at elevated temperature.

Other sources of hydrogen are ducts, O-ring seals, or hoses made of various rubbers or plastic materials, residues from cleaning processes, and occasionally, residues from chemical plating processes such as nickel plating. The

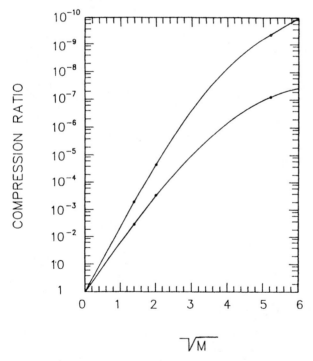

Figure 7.27 Typical maximum compression ratios, depending on molecular weight of the pumped gas, for conventional pumps. (Data by Pfeiffer).

following example may serve as an illustration. If two turbopumps are connected in series and an ionization gauge is placed in the duct between the pumps, when the filament of the gauge is lit, the inlet pressure in the vacuum chamber will immediately rise. The ionization gauge will produce enough hydrogen due to reactions at the hot filament to increase the partial pressure of hydrogen. Despite the relatively high speed for hydrogen at the second turbopump (compared to the mechanical backing pump), hydrogen will backdiffuse through the upstream turbopump.

An interesting phenomenon has been observed regarding the compression ratio for hydrogen when other gases are present in the system. If some other gas (e.g, nitrogen or argon) is deliberately introduced into the duct between two turbopumps connected in series, the compression ratio for hydrogen will increase. The same phenomenon can be also observed in vapor jet pumps. The presence of the intermediate gas (nitrogen or argon, for example) enhances the pumping action for hydrogen. Apparently, the intermediate collisions improve the capture probability for the lighter gas. Thus the variations in total pressure can affect the hydrogen pumping mechanism. Awareness of this interaction between gases is helpful in dealing with some applications, for example in contra-flow leak detection, where similar effects can be noticed when pumping helium.

The maximum compression ratio is maintained throughout the molecular flow regime: that is, as long as the discharge pressure is low and the inlet pressure is not affected by the ultimate pressure limitations. The decay of the compression ratio is illustrated in Figure 7.28, where the variation in compression ratio is shown as the pump is approaching a high discharge pressure. A simple interpretation of this graph is that the maximum pressure difference this pump is capable of maintaining is approximately 1 torr. It may be said, in analogy with vapor jet pumps, that the tolerable forepressure of this turbopump is 1 torr.

The data in Figure 7.28 are obtained by deliberately increasing the discharge pressure and observing the effect on inlet pressure. The discharge pressure can be increased either by throttling the backing pump or by introducing a gas leak into the foreline. Maximum obtainable pressure difference is a basic characteristic of all compressors (or pumps) and the graph of Figure 7.28 is entirely analogous to Figure 6.17. The difference in presenting the data is due simply to the individual preferences or established traditions in technical literature. For a complete characterization, the curves for each gas in Figure 7.28 should be obtained with varying gas flow. The results could be represented in a plot of pressure ratio versus flow rate (or throughput) rather than as inlet pressure versus flow rate, as in Figure 7.23.

The discharge pressure of turbopumps (tolerable forepressure) can be increased by changes in conventional designs. There are several avenues possible,

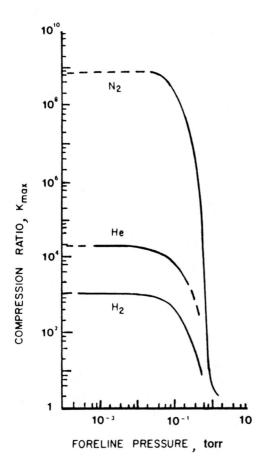

Figure 7.28 Decay of the maximum compression ratio at higher foreline pressures (for conventional pumps).

such as increasing the number of stages, increasing the motor power, and changing blade design. The most promising avenue is to use combination designs. For example, beginning stages can be of turbomolecular type and the later stages of turbodrag type, mounted on the same rotor. It is possible that any type of device belonging to the general class of turbomachinery equipment can be used in some combination fitted on the same rotor. Axial-flow stages can be followed by centrifugal stages and then by regenerative blower impellers. The latter can be designed to produce the highest possible compression with very little volume flow because of the enormous volume reduction produced by high-vacuum pumps.

The advantage of high discharge pressure is the possibility of using entirely oil-less backing pumps. Above about 10 torr, two-stage diaphragm pumps

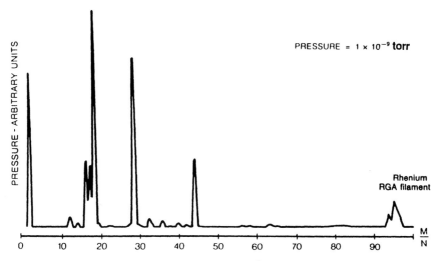

Figure 7.29 Residual gas spectrum in an unbaked small system pumped by a turbopump.

become feasible. In concept, it should be possible to design turbopumps that will discharge directly to atmosphere, but this may not be practical. One advantage of such pump would be keeping the bearings with their lubrication arrangements in the atmosphere, isolated from the gas flow ducts.

7.3.6 Hybrid and Compound Pumps

The first attempts to describe the turbomolecular pumping mechanism were based on certain assumptions: 1) Both upstream and downstream regions had Maxwellian molecular speed distributionss. It was noted, however, that small departures from such distributions do not greatly affect the results. 2) The flow was molecular throughout the pump with cosine reflection trajectories. 3) The temperature remained constant. 4) Steady-state flow prevailed. 5) Blades had an infinite length. The latter implied, as noted in the study, that the molecular collisions with the walls of the pump were neglected.. Also, the tangential (centrifugal) trajectories were neglected. It also implied that the blade velocities at the root and at the tip of the blade were equal, which is not the case in practice, particularly in small pumps. In addition, the differences in the blade distances at the root and the top were neglected. Most of these are usually minor issues, as proven by experimental results. A rotor made according to these principles is shown in Figure 7.30.

All studies include another important assumption which apparently was perceived to be so axiomatic that it was not even mentioned. The rotors and sta-

Figure 7.30 Bladed rotor disks and an assembled rotor of a conventional turbomolecular pump (Varian V-200).

tors were almost always the same size and had the same number of blades and the same mirror-image geometry. This was inherited from axial compressor examples and was often claimed to be the best arrangement. Because compression ratios in turbomolecular pumps are nearly an order of magnitude greater than in axial compressors, the a priori similarity of rotors and stators may warrant further study. This is especially important at the intermediate pressure range in hybrid pumps.

In recent years, compound pumps have appeared on the market that extend the permissible discharge pressure values to 20 torr or higher. This allows the use of oil-free backing pumps to avoid the possibility of backstreaming from the ordinary oil-sealed mechanical pumps. The designers of such pumps sought inspiration by looking into the history of molecular pumps and essentially attaching half of a Holweck pump (a cylindrical drum with a helical pumping channel) to the end of the conventional turbomolecular pump with the normal number of axial fan-type stages. This makes the pump nearly twice as long and more costly. A more advanced design is shown in Figure 7.31a, which also features magnetic bearings. Another example of a turbo-drag pump is shown in Figure 7.31b, which resembles the Holweck-type molecular drag pump of Figure 7.5 but has an additional bladed section at the inlet and intermediate pumping (or backing) ports.

Recent emphasis on oil-free vacuum pumping systems has revived the interest in pumps that can tolerate higher discharge pressures. When exit pressures reach 10 mbar (or higher), the well-established molecular flow analysis cannot be used. It appears that the existing turbomolecular and turbodrag pumps, which take their basic design from historical models, are not optimized in regard to their speed and throughput performance. Following a concept of bulk pumped gas velocity variation, which can be used for assessing the gas density (despite the existence of molecular flow through parts of the pump), and optimized pumping speed and compression staging arrangement was developed. It demonstrates an improvement in compression ratios as much as 40 times higher for light gases (helium and hydrogen) without increasing the size and power, compared to a conventional pump. Turbine-type pumps, unlike oil-vapor jet and high-vacuum cryopumps, permit practical engineering solutions for expanding the pressure regime to an almost arbitrarily high pressure. This makes it possible to use oil-free backing pumps at intermediate pressures of 10 to 100 torr using relatively small changes in design.

Usually, in high-vacuum technology flow descriptions, the existence of density distributions are not considered, although there are some exceptions. In terms of mathematical modeling, molecular density gradients can be used to express gas flow behavior and, at least in cases of simple geometry, give correct results. In this manner, one can speak of bulk flow, disregarding individual molecules. This simple idea can be carried into a discussion of the concept of conductance. If the numerical value of conductance (or pumping speed) is divided by the cross-sectional area of the conduit, we obtain a value for bulk velocity of flow, for a given size and geometric configuration, which is constant in the free-molecule domain.

Applying this to the conditions existing at the inlet of a typical modern turbomolecular pump, we obtain approximately 50 m/s. Considering that the pressure ratio in a turbomolecular pump, even without the additional drag stages, usually exceeds one million, we may ask what is the bulk velocity at the exit stages? Ideally, due to great compression, every stage of a turbine pump should have a different geometry but this is regarded to be too expensive. Typically, in addition to changing the angles of inclination of the blades, their length is changed one or two times. This results in having too much pumping speed in the downstream stages of the pump where the bulk velocity is nearly zero. In conventional turbomolecular pumps, which are backed by dual-stage oil-sealed mechanical pumps, this excess pumping speed may be acceptable. However, when the discharge pressure is increased to perhaps 0.5 torr, significant effects can be noticed that are associated with a waste of power (producing heating) and with the opportunity for backdiffusion of lighter gases.

The power required for compression in an isothermal process, $W = Q \ln P_2/P_1$ is only a minor part of the total power normally used in a high-vacuum

306 CHAPTER 7

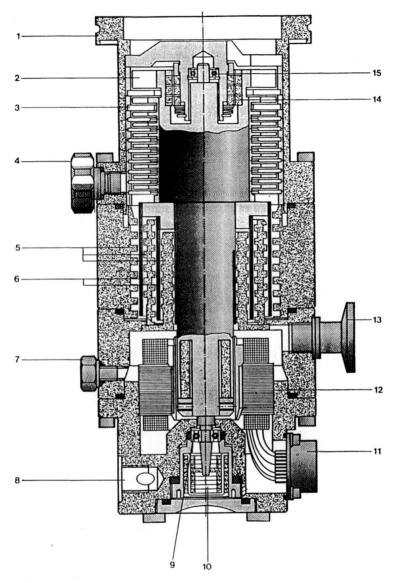

Figure 7.31a A compound turbomolecular pump with 10 turbine-type stages and 5 Holweck-type drag channel sections (Pfeiffer Co.). 1, High-vacuum connection; 2, rotating turbo disk; 3, stator disk; 4, venting connection; 5, molecular stator; 6, molecular rotor; 7, sealing gas connection; 8, cooling water connection; 9, emergency bearing; 10, pump lubrication reservoir; 11, electrical connection; 12, motor; 13, roughing vacuum connection; 14, permanent magnetic bearing; 15, emergency bearing.

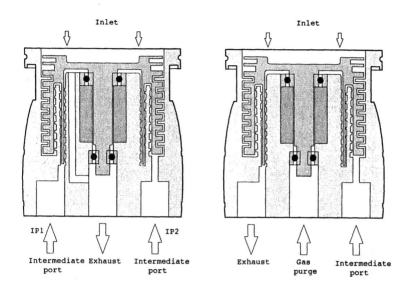

Figure 7.31b Cross section of a modern turbo-drag molecular pump combining Holweck-type drag stages with a turbine-type bladed section at the inlet. (Courtesy of Alcatel Vacuum Technology.)

pump. High-vacuum pumps are generally very inefficient. In turbopumps, power is consumed in bearings and gas friction at the discharge end of the pump. In addition, in most compressors (particularly valveless compressors such as turbopumps), some power is required to maintain the pressure difference across the pump, even at zero flow.

The total power may be expressed as

$$W = Q \ln (P_2/P_1) + S (P_2 - P_1) + W_f \qquad (7.13)$$

where Q is throughput (mass flow), P_1 and P_2 are inlet and exhaust pressures, S is the pumping speed, and W_f may be called wasted power.

Therefore, the greater the size of the rotor/stator stage, the greater the power, which at nearly stagnant flow conditions at the exit stages is wasted on repumping the gas, which tends to leak back into the preceding stator. The use of smaller pumping stages tends to reduce this power waste. In addition, the increased bulk velocity in smaller pumping channels provides a greater impedance to backdiffusion of lighter gases. It has been demonstrated that gas interactions occur in turbomolecular pumps. When argon was introduced in the middle of the pump, the hydrogen compression ratio increased. The higher bulk velocities should have an effect on such interactions. Hydrogen and helium pressure ratios are usually measured with more or less pure gases. It may be

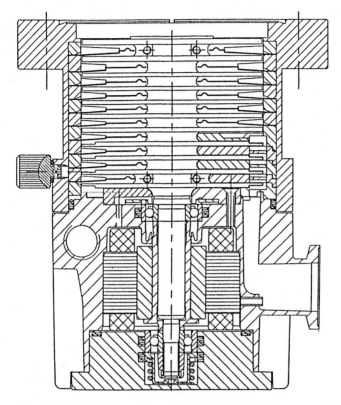

Figure 7.32 A schematic cross section of a hybrid pump with 7 turbine-type stages and 4 drag channel disks (Varian V-60-10), right, and the old design on the left.

of interest to also measure it using small quantities of the light gas in the presence of other, heavier gases.

Following these concepts, pumps were designed that produce a tenfold increase of the compression ratio for air and light gases without an increase in size, cost, or power, and without significant sacrifice in any other performance aspect of the pump. The compression ratio for hydrogen is over 10^5 and, for helium, over 10^6 using only 11 stages (11 rotors and 10 stators). The last three stages are essentially molecular drag disks resembling the Gaede pump as shown in Figure 7.32. Note that the last stator in the turbomolecular section is absent. The blade length is changed three times and the blade angle five times.

The photographs in Figure 7.33 show the hybrid rotors of a 250 L/s pump in which the last three (exit) stages are modified (compared to Figure 7.30), by omitting turbine-type blades and placing the uncut disk rotors into molecular drag channels. The lower photograph shows the assembly including the

TURBOMOLECULAR PUMPS 309

Figure 7.33 Hybrid pump rotors (Varian V-250 and V-60-10) and an assembled view of the V-250 pump.

stators, which are inserted between each rotor in pairs of half-circle segments and held in place by spacers at the pump wall. The resulting improvement in compression ratio compared to the conventional 250 L/s pump is shown in Figure 7.34 (for helium) and in Figure 7.35 (for nitrogen). In the latter case, the conventional pump had an additional rotor bladed so that the maximum value of the compression ratio was the same but the extension to the higher discharge pressure was more significant.

Figure 7.36 shows compression ratios for different gases of a 550 L/s hybrid pump, which are nearly two decades higher compared to conventional pumps of that size. The figure includes typical pumping speed curve for nitrogen in a 550 L/s hybrid pump, which produces nearly 2.5 to 3 times greater maximum gas flow (and uses significantly lower power compared to a typical vapor jet pump of similar size). In larger hybrid pumps (500 to 1000 L/s), it is relatively easier to achieve higher compression ratios, near 10^6 for hydrogen, and more than 10^7 for helium, which is more than 1000 times greater than older conventional pumps.

7.3.7 Operation and Maintenance

Turbopump routine maintenance consists of replacing or adding oil or grease at specified intervals and monitoring noise and vibration levels, which may indicate bearing problems. Typically, bearing life is 2 to 3 years of continu-

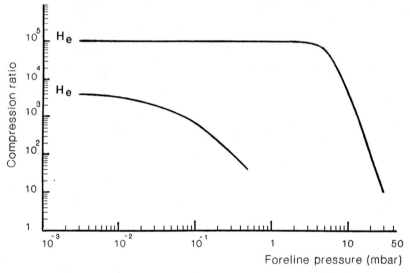

Figure 7.34 Compression ratios of two pumps for helium depending on the pressure at the pump exit: upper curve, hybrid 250 L/s pump (V-250); bottom curve, conventional 200 L/s pump (V-200).

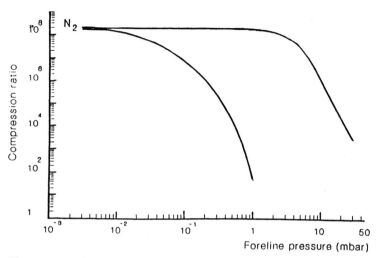

Figure 7.35 Compression ratios for nitrogen of two pumps from Figure 7.34.

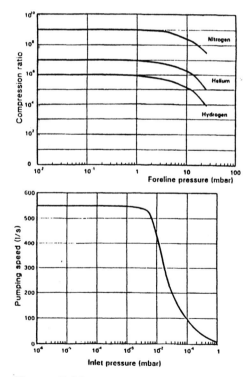

Figure 7.36 Pumping speed and compression ratios of a 550 L/s hybrid pump (V-550).

ous operation. It is usually better to have bearings replaced by manufacturers, at which time the rotor balance can also be checked. Rotor balance is a critical parameter that determines bearing life.

Generally, due to the extremely high compression ratios for larger gas molecules, turbomolecular pumps provide a clean vacuum environment, free of hydrocarbons. This is true for a turbopump operating at speeds between 60 and 100% of rated speed. Figure 7.29 shows an example of a residual gas spectrum obtained in an unbaked system pumped by a turbopump. Turbopumps can be mildly baked and they can produce 10^{-10} torr pressure using metal gasket flanges at the inlet and throughout the system. In a baked system the residual gas spectrum will usually show hydrogen as the most prominent peak.

The presence of lubricating oil or grease in the bearings of turbopumps requires attention regarding the possibility of backstreaming. The conditions are not as difficult as with vapor jet pumps. The oil is properly in the foreline rather than in the inlet, and it is not as hot. However, the oil is lighter (having a higher vapor pressure) and it can more easily migrate if certain precautions are not taken, although it is easier to remove the oil contamination from the vacuum chamber by mild baking.

Oil-lubricated turbopumps use a recirculating system that provides a steady flow of oil through the bearing assemblies during operation. When the pump is stopped, some residual oil remains in the bearings while the remainder is drained by gravity into the oil sump at the bottom of the pump.

Leaving a turbopump under vacuum when it is not running permits backing-pump oil and lubrication oil to migrate into the vacuum chamber. Proper stopping and venting procedures are especially important in a valveless system. Turbopumps do not have to be isolated by valves as are vapor jet pumps, where the hot oil cannot be exposed to atmosphere. Valves, therefore, are often omitted. The absence of valves permits commission of errors in overloading and in venting through the foreline.

When power to a turbopump is turned off and the pump gradually decelerates, the compression ratio decreases. Below 60% speed, hydrocarbons, present in the foreline, may migrate through the turbine blades and ultimately into the vacuum chamber, causing contamination.

This potential hazard can be prevented by proper venting methods. It is important to vent before the turbopump speed falls below 60% of its rated speed. A turbopump may be vented immediately after power is removed. This method is common and acceptable for many applications; however, there are good reasons to delay venting:

Coast through brief power failures,
Allow time for valves to close,
Avoid the high cost of inadvertent venting, and
Avoid the possibility of mechanical failure.

The oil must be changed every 6 months or every time it becomes opaque or dark brown. A typical grease-lubricated pump must be regreased every 6 months or every 3000 h of operation (whichever comes first). However, in some continuous heavy-duty applications, a shorter interval may be required. Lubrication for grease-lubricated turbopumps is provided by grease, which is packed in the bearing assemblies. The grease contains two parts: first, a thickener, which acts as "sponge" to retain the second part, the lubricating base oil. Pumps with grease-lubricated, ceramic ball bearings can be operated without any maintenance for more than two years.

The base oil portion of the grease requires periodic replenishment, as small quantities are continuously lost through evaporation whether or not the pump is operating. The thickener, which is the oil sponge or retainer part of the grease, does not require replenishment. It remains in place within the assembly for the life of the bearing. At specified intervals, relubrication of the bearings is necessary. The lubricant supplied is an oil, formulated for properties required for the specific application. No substitute may be used, nor should it be confused with oil formulated for, and used in, oil-lubricated pumps.

In recent years magnetic bearings have been introduced in some turbopumps. Because the rotor is relatively light (usually made of aluminum) the entire rotor assembly can be magnetically suspended as well as restrained in radial displacement. Regular bearings are used during starting and stopping to support the rotor. The advantage of magnetic bearings is the absence of vibrations, long life, and if the support bearings are dry, the absence of the possibility of backstreaming. Magnetic bearings should be used only when prolonged uninterrupted operation is required, because starting and stopping events produce a certain amount of damage and cannot be repeated more than several times.

Generally, it is not a good idea to expose an operating turbopump suddenly to air. Even if a pump can survive this, the sudden exposure may produce damage to the bearings which may not be immediately apparent but may shorten the bearing life. In larger pumps a very sudden air release may even bend the blades, producing catastrophic results.

Venting time depends on pump size and can range from several seconds to a few minutes. Once vented, the minimum time to leave the vent valve open is the time required to reach a pressure of at least 375 torr (500 mbar) inside the turbopump. This pressure is required to keep mechanical pump oil from backstreaming and contaminating the turbopump.

Special valves are available for proper venting of a turbopump to the atmosphere. The valve can be adjusted to go through the vent procedure automatically in the event of a power failure. Inadvertent venting can be prevented with an adjustable delay-time control in case of a brief power interruption.

In a valveless system the turbopump should be turned on when the chamber pressure is approximately 1 torr. This depends on the system volume and the size of the turbopump. To avoid backstreaming, the timing should be such that about two-thirds of the rotational speed is attained when the chamber pressure is at a few hundred millitorr.

Although it is usually not emphasized in turbopump operation discussions, it is a good idea to keep the discharge pressure below 0.5 torr in conventional turbopumps to avoid the deterioration of the compression ratio and the possibility of increased backstreaming. In general, the inlet pressure should not exceed a few millitorr and the discharge pressure a few hundred millitorr. Some pumps have very small motors and may be overloaded unless the discharge pressure is below 20 mtorr.

The temperature of the cooling water and air must be kept within certain limits to keep the lubricating oil or grease in the proper viscosity range. For example, in grease-lubricated pumps, air cooling can range from 5 to 30°C, depending on inlet pressure. When inlet pressure is near 10^{-3} torr, air cooling can be as high as 30°C, but with room-temperature cooling air, the pressure should not exceed 30 mtorr. With water cooling, the temperature should not be (typically) below 10°C unless design adjustments are made in the cooling system.

To operate turbopumps in systems that use corrosive gases, the main concern is to protect the bearings and motor areas. One simple method of protection is to provide a dry air or nitrogen bleed in the motor section so that the pressure in the motor section is somewhat higher than the foreline pressure. The motor section can be separated from the rest of the pump by a wall, leaving only an opening for the shaft. The flow of the bleed gas through this opening will prevent the entrance of corrosive gases to the motor section.

Rotor blades of an operating turbopump must not be allowed to contact any solid (or even liquid) material. Protective wire mesh screens are often placed at the inlet to the pump. Such screens typically reduce the pumping speed by about 15%, measured at the inlet. The net effect of the screen in a system is lower if there are other ducts present between the pump and the system.

7.4 SUMMARY OF PROPERTIES

High-vacuum pumps that function primarily at inlet pressures in the molecular flow range have many similarities in basic performance—often including similar initial evacuation and discharge pressure requirements. It may not always be easy to choose among vapor jet pumps, turbomolecular pumps, and cryopumps. The choice involves considerations of initial cost, operational cost, maintenance cost, reliability, and a review of certain properties that may be more suitable for a particular application. The general properties of turbo-pumps are reviewed in the following sections.

7.4.1 Advantages of Turbopumps

Roughing Valves May Not Be Necessary. This means that the initial evacuation can be accomplished through the turbopump by a roughing pump that may or may not also be used as a backing pump. The omission of a separate roughing line is possible in small systems. This represents a saving and provides some simplicity in system operation. However, when the volume of the vacuum chamber is large compared to the turbopump, the roughing time will be increased because the turbopump may have a limiting conductance.

High-Vacuum Valve May Be Omitted. Again, in small systems and undemanding applications, this is entirely feasible. Operating turbopumps can be exposed to the atmosphere with a proper venting procedure. The need for isolation of hot oil (as in vapor jet pumps) and isolation to avoid saturation (as in cryopumps) do not exist.

No Contaminating Motive Fluid. When operating properly, the lubricating medium (oil or grease) is present only at the foreline. The possibility of backstreaming is more indirect and associated primatily with malfunctioning.

Less Chance of Backstreaming Accidents. The oil reservoir is not in the direct path of inrushing air if the pump is incorrectly air-released from the foreline. Overloading by exceeding the maximum mass flow capacity or via the discharge pressure does not produce gross backstreaming events.

Mounting in Any Position. Pumps with magnetic bearings and pumps with grease-lubricated, ceramic ball bearings can be mounted on the vacuum chamber at any angle, although a vertical position is usually better for longevity.

Quick Startup and Shutdown. There is no need to wait for heating or cooling. Small and medium-sized turbopumps can attain operating speed in a few minutes. When properly vented they can also be stopped in a few minutes.

Less Need for Traps. In some vapor jet pump applications liquid-nitrogen traps are necessary to suppress the vapor pressure of the pumpling fluid at the inlet of the pump. In turbopumped systems it is easier to achieve inlet pressures below 1×10^{-8} torr without baking and without liquid-nitrogen traps. Cryogenic traps protect the system in case of minor errors in operation, and they of course increase the pumping speed for condensable vapors such as water vapor.

No High Voltages. Turbopumps operate at rather low voltages. The high rotational speed is obtained by using high-frequency motors. Currents are also low because motor power requirements are comparatively low.

Low Operating Expenses. The lower power requirement for unit mass flow rate and less dependence on liquid nitrogen produce a lower operating cost.

Pumps All Gases Effectively. There are no saturation effects as in cryopumps, involving light gases such as helium and hydrogen, although hydrogen may provide an ultimate pressure limitation. Corrosive gases can be handled with special design and care. It is easier to prevent chemical reactions with lubricating oil than it is to prevent reactions with pumping fluid in vapor jet pumps.

More Predictable Pumping Performance. The absence of the pumping fluid eliminates the need for monitoring its condition in regard to the possibility of overheating.

7.4.2 Disadvantages of Turbopumps

High Initial Cost. Considering the precision machining and balancing requirements, high-precision, high-speed bearings, and relatively high stresses developed in the rotor; turbopumps can be classified as precision machinery devices. They can cost as much as five times more than vapor jet pumps of comparable speed, although this comparison is not as severe when traps and valves are added to the vapor jet pump system.

Moving Parts Subject to Wear. This primarily concerns the bearings. Bearing life is not easy to predict because it depends on conditions of use. These include process cleanliness, proper maintenance, the number of accidental air exposures, and other factors. Typically, bearings should be replaced every 2 to 3 years. Generally, it is better to have bearings replaced by the manufacturer. Therefore, maintenance cost must be taken into account. If the pump is left unattended, serious damage can result if bearing maintenance is not performed. Similarly, a very sudden exposure to air or "ingestion" of small objects can produce a total loss of the pump.

Proper Operation Is Important. For proper operation turbopumps need careful handling and attention. To prevent damage, to prevent reduced bearing life, and to prevent operational difficulties, starting and stopping must be done with some care. Noise and vibration levels should be monitored. In some systems isolation valves may be necessary to avoid mechanical damage.

Not Available in Large Sizes. The largest turbopumps yet built have a pumping speed of 9000 L/s. There is no intrinsic reason for not making larger pumps except perhaps the forbidding expense. Very large multistaged axial-flow compressors are made for aircraft engines and for air-moving equipment.

Pumping Limitations with Light Gases. The compression ratio limit for hydrogen (for some pumps, as low as 300) may produce special problems in system design.

The success of turbomolecular pumps is associated with a few clear basic advantages compared to earlier vapor jet pumps. One of the most important issues is backstreaming. In vapor jet pumps, the condensed pumping fluid appears at the inlet of the pump, where it can evaporate into the vacuum chamber even if gross backstreaming events due to overloading are prevented. In turbopumps, the lubrication system is usually in the exit side and, at least when the pump is running, the backstreaming is minimal especially in grease-lubricated pumps. Turbopumps respond better to transient overloads because there is sufficient storage of rotational energy to maintain the speed. In vapor jet pumps, it is possible to overload the inlet stage (and cause excessive backstreaming) while the lower stages are still functioning properly. In turbopumps, the inlet stages may be overloaded without any adverse effects because the stages are mounted on the same shaft. Grease-lubricated or magnetic-bearing turbopumps can be oriented to any position, which is rather difficult to do with vapor jet pumps. Perhaps the most important feature of a turbopump is that it can be operated by a push-button initiation without a waiting period for heating or cooling as required of vapor jet pumps and cryopumps. In general, this push-button, on-off operation feature makes the system design and control simpler. Finally, turbopumps can be exposed to atmospheric air without harm, and as a result, they usually do not need isolation valves.

REFERENCES

1. W. Becker, Vak.-Tech., 7, 149 (1958).
2. M. Hablanian, in *Advances in Vacuum Science and Technology*, Proc. 1st. Int. Cong. Vac. Tech., Namur, Belgium, 1958, E. Thomas, ed., Pergamon Press, Elmsford, N.Y., 1960.
3. C. H. Kruger and A. H. Shapiro, in *Seventh National Symposium on Vacuum Technology Transactions*, C. R. Meissner, ed., Pergamon Press, Elmsford, N.Y., 1961.
4. G. Osterstrom, in *Methods of Experimental Physics*, Vol. 14, *Vacuum Physics and Technology*, G. L. Weissler and R. W. Carlson, eds., Academic Press, New York, 1979.
5. D. J. Hucknall and D. G. Goetz, *Vacuum*, 37(8/9), 579 (1987).
6. C. M. Van Atta, *Vacuum Science and Engineering*, McGraw-Hill, New York, 1965.
7. L. N. Rozanov, *Vakuumnaya Tekhnika*, Vysshaya Shkola, Moscow, 1990.
8. I. Rubet, *Le Vide*, 123, 227, Mai/Jun 1966.
9. L. Maurice, *Jpn. J. Appl. Phys.*, Supl. 2, Pt. 1, 21, 1974.
10. M. H. Hablanian, *Le Vide*, Supl. 252, 19, (Mai-Jul 1990).
11. M. Masse et al., *J. Vac. Sci. Technol.*, A6 (4), 2518, (1988).
12. M. V. Kurepa and C. B. Lucas, *J. Appl. Phys.*, 52 (2), 664, (1981).
13. F. J. Schittko and C. S. Schmidt, *Vak. Technik.*, 23, 8, 243 (1974) and 24, 4, 110, (1975).
14. G. Levi, *J. Vac. Sci. Technol.*, A10 (4), 2619, 1992.
15. J. G. Chu and Z. Y. Hua, *J. Vac. Sci. Technol.*, 20 (4) 1101, 1982.
16. J. A. Basford, *J. Vac. Sci. Technol.*, A10 (4) 2623, (1992).
17. M. H. Hablanian, *J. Vac. Sci. Technol.*, A (4), 2629, (1992).
18. M. De Simon, *Vacuum*, 41, 7-9, 1990.

19. T. Sawada et al., *Sci. Papers. Inst. Phys. & Chem. Res. Jpn.*, 62, 2, 49 (1968) and *Bull. J.S.M.E.*, 14, 67, 48 (1971), 16, 92, 313, (1973), 16, 96, 993, (1973).
20. T. Sawada, *Jpn. Inst. Phys. & Chem. Res.*, 70, 4, 79, (1976) and *Bull. J.S.M.E.*, 22, 169, 974 (1979), 24, 195, 1666 (1981), 29, 252, 1770, (1986), *10th Intern. Conf. Fluid Seal* 409, (1984) and *Vacuum*, 41, 7–9, 1833, (1990).
21. J. Y. Tu, *Vacuum*, 37, (11/12) 831, (1987), 38, 1, 13, (1988), and *J. Vac. Sci. Technol.*, A6, 4, 2536, (1998).
22. F. Casaro and G. Levi, *J. Vac. Sci. Technol.*, A9 (3), 2058, (1991).
23. J. G. Chu, *J. Vac. Sci. Technol.*, A6 (3), 1202, (1988).
24. N. Liu and S. J. Pang, *Vacuum*, 41, 7–9, 2015, (1990).
25. W. A. Wilson at al., *Trans. ASME*, 1303, (Nov. 1955).
26. H. Enosawa et al., *J. Vac. Sci. Technol.*, A8 (3), 2768, (1990).
27. J. C. Helmer and G. Levi, *J. Vac. Sci. Technol.*, A13 (5) (1995).
28. G. Levi, M. De Simon, and J. C. Helmer, *Vacuum*, 46, 357, (1995).
29. T. Sawada, *Vacuum*, 44, 5–7, (1993).
30. T. Sawada, *Vacuum*, 4, 7–9, (1990).
31. K.-H. Bernhardt, *J. Vac. Sci. Technol.*, A1 (2) (1983).
32. F. Tazioukov et al., *Vakuum in Forschung und Praxis*, 7, 1, (1995).
33. G. Levi, *Vacuum*, 43, 5–7, (1992).
34. J. C. Helmer and G. Levi, *J. Vac. Sci. Technol.*, A13 (5) (1995).

8
Cryogenic Pumps

8.1 BASIC PRINCIPLES OF OPERATION

One of the simplest ways of creating vacuum is to lower the temperature of a gas until it condenses into a liquid or becomes a solid. There are, however, two practical difficulties in using this concept. First, it is not practical to cool the entire vacuum system to the cryogenic temperatures required. Second, it is very expensive to lower the temperature enough to condense and solidify light gases such as hydrogen, helium, and neon. The practical difficulty can be appreciated immediately from Figure 8.1, which shows the equilibrium vapor pressures of relevant gases at given temperatures. It is seen that at 20 K the three lightest gases are not removed from the gas phase and helium (partial pressure in dry air 4×10^{-3} torr) is not condensed even at 4 K, which is the temperature of liquid helium (at atmospheric pressure).

In industrial practice, instead of cooling the entire chamber it is sufficient to introduce in the chamber a cold surface that will condense the gas. Also, instead of relying on condensation alone, special porous materials are used to provide a large surface area to adsorb the gas more efficiently. The adsorption of gases on surfaces that are at a higher temperature than those shown in Figure 8.1 is called cryosorption. In the sense of the effect it produces, cryosorption may be thought of as a means of lowering the vapor pressure of the gas.

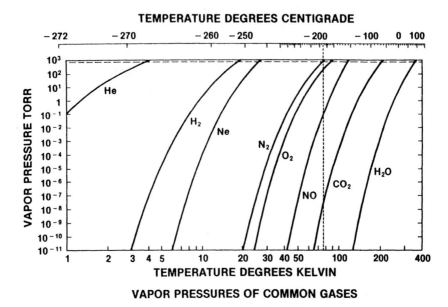

Figure 8.1 Vapor pressure curves of some common gases.

Both cryocondensation and cryosorption are not new in vacuum technology. For example, in helium liquefaction machines, cryopumping effects were clearly noticeable at the vacuum gauges used to measure the vacuum for thermal insulation of liquid-helium containers. Whenever the liquid-helium temperature was approached, the gauges would show a sharp pressure decrease (to the residual low-pressure reading of the triode ionization gauge). A small number of liquid helium cryopumps were made in the 1960s during the early stages of space simulation technology, but they have never become popular because of the expense and the difficulty of handling liquid helium. Typically, the guard vacuum could be maintained at 10^{-7} and 10^{-8} torr using liquid-nitrogen trapped vapor jet pumps. The inner vacuum chamber was pumped by a special vapor jet pump which penetrated the outer wall and had a double liquid-nitrogen trap. The inner vacuum space could be pumped to 10^{-10} or 10^{-11} torr before introduction of liquid helium. After the introduction of liquid helium, the gauges mounted on the liquid-nitrogen shroud (a nude UHV ionization gauge and a UHV cold cathode gauge) would indicate an abrupt drop in pressure to the minimum detectable levels or the ultimate residual readings. Presumably, the pressure was below 10^{-12} torr.

Another example of cryogenic pumping effects is the pumping of water vapor by liquid-nitrogen shrouds and traps. Traps are essentially cryopumps for water vapor and the vapors associated with the pumping fluid. In the early days of oil vapor jet pumps, pressure would drop approximately by a factor of 10 when the trap was cooled. This was partly due to relatively poor pumping speed, use of pumping fluids of higher vapor pressure, higher rate of breakdown of the pumping fluid, and poor degassing of the oil returning to the boiler. With modern vapor jet pumps the pressure typically drops by a factor of 2 or 3. This is due primarily to the increase of the pumping speed for water. The pumping speed is increased, because without liquid-nitrogen cooling the trap represents an impedance, and in addition it is placed closer to the vacuum chamber than the jets of the vapor jet pump, so that the overall conductance is increased. Any surface at liquid-nitrogen temperature is a very effective pump for water vapor, the pumping speed being approximately 14 L/s for each cm^2 of the surface, provided that the surface has a direct access to the incident gas. In other words, provided that the conductance to the surface is high enough not to produce a barrier to flow. For example, when estimating the pumping speed of a chevron-type cryobaffle, it is best to use the projected area at the inlet plane rather than the entire inner surface area of the elements.

Modern cryopumps are usually built as self-contained units with their own vacuum shell and an inlet flange, similar to vapor jet pumps and turbomolecular pumps. The cooling is achieved by the use of mechanical cryo-refrigerators at a temperature of 10 to 20 K and combining condensation and sorption.

8.2 CRYOSORPTION PUMPING

Sorption pumps were discussed briefly in Section 5.8.5 in association with rough vacuum pumps. Cryopumping or cryosorption mechanism can be employed for pumping effects at any pressure of interest to a high-vacuum technologist, from ultrahigh vacuum to atmospheric pressure. However, devices used near atmospheric pressure are very different from those used in the high-vacuum region.

Cryosorption pumps are not continuous throughput pumps. The pumped gas is not exhausted to the atmosphere, but kept inside the pump. The gas remains in the pump either as a thin film (one or a few monolayers thick) having a density of the liquid, or as a deposit resembling snow or frost. Because of saturation effects, the pump has to be regenerated or degassed occasionally. The amount of the pumped gas and the required period between regeneration govern the design at the extreme ends of the vacuum pressure scale.

The concept of cryopumping is not new. Near the beginning of this century, one of the common methods of creating vacuum was the use of activated charcoal cooled with liquid air. Later, when efficient oil-sealed mechanical pumps were developed, the use of sorption pumps ceased in routine vacuum pumping technology. The revival of interest began with the non-contaminating requirements associated with ionization and gettering pumps (Chapter 9) and the advent of the mechanically refrigerated cryopumps.

8.2.1 Adsorption and Cryosorption

As noted in Chapter 2, when a gas molecule collides with a surface, it does not generally rebound in a completely elastic manner. For the purpose of this discussion, we can assume that the molecule resides on the surface for a period of time (however short) and then may rebound back into the gas phase in a more-or-less arbitrary direction. It may also move on the surface before rebounding. The time of residence depends highly on the molecular weight and the temperature. It may range from fractions of seconds for hydrogen at room temperature to centuries for a large complex molecule at liquid-nitrogen temperature. It depends on the binding forces (related to the vapor pressure) and the points of contact to the surface and the number of such contacts. As a mnemonic device, a simple experiment can be made to illustrate this concept. If a "superball" made of highly elastic material is dropped on a hard surface, it bounces back almost as high as the original level from which it was dropped. If three or four such balls are held together by a thin string and the experiment is repeated, they will hardly bounce at all. A complex molecule needs synchronized interactions at several contact points in order to separate from the surface.

8.3 GASEOUS HELIUM CRYOPUMPS

8.3.1 Basic Construction

The pumping surfaces of cryopumps are arranged in three distinct variations. The first is a metal surface kept at temperatures of 50 to 80 K which can be associated with liquid-nitrogen temperature. This surface is attached to the first stage of the cryogenerator. Usually, it has a shape of a cylinder, the closed end of which is attached directly to the cold end of the refrigerator. The open end includes an optically opaque chevron baffle (Figure 8.2). The second section consists of a metallic surface attached to the second stage of the refrigerator, which is kept at 10 to 20 K. The third section is at the same temperature as the second but includes a layer of charcoal bonded to the inner surfaces of the second-stage cryoarrays. There are many different geometric arrangements possible (Figure 8.2 is an example).

The basic design consideration is to provide pumping surfaces for three different groups of gases, some of which are pumped as a result of condensation and some due to sorption. The first-stage chevrons are the place where primarily water vapor is pumped. The temperature of the first stage is more than sufficient to pump water vapor and hold it at extremely low equilibrium

CRYOGENIC PUMPS 323

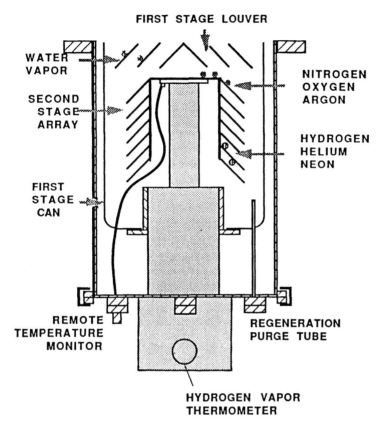

Figure 8.2 Schematic drawing of a typical cryopump.

partial pressure. In addition, any other gases that condense at the temperature will be pumped. The cylindrical enclosure of the first stage also serves as a thermal enclosure for the second stage. It reduces the radiation heat transfer to the second stage, thereby increasing its refrigeration capacity. The next group of gases, oxygen, nitrogen, argon, and other gases of similar molecular weight are pumped at the exposed metallic surface elements at the second stage. A review of the vapor pressure curves will indicate the equilibrium vapor pressures obtained for those gases at temperatures near or below 20 K. The third group, mainly hydrogen, helium, and neon, will not condense at 20 K and will not be pumped efficiently by the metal surface because the equilibrium vapor pressure for condensation will be too high. Not only will the capacity be limited to fractions of a monolayer, but the pumping speed will decrease as the surface coverage increases.

One way to improve this situation is to increase the surface area. This is achieved by the use of porous materials such as charcoal which have an estimated internal surface area approaching 1000 m^2/g. In addition to the very great surface area, the pore sizes approach the dimensions of the atoms and it is conceivable that the gas atoms are held in very small pores with much higher binding forces because the gas atom may interact with more than a few atoms in the solid.

In the conventional vertical arrangement, the cryoarray occupies only one-half of the pump body. The rest of the body length is necessary for locating the first-stage expander section and the motor, which operates the valves and the expander pistons. Cryopumps can be operated in any position, even upside down. The only concern may be the possibility of flaking particles from the pumped frost deposits. For convenience of attachment to a process chamber, pumps can be arranged with the expander unit at a right angle to the pump inlet flange (so-called low-profile cryopumps, Figure 8.3). In this way, if the pump is mounted underneath the chamber, only 25 cm of vertical space is required for a pump with 4000 L/s pumping speed (for water vapor).

The cold surfaces at the second stage contain charcoal and a "glue" (usually low-vapor-pressure epoxy) with which the charcoal is attached to the metal. However, at the operating temperatures of the cryopump, the vapor pressure of the epoxy is entirely insignificant. The cooling time of typical industrial cryopumps, starting from room temperature, depends on size and construction. The heat capacity of internal cooled parts should be minimized. Fortunately, the thermal conductivity of metals at cryogenic temperatures is high enough to permit relatively thin pumping panels and still minimize any temperature difference at the protruding edges of the panel.

The cool-down time of cryopumps with inlet diameters from 15 to 30 cm is typically 1 to 2 h, depending on the size of the cryoarrays and their construc-

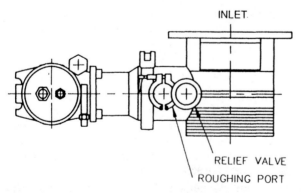

Figure 8.3 Cryopump with a lateral cold section.

tion, or, in other words, the heat capacity of the structure. A thermometer is usually provided for monitoring the operation of a cryopump. This can be a gas bulb thermometer with a conventional pressure gauge dial or a silicon diode sensor. The latter permits a remote location and convenient functional display to indicate when the cold temperature has been reached, when it is time to regenerate, and when room temperature has been reached during regeneration.

In a cryopump the only item exposed to the vacuum environment is a cold surface. All mechanisms, fluids, lubricants, and so on, are completely isolated from the vacuum space. This provides an inherent degree of cleanliness in regard to possible contaminants originating from the pump. This does not mean that cryopumps can be operated without certain care and attention. There exists the possibility of desorption of previously pumped gases if the temperature of the cryogenic arrays increases due to overloading and other mis-operation. However, in such cases it is the lighter gases that evolve first. Aside from hydrogen and helium, this may concern only nitrogen, argon, carbon monoxide, oxygen, and possibly methane, but no heavier hydrocarbons possibly associated with elastomer gaskets and other system contamination. Attention is also required during regeneration, which has to be performed in such a way as to prevent migration of desorbing gases back into the vacuum system.

Cryopumps can be made with all-metal gaskets using the well-established ultrahigh-vacuum techniques (see Chapter 11). Such pumps can achieve an ultimate pressure of 5×10^{-11} torr. the geometry of the cold surfaces (the cryoarray) can influence the shape of the pumping-speed curve and the optimum arrangement of pumping speed for various gases. Also, at higher pressures (such as used for sputtering applications) heat transfer due to convection may become significant. The array geometry can be made so that the formation of convection currents is inhibited.

Cryopumps, because of moving parts in the compressor and the expander, may transmit vibrations to the vacuum chamber. The degree of cryopump-induced vibration interference on process equipment will depend on system design, mass, and the location of the cryopump. Vibration spectra of cryopumps typically show peaks in the vicinity of 120 and 240 Hz (with a 60-Hz supply). There is also some low-frequency vibration associated with the reciprocation of the displacer in the expander (1.2 Hz at a 60-Hz line frequency). When attached to a medium-sized vacuum process chamber, the amplitudes are in the range of acceleration 25 to 50 mG respectively. The amplitude of the vibration is significantly reduced in pumps which use three-phase motors in the expander and the compressor. In this way the amplitude is reduced to 4 and 8 mG acceleration force at the two frequencies (Figure 8.4). Even the reduced amount, usually, does not have a sufficiently low level to permit direct use of cryopumps

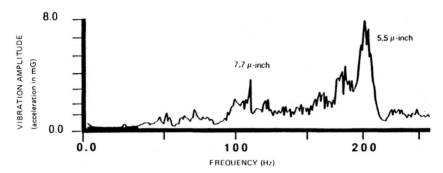

Figure 8.4 Vibration characteristics of a cryopump.

in applications such as scanning or transmission electron microscopes or electron beam lithography systems.

8.3.2 Expander Construction

The simplified schematic representation of the entire cryogenerator is shown in Figure 8.5, and the outline of the physical arrangement, in Figure 8.6. There are three basic components: the displacer (piston), the regenerator (heat exchanger) and the compressor with its surge volumes. Only a single stage is shown in Figure 8.5. For cryopumps, usually, the regenerator is placed inside the displacer, and the valves together with the piston drive motor are attached to the warm end of the displacer. In this way the system is reduced to two separate sections connected with flexible hoses. The cold end protrudes into the vacuum space. The drive motor assembly is hermetically sealed (from communication with atmosphere) and is submerged in the helium motive fluid.

The reciprocal motion of the displacer is mechanically synchronized with the valves, which open and close about once per second to channel the helium flow. At the beginning of a cycle the compressed helium is introduced into the cylinder by opening valve V_1. At this point the piston is at the bottom and valve V_2 is closed. Next, V_1 is closed, the piston is moved up, and the compressed helium passed through the regenerator into the lower part of the cylinder. While passing through the regenerator the helium is cooled because it expands. Then valve V_2 is opened, which expands the gas further and cools it more. Next, the piston is moved down and the expanded gas pumped by the compressor through the open V_2. Finally, V_2 is closed and the cycle repeated. The reciprocal motion of the piston can be obtained by a pneumatic cam drive synchronized with the proper sequence of valve operations. The heat exchange produced in the regenerator gradually cools one end of the cylinder to the desired low temperature. The regenerator is filed with a high-heat-capacity material (metal spheres

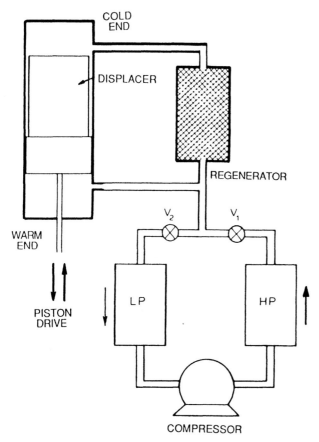

Figure 8.5 Cryogenerator diagram.

or disks). The cylinder is made of a very low heat conductivity material (reinforced composite resin) to prevent excessive heat transfer from the warm end of the cylinder to the cold end. Figure 8.7 shows a schematic arrangement of a two-stage expander section which produces two levels of cryogenic temperatures needed for practical design of cryogenic pumps.

8.3.3 Compressor Assembly

A compressor is required as part of the cryogeneration system to circulate helium through the expander assembly and to recompress it. The working fluid (helium gas) must be hermetically sealed to prevent leakage and frequent replenishment. Unfortunately, hermetically sealed compressors which are used for

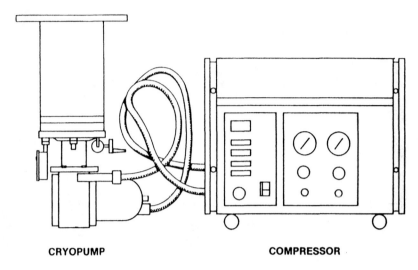

Figure 8.6 Cryopump system.

ordinary refrigerators, using Freon as the working fluid, cannot be used directly with helium. The major problem is oil separation. The lubrication oil permanently sealed inside the compression capsule cannot be permitted to migrate into the cold parts of the expander because it will freeze there. Completely oil-free compressors, such as perhaps diaphragm-type compressors, which could be used with helium, are typically not hermetically sealed and are too expensive for the purpose. Thus, for practical engineering reasons, the Freon-type compressors are adapted for helium service at cryogenic temperatures after considerable modifications and additions. These additions are necessary to extend the operation of the compressor to 2 to 3 years without major service.

The contents of a typical cryopump compressor module are shown in Figure 8.8. The compressor can be water cooled or air cooled. Generally, cryorefrigerators are not very efficient in regard to the overall energy expenditure. To obtain several watts of refrigeration capacity at 20 K, a 1-hp motor is used to drive the compressor. The heat developed by the motor and the heat developed during the compression of helium are removed by cooling the compressor capsule. The operating pressure is typically between 100 and 300 psig (approximately 7 to 20 atm). The efficiency as well as the cost of cryogenic refrigeration systems is lower for units of small capacity. They average 2% at 1 W, 5% at 10 W, to 10% at 100 W capacity. The refrigeration capacity diminishes as the temperature is decreased. A small two-stage system may reach

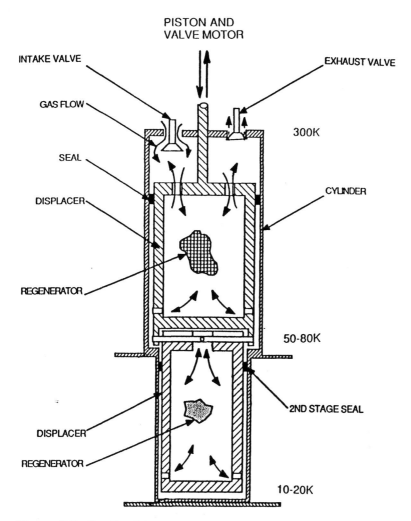

Figure 8.7 Details of the cryogenerator.

10 or 20 K with the capacity being near zero; it may be a few watts at 15 K and double the capacity at 25 K. Typical available refrigeration capacity for medium-sized cryopumps (15 to 20 cm inlet diameter) is 4 to 5 W at the second stage and 15 to 35 W at the first stage.

Figure 8.8 shows the oil recovery system and the pressure equalization, which permits the compressor to operate in the proper pressure regime regard-

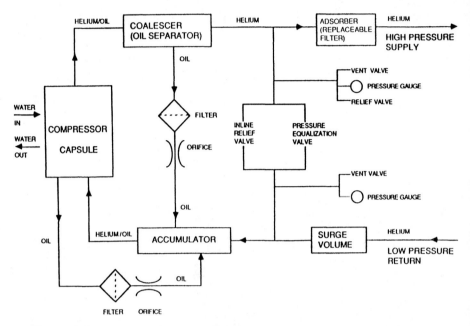

Figure 8.8 Diagram of the compressor section.

less of the cold-end temperature even if the compressor module is isolated from the expander section. The entire compressor module is packaged in a separate cabinet (approximately a cube with 50-cm sides) and connected to the expander with two hoses using quick-disconnect joints (Figure 8.6). The compressor module, then, consists of four main parts: the compressor, heat exchanger, oil separator, and oil adsorber (Figure 8.8). Oil is circulated in the low-pressure return circuit not only to lubricate the moving parts in the compressor (pistons, vanes, connecting rod bearings or shafts, etc.) but also as a medium for removal of heat. The helium is cooled to room temperature before it is returned to the expander section. The two hoses connecting the compressor module to the cryopump are essentially at room temperature. The basic system is analogous to the air conditioner, in which there are two separate heat exchange arrangements with two separate fans (even though they are attached to a common shaft). The outside circuit exchanges heat from the elevated temperature to the outside ambient air; the internal circuit exchanges heat from the room air to the refrigerated surfaces.

The efficient separation of oil from the helium returning to the cold section of the refrigerator is very important because even small amounts of oil producing a frozen film on the moving parts will interfere with the proper operation.

Typically, the helium leaving the oil separator contains approximately 1 ppm of oil. This final amount is removed in the adsorber, which has a replaceable filter. A typical replacement period is 1 year. Some newer versions of compressor modules are said to require filter replacement once in 3 years. Since filter replacement is a relatively simple chore, it is advisable to replace them frequently to avoid the possibility of need for major repairs. The adsorber-filter can be mounted in any part of the high-pressure side of the helium circuit and mounted in a convenient location for cartridge replacement.

The design of the pumping surfaces, their size, and their geometrical arrangement is made regarding certain common requirements encountered in typical applications. Pumping speed for various gases has to be arranged in such a way as to preclude rapid saturation for one gas while the pump still has left plenty of capacity for other gases. In other words, the saturation for various gases should occur more-or-less simultaneously. In some cases, special pumps are made for particular applications: for example, pumping of argon in sputtering installations. In addition to freedom from oil, the helium stream entering the cold refrigerator must be free of water vapor, which will also freeze on moving surfaces. The contamination of the working fluid (helium) may be manifested by a periodic temperature change, such as oscillation between 10 and 30 K with about 30-min cycles.

Careful processing of materials used in the entire helium circuit system is required to eliminate excessive amounts of water before the system is charged with helium. Oil must be degassed, parts must be baked before assembly, and the elastomer materials used in construction of the expander pistons and seals must be also degassed. It is best, therefore, not to attempt major service of the internal parts of the cryogeneration system without proper training and special facilities.

The electrical controls of the compressor module can be located remotely at the control panel of the vacuum system together with the controls for initiating and conducting the regeneration cycle. By using suitable pressure sensors the entire regeneration cycle can be performed automatically.

The convenience of separation between the compressor module and the cold end of the refrigeration systems also permits operation of two or more cryopumps from the same compressor, provided that it has sufficient capacity. Electrical control units that can control two compressor modules are available. Thus, in complex vacuum systems as many as four cryopumps can be operated from a single electrical control panel.

8.3.4 Pumping Speed

The pumping speed of a condensation or sorption pump depends on the arrival rate of the gas (if we restrict the discussion to molecular flow conditions), the capture probability (sticking coefficient), and the area of the cold surface. The arrival rate depends on the molecular weight and the temperature, as in any other high-vacuum pump. For cryosorption situations the pumping speed will also depend on the amount of previously pumped gas on the surface. In addition, the accumulated frost deposits can make the various gas passages in the pump narrower, thereby reducing their conductance. In actual use, the speed for a particular species of gas may depend not only on the amount of previously pumped gas but also on the sequence with which those gases have been introduced into the pump. For example, during the initial cooling of the cryopump, care must be taken to prevent a decrease in the hydrogen capacity of the pump. During cooling the pump should be isolated from the system (i.e., the high-vacuum valve should be closed). If the valve is left open while the pump is cooling, gases from the chamber may adsorb on the charcoal at the second stage before the metal surfaces of the pump become cold enough to freeze them and thus prevent their access to the charcoal. This will decrease the hydrogen capacity. The damage may be permanent if the charcoal pumping sites are occupied by hydrocarbon molecules. Rather high temperatures are required to drive them out, but cryopumps cannot be baked above 100 to 150°C because of the construction and the materials used in the expander section. One of the subtleties of cryopump design is the arrangement of thermal capacities such that the second stage remains warmer than the first stage until the first stage is cold enough to pump water vapor and other potential contaminants.

Pumping speed curves normally associated with vapor jet pumps and turbomolecular pumps cannot be supplied for cryopumps in the same definite way. The speed values given in specifications should be considered to be nominal, perhaps for a newly regenerated pump and for the particular gas alone. The pumping speed of a surface that fully captures gas molecules arriving at room temperature can be calculated to be, for hydrogen, 44.6 L/s for each square centimeter of the surface; for water vapor, 14.9 L/s $\cdot$ cm^2; and for nitrogen, 11.9 L/s $\cdot$ cm^2, provided that the access to the surface is not impeded by limitations in conductance. In industrial cryopumps, the full pumping speed for water vapor may be realized if measured directly at the inlet port, because the access to the first-stage chevron cryoarray is free and condensation efficiency is essentially 100%. The net speed for nitrogen is typically about 3 L/s $\cdot$ cm^2 and for hydrogen 5.6 L/s $\cdot$ cm^2. Nitrogen, oxygen, carbon monoxide, and argon have to pass through the first-stage chevron baffle, which reduces the conductance and allows about 25% of incident molecules to pass through. Hydrogen has still a longer path to travel before it reaches an inner charcoal surface; this is also true for helium and neon.

Thus hydrogen net speed is only about 12% of the incident rate. Figure 8.9 shows the pumping speed for a few different gases for a 20-cm-diameter pump. The increase in the pumping speed at the higher pressure cannot be fully uti-

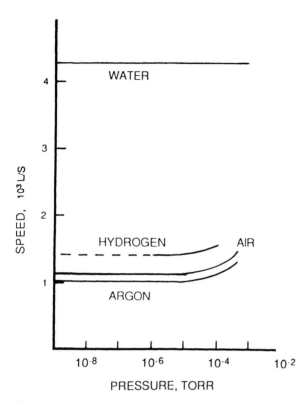

Figure 8.9 Pumping speeds for different gases.

lized in conventional cryopumps because of the resulting need for frequent regeneration.

The pumping speeds for water vapor, nitrogen, oxygen, and argon decrease only slightly in time. As frost deposits accumulate at the cold surfaces, the temperature changes very little. Water (ice or frost) can be allowed to accumulate on the chevrons until nearly half of the passage areas are blocked. At the outer second-stage panel, argon or nitrogen deposits can be 1 cm thick. The limit or the total hydrogen capacity specifications are usually based on the assumption that no other gases are adsorbed on the charcoal surfaces of the second stage and that this stage is at its normal low temperature. The limit is also based on having the equilibrium hydrogen pressure of 1×10^{-6} torr. For this reason the pumping speed for hydrogen in Figure 8.9 is shown as a dashed line at lower pressures.

If the desired hydrogen pressure is lower than 10^{-6} torr, the total capacity should be reduced accordingly. The effect is indicated in Figure 8.10, where

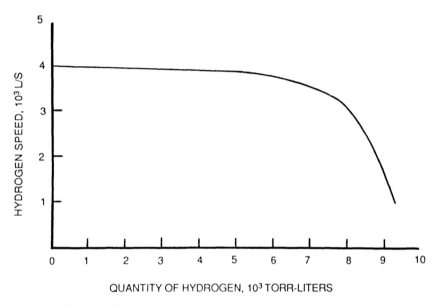

Figure 8.10 Decline of hydrogen speed with the pumped amount.

the pumping speed for hydrogen is shown, depending on the amount pumped previously. The capacity improves when the temperature is lower for a given desired equilibrium pressure. It can be appreciated that pumping performance, where the major pumping mechanism is due to cryosorption, is strongly dependent on temperature. Of most important concern is hydrogen, which is usually present in high-vacuum systems. Helium and neon, although they are present in air, are (or should be) removed almost completely during early stages of pumping. Figure 8.11 shows the decline in hydrogen speed due to argon pumped previously.

As noted before, the pumping-speed curves for cryopumping often show increased speed at the higher pressure. This is of significance for cryosorption pumps used for rough vacuum pumping. For high-vacuum cryopumps such curves may be misleading unless they are immediately qualified regarding saturation effects. For practical purposes, the basic short-term speed-pressure characteristics are very similar to those of vapor jet pumps and turbomolecular pumps, including, more or less, the values of impulse gas loads, for example during the crossover from rough to high vacuum pumping. The pumping speed for water vapor is nearly twice as high for cryopumps because of the efficient condensation of water vapor on the baffle placed immediately at the inlet of the pump. Yet a similar geometry for a liquid-nitrogen-cooled trap above a vapor jet pump (or turbopump) would produce the same pumping speed.

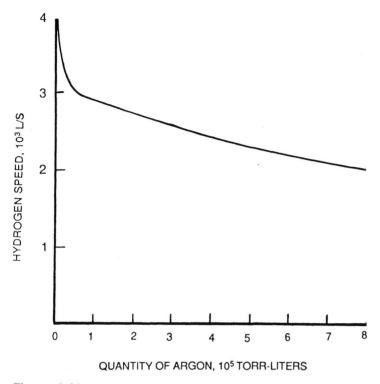

Figure 8.11 Decline of hydrogen speed after pumping argon.

8.3.5 Throughput and Capacity

An operating cryopump, like a vapor jet pump, cannot be exposed to air because of almost instant saturation and well as the high thermal load. After an accidental air exposure it will take 4 to 6 h to restore high-vacuum pressure. The pump has to be regenerated and cooled again. In vapor jet pumps, depending on the type of pumping fluid and the desired vacuum level, the damage may be minor or severe and require complete cleaning and replacement of the pumping fluid. In relatively undemanding applications, such as television picture tube evacuation, valveless vapor jet pump systems are used where the pump is exposed to atmospheric air with each process tube. This can be done with pumping fluids with high oxidation and thermal breakdown resistance (silicone oils), although it is not generally recommended practice. Such operation would be impossible with cryopumps. Cryopump systems need a high-vacuum isolation valve and, generally, a separate rough vacuum valve for the initial evacuation of the system.

Having gained some familiarity with three types of pumps capable of achieving and maintaining high vacuum, the reader may be concerned about the choice of pumps for particular applications. This is not a simple matter and it requires careful consideration of performance characteristics, including maintenance requirements. The cost (capital acquisition, operating, and service costs) is rarely the primary consideration, except, perhaps, for very high pumping speeds. In the case of capture pumps it is not sufficient to look at the cost per unit pumping speed, but the accumulated gas load (integrated throughput versus time) must be also considered. For example, both cost and power consumption per unit throughput should be considered in association with the required period between regeneration events. Figures 8.12 and 8.13 show comparisons between a single and a dual cryopump system and a vapor jet pump of a similar size. If the vapor jet pump system contains a liquid-nitrogen trap and a baffle, its cost will roughly double. These graphs indicate that at high gas loads, when the periods between regenerations became short, it is conceivable to use two independent pumps where one can be regenerated while the other is in operation. However, when cleanliness, fear of the possibility of contamination from the pumping fluid, and other performance characteristics are of paramount

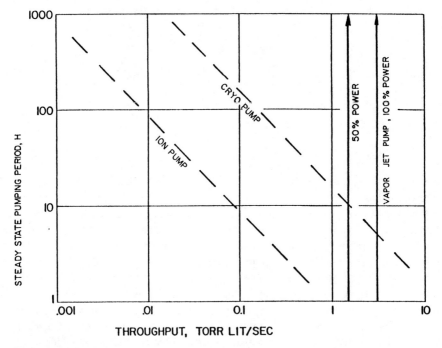

Figure 8.12 Pumping period depending on gas load.

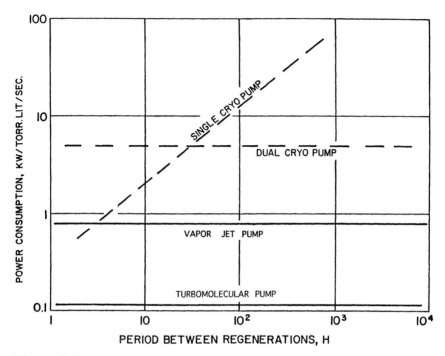

Figure 8.13 Power consumption per unit mass flow.

importance, the operational costs may be insignificant compared to the added value of processed product of good quality.

The high-vacuum cryopumps, despite their tolerance to high impulse gas loads, are intended primarily for operation at inlet pressures in the high-vacuum range (i.e., below about 1×10^{-3} torr). Partly this is due to saturation issues and the requirement of a reasonable time period between regenerations, and partly due to the pumping speed dependence on thermal parameters, such as the thermal conductivity of the gas. At pressure above 10^{-2} torr the degree of thermal insulation is diminished and the refrigeration capacity of the cryorefrigerator may be insufficient to maintain the required low temperature. In other words, depending on the refrigeration capacity, the effective pumping action ceases whenever the thermal load due to the pumped gas condensation and/or the thermal conduction by the gas exceeds the available refrigeration power. This is illustrated in Figure 8.14, showing the calculated value of nitrogen pumping at a liquid-helium-cooled surface with a 1-W refrigeration capacity. As indicated, very little refrigeration capacity is required to produce very high pumping speeds at low pressures.

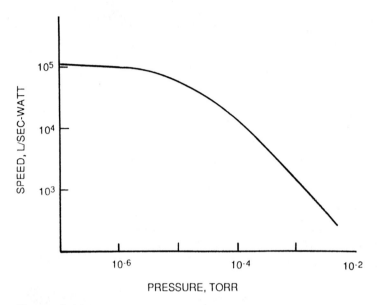

Figure 8.14 Theoretical pumping speed of nitrogen on a 4 K surface with 100% sticking probability.

It should be understood that the total gas pumping capacities given in specifications of cryopump performance are essentially nominal values. The total capacity values are sometimes associated with an operating pressure (typically 1 to 5 × 10^{-6} torr). This indicates that saturation effects will occur sooner if lower pressures are desired. Also, when a mixture of gases is pumped, it is not always possible to assume that the saturation times for each species are not interdependent. The basic reason for this is that there is simply a limited space for accommodating the solid (frost) deposits inside the pump. If two gases are pumped simultaneously on the same pumping surfaces, the volume of the deposit will perhaps be twice as great (assuming equal speed and partial pressure). Thus for estimating saturation time, total pressure should be used for gases that are pumped in a similar manner.

Nevertheless, at prevailing high-vacuum pressures, cryopumps can continue pumping a very long time before regeneration is required. For example, a 20-cm-diameter cryopump has a pumping speed for nitrogen of 1000 L/s. Its total capacity for nitrogen is 300,000 torr·L. At a continuous pressure of 5 × 10^{-6} torr we obtain

$$\frac{300{,}00 \text{ torr} \cdot \text{L}}{(1000 \text{ L/s}) (5 \times 10^{-6} \text{ torr})} = 6 \times 10^7 \text{ s}$$

which is equivalent to 16,700 h. Even if we allow a safety factor of 2 for the calculation, this represents 1 year of continuous operation. Correspondingly, at a pressure of 5×10^{-5} the operating time would be about 1 month. For periods when the pressure is not constant, for example, during the evacuation after opening the high-vacuum valve, the total gas load can be estimated by assessing the area under the pump-down curve and multiplying it by the pumping speed.

The maximum gas throughput given in cryopump specifications must be understood as the limiting value above which the pump operation is impeded. Temperature may rise suddenly, and previously pumped gases will be desorbed. The maximum throughput capacity is an important parameter of pump performance in regard to the ability of the pump to adsorb occasional high gas loads or sudden unexpected pressure bursts. However, it should not be associated with a continuous gas load without a clear understanding of the relationship between the amount of pumped gas and the period between regeneration steps. For example, Figure 8.12 shows the period between regenerations, depending on the continuous gas throughput. The values are for a typical 20-cm-diameter cryopump pumping air. The maximum throughput values for a vapor jet pump and an ion pump (see Chapter 9) of similar size are given for comparison.

8.3.6 Throttled Cryopumps

For some applications requiring higher process pressures, such as sputtering, it is necessary to control the pressure and the flow rate independently. Most high-vacuum pumps do not function above an inlet pressure of approximately 1×10^{-3} torr. If the process pressure has to be higher, typically 5 to 50 mtorr for sputtering, the high-vacuum pump must be throttled to avoid overloading. Unfortunately, in the case of vapor jet pumps and turbomolecular pumps, this requirement is often overlooked and pumps are operated in overload condition with some long-range detrimental effects. For the cryopumps, throttling becomes a necessity if the pump is to last a week of typical batch process operation before regeneration is required. The throttling apertures can be fixed or variable. The size and shape of the apertures, and their position and temperature, are chosen to optimize the performance in regard to the pumping speed for the sputtering gas (usually argon) and water vapor, which is the main residual gas during the later stages of evacuation of the process chamber as well as during the operation of the sputtering system. It is then desirable to reduce the access of argon to the second stage but keep the water vapor speed high. To achieve this, the disk containing the throttling aperture is attached to the frontal array, which is at the temperature of the first stage. The disk itself becomes the major pumping surface for water vapor. The high pumping speed for water vapor reduces the partial pressure levels of hydrogen and oxygen which form by dissociation of water during sputtering. Both hydrogen and oxygen can be contaminants in processing silicon wafers, for example.

In addition, restricting process gas access to the second stage helps keeping it at low temperature, which enhances the hydrogen pumping capacity. In pumps used for sputtering it is desirable to have a high refrigeration capacity in both stages, but especially at the first stage. The design must be optimized so that maximum hydrogen capacity is reached at approximately the same time as the other gas load limits.

The optimization of the performance of cryopumps in applications with throttling involves an interesting issue in the shape of the throttling apertures. In Figure 8.15 the performance of a pump with a single round hole as the restricting element is compared with a pump having multiple slots for throttling. This arrangement produces a more constant pumping speed in the area of interest. Thus the pump with the slots can be operated at higher pressures at the same throughput (or process gas flow rate). The higher restriction at higher pressures for the rectangular slots is due to the higher periphery of multiple slots, which affects the conductance in the pressure range exceeding the molecular flow conditions. Thus the conductance of the round orifice is lower at the molecualr flow range but higher at the higher pressures, and the rectangular slots permit the improvement of performance as shown in Figure 8.15.

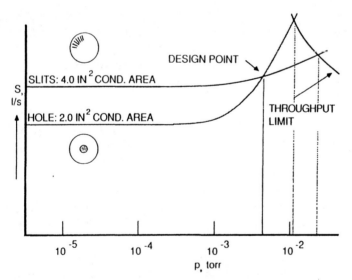

Figure 8.15 Throttling effect of rectangular slots.

8.3.7 Regeneration

Cryopumps do not have a continuously functioning exhaust port. Depending on the pressure at which they operate, cryopumps become saturated after a certain period of time. The gas deposits accumulated in the pump have to be removed periodically. The process of removal requires warming up the pumping surfaces to room temperature, which means that for a period of several hours the pump will not be functioning while it is being regenerated. The need for regeneration becomes evident when the pump is no longer able to produce or maintain expected pressures. Usually, the partial pressure of hydrogen will be first to rise and will begin to affect the total pressure measurement. The practical engineering problems in the design of gaseous helium cryorefrigerators limit the capacity of the cryosorption arrays at the second stage. Typically, the total hydrogen capacity is 50 to 100 times lower than, for example, the capacity for argon (10 to 30 torr · L compared to 1000 to 2000 torr · L). Even in an application when the process gas is argon, such as sputtering, the hydrogen saturation may occur before argon saturation, or it may occur nearly simultaneously. Significant amounts of hydrogen may be generated by dissociation of water vapor during sputtering.

If the pump has adequate refrigeration capacity the temperature of the second stage may not rise even though the pumping speed has been seriously diminished. In normal high-vacuum applications, the saturation will be noticeable due to the increase in evacuation time. Somewhat arbitrarily, regeneration may be advisable when the normal pumping time to operating pressure has doubled. In sputtering applications, if the argon saturation is approaching, the pressure decay during evacuation will become erratic.

The period between regenerations may be estimated by integrating the pressure at any time for the significant gases, but this requires special electronic instrumentation and knowledge of partial pressures. If the hydrogen partial pressure remains more or less constant, its throughput can be estimated by multiplying the pressure and pumping speed ($Q = Sp$). The regeneration time can then be obtained by dividing the total hydrogen capacity value (from pump specifications) by the estimated throughput. For example, if the hydrogen pressure is 2×10^{-6} torr, the pumping speed 1200 L/s, and the total hydrogen capacity 700 torr · L, the regeneration time

$$T = \frac{7000 \text{ torr} \cdot \text{L}}{2 \times 10^{-6} \text{ torr} \times 1200 \text{ L/s}} = 2.9 \times 10^6 \text{s} = 800 \text{ h}$$

Generally, it is not advisable to wait until erratic performance is noticed. In routine operations regeneration should be performed after about 60% of the expected maximum time of satisfactory operation has elapsed. It is often convenient to regenerate during weekends or any other scheduled maintenance intervals.

The regeneration procedure must be performed with care and adherence to established technique. In addition to general removal of adsorbed gases, care must be exercised not to contaminate the second-stage cryosorption beds with heavy gases (hydrocarbons) previously pumped at the first stage. This is accomplished by a properly located and properly operated gas sweep (usually nitrogen).

The process of regeneration consists of four steps. First, the pump is warmed while purging with dry, preferably heated nitrogen with a maximum temperature of 80 to 100°C, depending on the manufacturer's instructions. The proper location of the gas purge line and its flow rate and temperature will ensure that the charcoal at the second-stage cryoarray will not become contaminated by water vapor desorbing from the first stage and by hydrocarbons which are nearly impossible to remove. This is done by directing the purge gas flow from the second stage toward the first stage. The heated gas will keep the temperature of the second stage higher than the first, which will help to minimize the condensation at the second stage. The purge gas flow should be maintained until the cryosorption array reaches room temperature. For thorough degassing it may be maintained for another hour.

The desorbed gases are exhausted through the pressure relief valve (Figure 8.16). If dangerous gases are present, the venting lines must be provided. The use of nitrogen purge gas helps to dilute potentially explosive or toxic gas mixtures which may have been pumped during the operation of the pump.

The second step is evacuation, to remove the remaining gases and to create the vacuum that is necessary for thermal insulation before cryocooling is initiated. The pressure after this initial evacuation should be 25 mtorr or less. Great care should be exercised not to contaminate the pump if oil-sealed mechanical vacuum pumps are used for this purpose. The evacuation line must include an efficient trap to prevent backstreaming of oil vapors or their fractions reaching the cryopump. The molecular sieve (zeolite) traps must frequently be regenerated themselves of their replaceable filters changed. Regeneration of mechanical pump sorption traps consists of heating the trap to about 250°C while pumping with a purge gas flow at about 2 torr pressure. The location of purge inlets relative to trap isolation valves is important to make sure that during the desorption of the trap, contaminants do not travel upstream. Traps should be regenerated at least as often as the cryopump. The automatic regeneration control units include this step prior to the regeneration of the cryopump.

In recent years oil-free vacuum pumps have become available that produce adequate ultimate pressure for regeneration of cryopumps. They can greatly simplify the regeneration procedure by eliminating any concern about hydro-

CRYOGENIC PUMPS

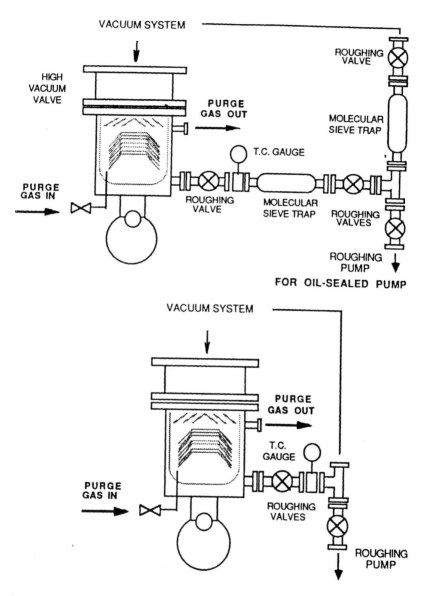

Figure 8.16 Arrangements for cryopump regeneration.

carbon contamination. The third step is of precautionary nature. A time rate of rise of pressure of an isolated cryopump should be performed to ensure that residual gases (usually water vapor) have been removed, and for reassurance that there are no external leaks. For conventional cryopumps (15 to 30 cm diameter) a typical pressure increase should be no more than about 1 mtorr/min.

The fourth step is cooling the pump to the operating temperature. The rough vacuum valve should be closed soon after cooling begins (no later than 15 min if oil-sealed pumps are used). Regeneration can also be carried out without the gas purge, but the procedure will be much longer as indicated in Figure 8.17.

The newest technical refinements for control of operation and regeneration of cryopumps have simplified the diagnosis of pump condition, the prediction of regeneration requirement, and permit automatic control of the regeneration process. A control module attached to the cryopump can include local and remote diagnostic displays and operation commands. It may be linked to a computer to enable the operator to perform all input functions, provide faster and more thorough regeneration, monitor temperature, pressure, and other functions such as defects in the purge gas flow and warm-up heater temperature, signal the completion of regeneration, and assure proper procedures.

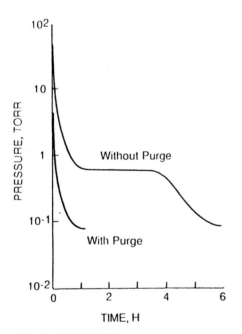

Figure 8.17 Regeneration with and without purge gas.

In addition to the refinement in monitoring and control, improved regeneration procedures have been made. For example, in some applications, the hydrogen pumping capacity is exhausted while the ability to pump other gases is still adequate. In such cases, the pump can be partially regenerated, which may take less than an hour compared to more than three hours for complete regeneration. Manufacturers usually provide hydrogen pumping data (from measurements with pure hydrogen), but the hydrogen pumping capacity may be reduced if other gases are, or have been, pumped (for example, due to a small rise in temperature). Partial regeneration is achieved by careful heating of only the second stage to emit hydrogen. Care must be exercised not to raise the pressure to the level where heat transfer by gas conduction may increase the temperature of the first stage.

In conclusion, it is only necessary to review the vapor pressure curves in Figure 8.1 to appreciate that cryopumping of hydrogen, helium, and neon at temperatures of 15 to 20 K is a considerable engineering achievement. Nothing comes easy in high-vacuum technology. All pumping methods have certain disadvantages and proper technique is required to use them successfully. As ultrahigh vacuum is approached, careful design and operational techniques become more and more important.

8.4 LARGE CRYOPUMPS

For some design requirements it is convenient to combine closed-loop gaseous helium cryopumps with liquid-nitrogen-cooled cryosurfaces. An example of such a combination is shown in Figure 8.18. Such arrangements are entirely practical whenever liquid nitrogen is plentiful and when high pumping speeds for water vapor is required or when the water vapor is the primary gas load in the system. A pump of roughly 1 m diameter can have a 100,000-L/s pumping speed for water vapor and other gases condensable at liquid-nitrogen temperature to the degree required to produce partial pressures in the high-vacuum region. The internal cryoarray of such pumps can be constructed according to the requirements. Because the first-stage cooling is not necessary for pumping action, somewhat more refrigeration capacity can be made available at the second stage. The actual geometry of the liquid nitrogen-cooled surfaces may be varied and adapted to the system needs. The only general requirement is that the inner (mechanically refrigerated) 20 K surfaces should be shielded or surrounded by a liquid-nitrogen-cooled enclosure.

8.5 WATER VAPOR PUMPS

Closed circuit helium mechanical cryogenerators can be used for pumping primarily water vapor. A single stage refrigerator is sufficient for this purpose,

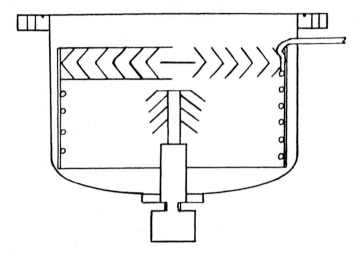

Figure 8.18 Large cryopump with liquid nitrogen.

providing enough cooling capacity to keep a cryopanel at a temperature near 107°K. A typical system is shown in Figure 8.19.

The cold surface consists of an open-ended cylinder placed in close proximity of the outer wall and connected to the cryopump. This provides a reasonably high conductance for other noncondensable gases, which can pass through the cylinder, while producing a reasonably high probability of capturing water molecules. For example, the water vapor pumping speed for a 150 mm diameter cylinder is 2,500 L/s and the conductance for nitrogen (or air) 1,000 L/s. Figure 20 shows a water vapor pump attached directly to a turbopump. This may decrease the speed of the turbopump (due to a reduction in conductance), but it substantially increases the pumping speed for water vapor. Such an arrangement can typically increase the water vapor pumping speed by 4 or 5 times when the cryopanel is cooled.

8.5.1 Polycold Pumps

In some industrial vacuum processes, where large amounts of water vapor must be removed, surfaces cooled by liquid nitrogen are placed inside the vacuum chamber. The simplest and most convenient arrangement consists of a coiled tube filled with liquid. The cooling may be started soon after the initial evacuation is completed, and the coils are defrosted each time before the chamber is exposed to atmosphere.

However, for pumping primarily water vapor, liquid-nitrogen temperature is not required. Effective cryopumping speed for water vapor can be obtained

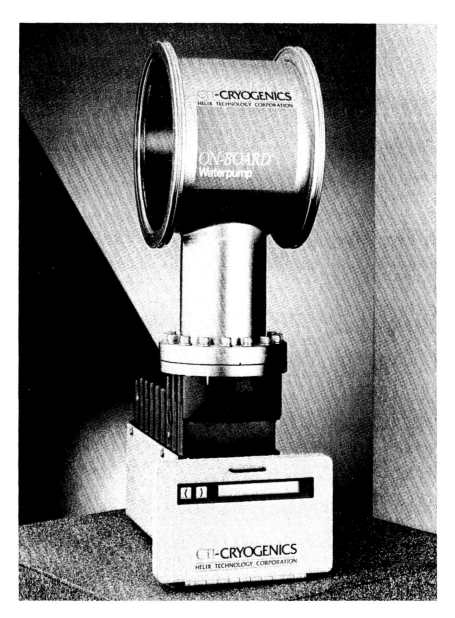

Figure 8.19 An open-ended cylindrical cryopanel attached to a single-stage cryopump. (Courtesy of CTI-Cryogenics.)

Figure 8.20 A water vapor cryopump (attached directly to a turbomolecular pump. (Courtesy of CTI-Cryogenics.)

at −110°C, and if the desired ultimate pressure is in the 10^{-8} range, −130 to −140°C is sufficient. There exists a special closed-cycle refrigeration system (Polycold Systems) which uses a single-compressor, mixed-component cascade circuit capable of reaching such temperatures.

Polycold cryogenerators use a compound freezing method which employs several different coolant fluids with only one compressor. The differences of boiling points of the different coolants is utilized for repeated cycles of cooling and vapor separation, extracting part of the mixture at lower boiling points to be used in the next cycle of refrigeration. The coolant stream is passed through several stages of heat exchange, with each successive fluid fraction consisting of lower boiling point components as the temperature decreases (as indicated in Figure 8.21). The arrangement is similar to ordinary cascade refrigeration systems. The advantages of the mixed refrigerant cooling process are: simplicity of equipment and compressor operation, versatility of refrigerant selection for varying applications, lower pressure ratios, lower compressor discharge fluid temperatures, greater reliability, and reduced maintenance.

The Polycold pump can be used to cool baffles and traps as well as nonisolated cryopumping tubes. With careful design, cooling and heating time can be reduced to a few minutes. for fast defrosting a twinned tube is used with one side used for cooling, the other for heating. The refrigerant flows through the cooling tube. Compressed gas, stripped of high-vapor-pressure components and after oil residues from the compressor are removed, is indirectly heated using compressor discharge vapors and passed through the other tube for the defrosting cycle.

This system uses less overall energy per unit amount of water vapor throughput (torr · L/s) because the temperatures are not as low compared to those obtained with helium refrigerators. Its cost per unit throughput is also lower. Nominal pumping speeds of such systems for water vapor can be as high as 100,000 L/s.

As can be expected, the pumping speed depends on the access of the pumped gas (water vapor) to the cold surface. Usually, the location of the cold surface in the vacuum chamber precludes the possibility of free, unimpeded access. Therefore, in estimating the extent of the pumping surface, it is often wiser to take the projected area rather than the total refrigerated surface area exposed in vacuum. The maximum possible pumping speed of an adequately cold surface for water vapor (at room temperature) is 14.7 L/s · cm^2, assuming a 100% condensation coefficient. In technical bulletins, the nominal pumping speed for an accessible surface is said to be near 9 L/s · cm^2 in molecular flow conditions. The high pumping speed level occurs in the pressure range 10^{-7} to 10^{-3} torr.

In transition and viscous flow, typically at pressures above 10 mtorr, an additional access impedance develops because the condensing gas must diffuse through the more or less stationary component in order to reach the cold surface. Therefore, the speed will be affected by pressure, the length of the diffusion path, and the pattern of flow. D. Missimer (Ref. 19) estimated actual

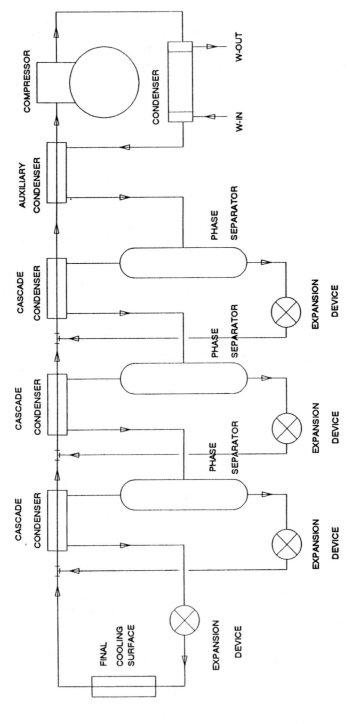

Figure 8.21 A schematic view of the Polycold cascade refrigerator. (Courtesy of Polycold Systems.)

pumping speeds in relatively large chambers for some geometric configurations at pressures from 10 to 100 mtorr and with various air-water vapor mixtures. The results show reductions of pumping speed to values at low at 1 to 10% of the possible maximum of 14.7 L/s · cm².

The shape and location of refrigerated pumping surfaces and the need and manner of periodic defrosting should be considered in the design and operation of a vacuum system. Generally, it is better to not permit a heavy buildup of condensed deposits. It is best to defrost the cold surface often, perhaps during each process cycle, even if the cold surface is located between the high-vacuum valve and the throughput pump (vapor jet, turbomolecular, or Roots-type). It is good to remember that 1 cm³ of ice will become more than one million L/s of water vapor at $1 \cdot 10^{-3}$ torr.

REFERENCES

1. B. A. Hands, *Vacuum*, *37*(8/9), 621 (1988).
2. R. A. Haefer, *Kryo-Vakuumtechnik*, Springer-Verlag, Berlin, 1981.
3. B. A. Hands, *Vacuum*, *26*, 11 (1976).
4. K. M. Welch and C. Flegal, *Ind. Res. Devl.*, *83* (Mar. 1978).
5. W. G. Baechler, *Vacuum*, *37*(1.2), 21 (1987).
6. L. T. Lamont, Jr., in *Methods of Experimental Physics*, Vol. 14, *Vacuum Physics and Technology*, G. L. Weissler and R. W. Carlson, eds., Academic Press, New York, 1979, p. 231.
7. J. F. O'Hanlon, *A User's Guide to Vacuum Technology*, John Wiley & Sons, 1980, p. 221.
8. F. Turner, Cryosorption pumping, *11th Vacuum Technology Seminar*, 1973, Varian VR-76.
9. J. P. Hobson, *J. Vac. Sci. Technol.*, *10*(1), 73 (1973).
10. S. W. Van Sciver, *Helium Cryogenics*, Plenum Press, New York, 1986.
11. D. J. Missimer, *31st Conf. Soc. Vac. Coaters*, May 1988.
12. J. F. Peterson and H. A. Steinherz, *Solid State Technol* (Dec. 1981).
13. K. M. Welch and R. C. Longsworth, *An Introduction to the Elements of Cryopumping*, American Vacuum Society, New York, 1980.
14. K. M. Welch, *Capture Pumping Technology*, Pergamon Press Ltd., Oxford, 1991.
15. J. Hengevoss and E. A. Trendelenburg, *Vacuum*, *17*,9, 495, (1967).
16. M. M. Eisenstadt, *J.Vac.Sci. & Technol.*, *7*,4, 479, (1970).
17. S. B. Nesterov and A. P. Kryukov, *Vac. Phys. Technol.*, *1*,1, (1993), and in Russian *Vac. Techni. Technol.*, 4,4, (1994).
18. J. P. Hobson, *J.Vac.Sci. & Technol.*, *10*,1 73, (1973).
19. D. J. Missimer, *Proc.Soc.Vac. Coaters*, *39*, (1996).

9
Gettering and Ion Pumping

Ultrahigh-vacuum techniques and system components have evolved together with special pumping methods uniquely adapted to the requirements of extreme cleanliness, bakeability, and pumping speed. The principal pumping mechanisms employed are chemical transformation, whereby gases are chemically combined into solid compounds having very low vapor pressure, and a method involving an ionization step that permits acceleration of the atoms or molecules in an electric field to drive the ions directly into a solid surface, where they remained captured.

9.1 GETTERING PUMPS

At high-vacuum conditions a surface can hold large quantities of gases compared to the amount of gas present in the space. This produces a pumping action either by physisorption or gettering, which refers to a chemical combination between the surface and the pumped gas. Many chemically active materials can be used for gettering. In vacuum systems the material commonly used is titanium because it is chemically reative with most gases when it is deposited on a surface as a pure metallic film, but it is rather inert in bulk form because of the tenacious oxide film covering its surface.

To produce a pumping action, all that is needed is a source of titanium and a means of producing a fresh (unoxidized) layer preferably on a large surface. Various forms of heating can be used to deposit a titanium film by evaporation or sublimation. The fresh titanium, deposited on surfaces surrounding the source, forms stable, solid compounds with chemically active gas atoms or molecules that strike the surface. This capture process can be continuous if new layers of titanium are constantly produced. Figure 9.1 shows a typical arrangement.

The pumping speed versus pressure curve for any titanium pump is shown in Figure 9.2. Pumping speed is maximum when a complete film of titanium is maintained on the pumping surface. Pumping speed, then, is maximum at low pressure. The number of gas molecules striking the titanium surface decreases with decreasing pressure. Therefore, below some pressure point a particular titanium source is able to supply titanium fast enough to sustain a complete film of fresh titanium. In this pressure range, the pumping speed is limited only by the intrinsic speed S_j of the titanium surface area and the gas conductance from the chamber into the pump.

It should be noted that at low pressure, periodic sublimation is employed to allow greater saturation of the titanium film before more titanium is deposited. The life of a titanium source is finite, but with judicious use (to be discussed later) it can be made to serve for a relatively long period. When pressure increases, and hence the density of gas molecules per unit volume is greater, pumping speed decreases and is dependent on the rate of continuous titanium sublimation. At higher pressures, the impinging gas molecules combine with the titanium virtually as fast as it is deposited and a complete film of fresh titanium is never formed. Consequently, the probability of capture for gas molecules decreases. Throughput is constant in this pressure region because the pumping action is limited by the rate of titanium deposition.

As in the case of cryopumps, the pumping speeds of gettering pumps cannot be defined precisely because it depends somewhat on the mixture of gases,

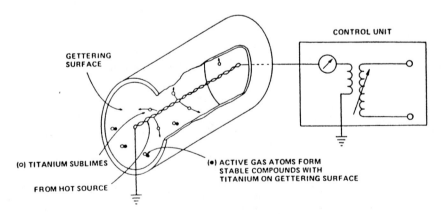

Figure 9.1 Typical arrangement of a titanium gettering pump.

GETTERING AND ION PUMPING

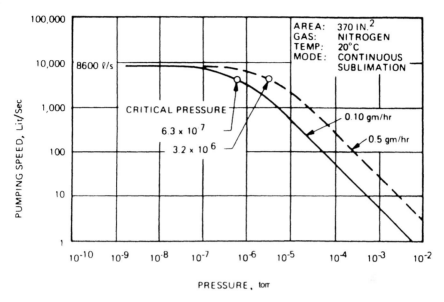

Figure 9.2 Pumping speed characteristic of a gettering pump.

the presence of gases that are not chemically pumped, the temperature, and the previous history of a surface. The following method can be used for computing an approximate pumping speed. Surface area, speed for specific gases, and the conductance of the titanium sublimation pump inlet aperture are the variables considered (see Table 9.1). The area-limited speed for specific gas is

$$S_A = AS_j \quad \text{liters/second}$$

where A is the area of gettering surface in square centimeters and S_j is the intrinsic speed per square centimeter of titanium surface for a specific gas, in

Table 9.1 Typical[a] Pumping Speeds Per Square Centimeter Titanium Sublimation Surface for Various Gases (L/s)

Surface temperature (°C)	H_2	N_2	O_2	CO	CO_2	H_2O	CH_4	A	He
20	20	30	60	60	50	20	0	0	0
-195	65	65	70	70	60	90	0	0	0

[a]Actual speeds vary with the condition of the deposits; usable speed is reduced by the conductance between the pumping area and the process area in the vacuum chamber.

liters/second. To use titanium economically, it is essential to determine the time required to deposit a complete but nonwasteful film and then cycle power to the filament accordingly.

Figure 9.3 shows typical pressure rise characteristics for a system in which the titanium sublimation pump yields a pressure reduction of a factor of 10 before saturation of titanium causes the pressure to rise. At successively lower pressures, saturation occurs after a longer period of time because the number of reacting gas molecules is decreased proportionally.

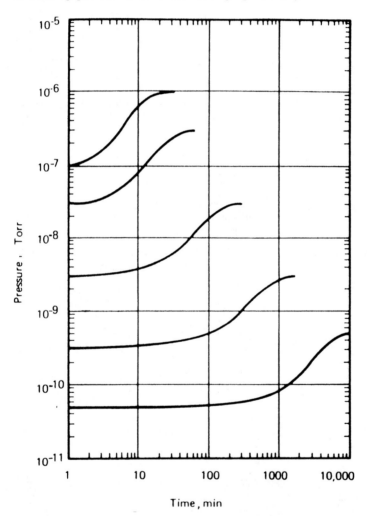

Figure 9.3 Pressure history of a gettering pump at various pressure levels.

Since the mechanism of titanium pumping is chemical combination, the theoretical maximum throughput can be found by using the appropriate chemical formula and the rate of titanium sublimation in atoms per second. To use as an example pumping nitrogen (compound: TiN), the typical sublimation rate from a titanium source is 3×10^{17} atoms/s. since one titanium atom can combine with one nitrogen atom, the theoretical maximum nitrogen throughput is

$$A = 3 \times 10^{17} \text{ atoms/s}$$

However, nitrogen forms molecules with two atoms. Therefore, the throughput in molecules per second is 1.5×10^{17} mol./s, which is converted to the normal units for throughput by using Avogadro's number:

$$Q = \frac{1.5 \times 10^{17} \text{mol./s}}{6.023 \times 10^{23} \text{mol.}} (22.4 \text{ L} \times 760 \text{ torr})$$

$$= 0.0042 \frac{\text{torr} \cdot \text{L}}{\text{s}}$$

There are 6.023×10^{23} molecules in a mole of gas that at atmospheric pressure (760 torr) has a volume of 22.4 L.

Titanium evaporation can be of great assistance in vacuum systems by providing high pumping speeds at relatively high pressures. Typical pumping speeds for pumps of different diameters are shown in Figure 9.4.

In Figure 9.1 the source of titanium is shown in the form of a wire. In practice, there are several ways to produce a mechanically stable evaporating surface. For example, molybdenum-titanium alloy wire will retain its shape at an elevated temperature while preferentially evaporating titanium because it has a higher vapor pressure compared to molybdenum. One or more filaments can be used. If one of the filaments has been depleted, the other can be used by external switching. To provide a larger storage of titanium, a hollow titanium spheroid piece can be heated by an internal heater (Figure 9.5). The titanium piece is removable and should be replaced before it gets too thin to retain its shape.

A titanium sublimation control unit provides a regulated power source for operating either a filament sublimation cartridge or a radiation-heated titanium sphere. Essentially, it is a current-regulated power supply (low-voltage ac power) which is applied directly to the filaments of the sublimation cartridge or to the heaters inside the Ti-Ball sphere. Power is adjustable according to the desired evaporation rate.

In the manual mode it can provide 0 to 50 A either continuously or periodically (a maximum of 10 V). In the automatic mode the current is supplied for a period of a few minutes and then switched off or reduced to a standby

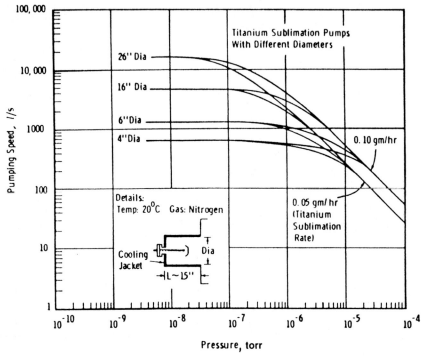

Figure 9.4 Pumping speeds of pumps of various sizes.

value for a period that depends on the pressure in the vacuum system. This period is adjusted according to the signal received from the pressure gauge. Often, a life indicator is used to show the amount of titanium expended as measured through an ampere-hour integrator.

When a new filament is put in operation, it must be degassed to reduce the initial burst of gas, which is released when the sublimator is heated. For pressures above 5×10^{-6} torr, it is recommended to maintain sublimation continuously. Manufacturers usually recommend a certain waiting period between evaporations, in the manual mode, according to a system pressure. However, the optimum "off" period may vary for different systems because it depends on the geometric relationships and the size of the surface receiving the evaporating gettering material. Typically, a monolayer of titanium is evaporated during 2 min of sublimation time. Typical sublimation rates for a filament and a Varian Mini-Ti-Ball are shown in Figure 9.6 and 9.7.

Chemical gettering is not possible with inert gases and it is almost useless with a gas such as methane (Table 9.1). A technique has been developed to

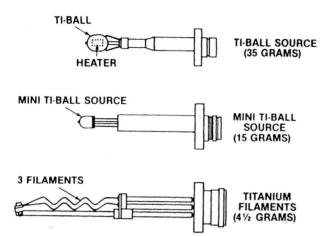

Figure 9.5 Various titanium evaporators.

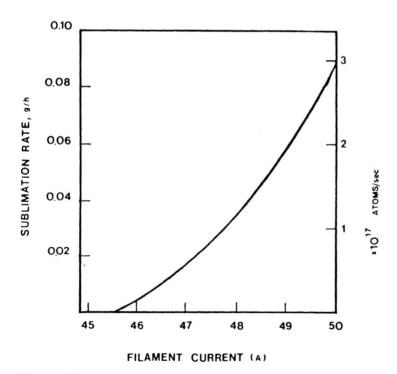

Figure 9.6 Titanium emission rate from a filament source.

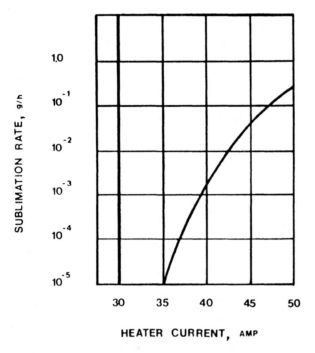

Figure 9.7 Titanium emission from Ti-Ball source.

pump inert gases which includes an ionization process. Usually, the ionization is produced as a result of a collision with electrons. The electrons may be produced either by cold cathode field emission or thermionic emission from hot filaments. The electrons and then the ions are accelerated through an electric field of a few thousand volts. the velocities obtained are sufficient to drive the ions into a solid metal surface. They are "buried" a few atomic layers deep and they remain captured in the surface more or less at room temperature, and if the diffusion rates are low, they will not reappear in the gas phase for a long time.

9.2 SPUTTER-ION PUMPS

Pumps utilizing chemical and ionization pumping effects can generally be called ion-getter pumps. Early designs (after 1955) had a variety of arrangements for electron sources and for titanium evaporation. Later designs, which are used almost exclusively now, are called sputter-ion pumps because the supply of a fresh titanium film is produced by a process called sputtering. The basic structure consists of two electrodes, anode and cathode, and a magnet (Figure 9.8).

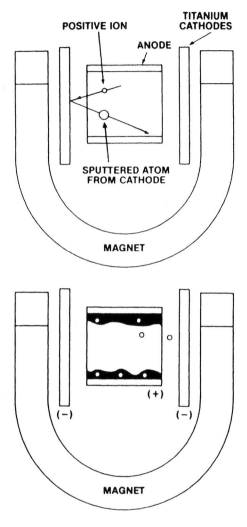

Figure 9.8. Basic cell of an ion-getter pump.

The anode is usually cylindrical and made of stainless steel; the cathode plates positioned on both sides of the anode tube are made of titanium, which serves as gettering material. The magnetic field is oriented along the axis of the anode. The basic cell resembles a cold cathode ionization gauge (see Chapter 12). Electrons are emitted from the cathode due to the action of an electric field, typically 5000 to 7000 V applied between electrodes. The presence of the magnetic field produces long, more-or-less helical trajectories for the electrons.

The long travel path of electrons before reaching the anode improves the chances of collision with the gas molecules inside the pumping module. The common result of a collision with the electron is the creation of a positive ion. In other words, when the electron collides with a gas molecule (with the velocity or energy attained due to the high-voltage field), it tends to knock out another electron from the molecule. The positive ions travel to the cathode due to the action of the same electric field. The collision with the solid surface, when the velocity or energy is adequate, produces a phenomenon called sputtering (i.e., ejection of titanium atoms from the cathode surface). The fresh titanium film covers various surfaces in the pump, where it reacts with gas molecules as illustrated in Figure 9.8 (black layer inside the anode). Some of the ionized molecules or atoms impact the cathode surface with high enough force to penetrate the solid and remain buried.

The detail pumping mechanisms in sputter ion pumps are more complex than outlined here. It may be appreciated that different gases are pumped in a different way and at different pumping speeds. Also, there exists the possibility of previously pumped gases to be reemitted: for example, when the cathode surface is eroded, as illustrated in Figure 9.9. Sputter-ion pumps are generally associated with ultrahigh vacuum, although they can be used successfully in the high-vacuum region. Also, other pumps can produce an ultrahigh vacuum. The unique advantage of sputter-ion pumps, aside from their relative cleanliness, is their complete isolation from the atmospheric environment. Like cryopumps, they do not have an exhaust. They accumulate the pumped gases inside the

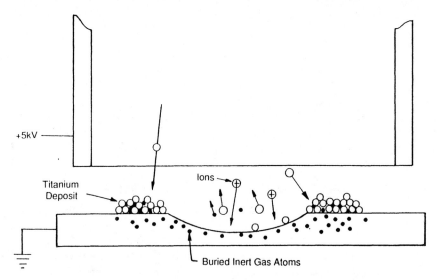

Figure 9.9 Cathode pumping effects and erosion.

GETTERING AND ION PUMPING

pump. Unlike cryopumps, they are not regenerated periodically. They are used until the cathode is exhausted. Naturally, their life depends on the amount of pumped gas, and therefore they should not be used at higher pressures for extended periods.

9.3 BASIC PERFORMANCE OF SPUTTER-ION PUMPS

The general pumping speed characteristics of sputter-ion pumps are similar to other vacuum pumps. There is a more-or-less constant-speed region at intermediate pressures, the net pumping speed is reduced to zero at low pressure, and it decreases to very low values at higher pressures. Due to practical engineering considerations sputter-ion pumps are not designed for high continuous-gas-throughput conditions. Compared to other high-vacuum pumps of equal pumping speed, sputter-ion pumps have approximately 100 times lower maximum throughput values. the pumping speed begins to decrease near 1×10^{-5} torr rather than near 1×10^{-3} torr. A typical pumping speed characteristic is shown in Figure 9.10. The ultimate pressure of an isolated pump is the 10^{-12} torr range. There is not single, simple, specific reason for the particular ultimate pressure. Presumably, it is due partly to the equilibrium between the residual gas evolution inside the pump and the pumping speed. This is similar to other pumps. Partly, however, it may be due to the tendency for the electric discharge in the gas to extinguish at very low pressures.

The pumping speed of ion-getter pumps depends, as in cryopumps, on the arrival rate of the gas molecule (i.e., their molecular weight and temperature)

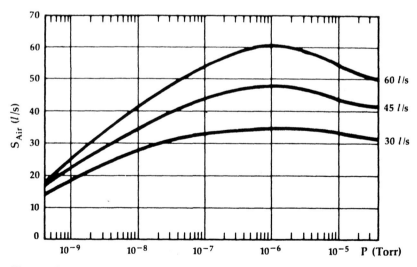

Figure 9.10 Typical pumping speeds of three sputter-ion pumps.

and the sticking coefficient or the probability of capture upon arrival at the surface. The pumping speed per unit area, S (liters per second per cm^2), can be expressed as

$$S = 3.64 \, s \sqrt{\frac{T}{M}}$$

where s is the sticking coefficient, T the absolute temperature (K), and M the molecular weight. This equation is not very useful unless the sticking coefficient is known. Unfortunately, it can vary within a factor of 2 or more. The initial sticking coefficient for nitrogen on titanium surface has been quoted in the technical literature at values ranging from 0.1 to 0.7. For sputter-ion pumps an additional variable is the efficiency of the sputtering process or the sputtering yield.

At the present time there are five varieties of industrial sputter-ion pumps: the diode pump (Figure 9.8), in which the cathodes are attached to the body of the pump and a high (positive) voltage is applied to the anode; the triode pump, where the cathodes are separated from the pump body and a high (negative) voltage is applied to the cathode (Figure 9.11); "differential" pumps, where one of the cathodes is made of a metal other than titanium; StarCell pumps (Varian Associates), which are triode pumps with a distinctive star-shaped configuration; and pumps that include a nonevaporative gettering module. The latter provides a high speed for hydrogen and consists of a corrugated strip of constantan into which is sintered a special alloy (zirconium, vanadium, iron). When heated this alloy undergoes a transformation producing a very large gettering surface.

In summary, the pumping action in sputter-ion pumps is thought to be produced by the following processes:

1. Ion burial and entrapment within the metal lattice a few atomic layers under the surface of the cathode.
2. Gettering of chemically active gases at the cathode and elsewhere by sputtered deposits.
3. Diffusion of hydrogen into the cathode material.
4. Dissociation of complex molecules into simpler fractions which are then pumped by one of the mechanisms.
5. Production of neutral atoms of high velocity (or energy) by neutralization of ions and scattering from the cathode surface. The neutral atoms can subsequently be pumped by the burial process in areas of deposited material on the anode, cathode, or pump envelope.

For practical designs of sputter-ion pumps the anode cells are between 15 and 25 mm. This is suitable for operations with magnetic fields of 1 to 1.5 kG.

GETTERING AND ION PUMPING

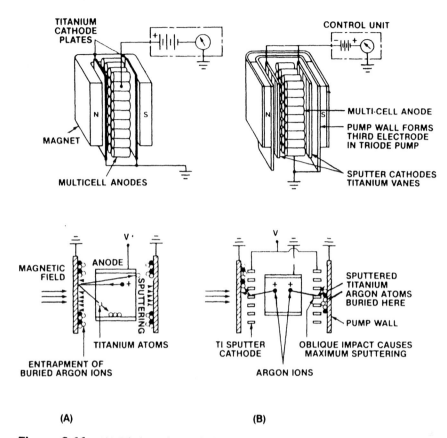

Figure 9.11 (A) Diode and (B) triode ion pumps.

A voltage of at least 3 kV is necessary to start the pump reliably. The pump current increases with the voltage until 6 or 7 kV is reached, which is a common range of operation. These conditions produce pump currents between 5 and 30 A/torr; for the cell, the pumping speed is between 0.3 to 2 L/s. Pumps are made normally up to 200 L/s but can be made larger.

9.4 PUMP TYPES AND PERFORMANCE WITH DIFFERENT GASES

Ion-getter pumps are capable of pumping any gas, including noble gases and hydrocarbons. Due to the differences in molecular weight and chemical activity, the pumping mechanisms and the associated pumping speeds are different for different gases. The ionized molecules bombarding the cathode produce different rates of sputtering depending on their molecular weight and the angle of approach. Small angles (grazing the surface) produce a higher sputtering rate. For light gases such as hydrogen the sputtering rate is too low to obtain pumping speeds comparable to those for nitrogen and oxygen.

Titanium is chosen for the chemical gettering action because it is a highly reactive element that forms stable compounds with many different gases. Ironically, titanium in bulk form is very corrosion resistant at normal temperatures because it has a tightly packed oxide layer which protects its surface. Getterable gases are oxygen, nitrogen, CO, CO_2, and many others.

In addition to sputtering, a fraction of impinging ions can remain buried in the cathode. However, as noted before, the depth of penetration is only a few atomic layers. Because the cathode is continuously eroded, the pumping speed for getterable gases by this mechanism is negligibly small. Hydrogen, in principle, is also getterable, but by itself, does not produce a sufficient sputtering yield. When hydrogen is pumped together with other gases, an important fraction of hydrogen is pumped at the anode. But the major mechanism of hydrogen pumping is due to the diffusion into the cathode material. It diffuses into the cathode to depths much beyond the range of ion implantation and therefore can be pumped in a stable manner and high amounts. The pumping speed for hydrogen is typically two times higher than the speed for nitrogen. Cathode structure should be resistant to, or should accommodate, distortions because of the presence of large amounts of hydrogen in the lattice. If the temperature increases, the diffusion rate is increased but a fraction of pumped hydrogen may be released because the binding energy of titanium hydride is lower than that for nitride and oxide.

Inert gases cannot be pumped by chemical reactions. As noted before, they are pumped primarily by burying into surfaces in various parts of the pump after impinging either as ions or as neutral atoms at high velocity (or energy). They can then be covered by layers of sputtering titanium films. The inert gas atoms implanted at the cathode may be released as the cathode erodes due to continuing ion bombardment. Early diode pumps, in association with this, had a rather low speed for inert gases and tended periodically to release previously pumped gas (called an argon instability)

One way to overcome this problem is to increase the portions of the atom that rebound from the cathode surface as neutrals and then are implanted into the anode and the walls of the pump, where they are held more or less permanently. The probability of this reflection depends on the masses of the impinging ions and the cathode material as well on the incidence angle. The higher the mass of the target material, the more likely it is that the ion will rebound and will also retain more energy after reflection. The use of heavier metals, such as molybdenum or tantalum as an auxiliary cathode material, then provides the necessary improvement. If both cathodes were made of the heavier metal, the sputtering yield would be reduced. Pumps with such differentiated cathodes produce a very substantial improvement in the pumping speed of inert gases and also improve its stability (Table 9.2), although the speed for getterable gases may be somewhat reduced.

A more effective approach is to provide a more favorable incidence angle for the ions. One way to achieve this is to slot the cathode surface, providing shallow angles for incident ions, and to increase both the reflection probability and the sputtering yield. Such pumps produce inert gas speeds that are 5 to 10% of the pumping speed for air, and the pumping speed for the inert gases is stable.

The next logical development is to make the cathode in the form of thin parallel strips with some space between them. This grid structure is partly transparent for ions and allows the pump wall (behind the cathode) to be available as a pumping surface. The wall can be considered as an auxiliary anode. This type of pump is called a "triode pump" (Figure 9.11), as expected, the pump walls are kept at a ground potential and the high negative voltage is applied to the cathodes. In a triode pump the ions that retain their energy after being neutralized at the cathode may travel to the pump wall and be buried there or reflect and be pump at the anode. The pumping speed for the inert gases for triode pumps is 20 to 25% of speed for air and it is very stable even after a long period of pumping. The pumping speed for hydrogen is somewhat lower than that of diode pumps with titanium cathodes. Also, distortion of cathode

Table 9.2 Relative Pumping Speeds to Various Gases[a]

Gas	Diode pump (3.5 kV)	Triode pump (5.2 kV)	Diode pump with slotted cathode (7.3 kV)	Diode pump with Mo and Ti cathodes
H_2	2.5-3.0	2.0	2.7	
CH_4	2.7	0.9-1.05		
NH_3	1.7			
H_2O	1.3		1.0	
Air	1.0	1.0	1.0	1.0
N_2	0.98	0.95	1.0	
CO	0.86		1.0	
CO_2	0.82	1.0	1.0	
O_2	0.55	0.57	0.57	
He	0.11	0.3	0.1	0.25
Ar	0.01	0.3	0.06	0.2

[a]Taken from manufacturers' catalogs. Air speed taken as unity.

strips can occur after pumping large amounts of hydrogen or when pumping at the high end of the pressure range.

To overcome problems with cathode distortions the cathode has been designed with radial strips or spokes located in alighment with each anode cell (Figure 9.12). This avoids the tendency for distortions because the spokes are attached only on one end. They are free at the center and are symmetric relative to the ion activity in the pumping cell. Because of the shape these pumps are known under the trademark StarCell (Varian Associates). An additional important advantage is obtained from the symmetric pattern of cathode erosion, which begins at the center and progresses toward the periphery. In this way more titanium can be sputtered before the cathode structure loses its integrity. In a diode pump, the ion bombardment eventually produces a hole in the cathode plate after which the pumping action is nearly lost. In a triode pump the end of the cathode life occurs when a strip brakes, which often causes a short circuit. Due to the gradual symmetric erosion in the StarCell pump its life is increases by nearly a factor of 2 compared to a standard triode pump and by 60% compared to a diode pump, so that the pump can operate continuously at 1×10^{-5} torr for a period of 1 year. Pumping speeds of StarCell pumps for different gases compare as follows: for oxygen, nitrogen, water vapor, and methane, nearly the same; for carbon dioxide, 10% higher; for hydrogen, twice as high as nitrogen; for argon, 22%; and for helium, 30% of nitrogen. The most recent development in ion pump design is the addition of a nonevaporating getter strip inside the pump. This increases the pumping speed and the maximum throughput capability of the pump.

Figure 9.12 Cathode structure of a StarCell pump.

9.5 OPERATION OF ION PUMPS

Sputter-ion pumps are normally shipped to the user after evacuation, degassing, and sealing under vacuum. This keeps the pump surfaces clean, unsaturated with adsorbed gases, and makes it easier to start the pump. The pump inlet is covered and sealed by a flanged plate, which includes an evacuation tube made of copper. After the final processing this tube is sealed usually by a pinch-off tool that produces a leaktight cold weld as it cuts off the tube. When constructing a new vacuum system the pump should be opened only after everything is completed and the system is ready to be evacuated. The exposure of the pump to atmosphere should be minimized.

Ion pump systems normally have a high-vacuum valve which can isolate the pump when the system is exposed to air. If ultimate pressure in the 10^{-9} torr range is adequate, baking may not be mandatory and the system may have two Viton O-rings, one under the bell jar (Figure 9.13) and one in the valve plate. For lower pressures the chamber and all seals must be made of metal and the entire system surrounded by a bake-out oven.

When pumping speeds of ion pumps are measured or specified it should be clearly understood whether the measurements are made with a new, processed pump or after a certain amount of gas has been pumped. The pumping speed of new ion pumps decreases during operation until a stable level is reached. The

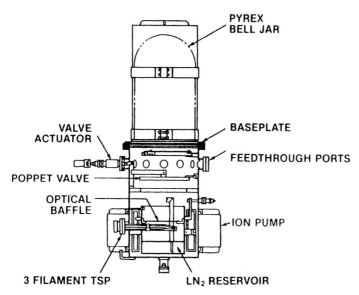

Figure 9.13 Ultrahigh-vacuum system.

initial pumping speed fo rnonsaturated pumps (StarCell types) is shown in Figure 9.14. This should be compared with Figure 9.10, which shows the saturated pumping speeds for the same pumps. The higher pumping speeds at pressures above 10^{-7} torr (Figure 9.14) may or may not be utilized, depending on the operation of the system. For example, to saturate a 60-L/s StarCell pump 2.5 torr · L of gas is normally required.

Sputter-ion pumps are sometimes difficult to start if the pressure is too high or too low when the high voltage is applied. At the higher pressures, near 10^{-2} torr, particularly if the pump has been exposed to the atmosphere for a few days, the initiation of the electric discharge may produce enough current drain from the power supply to increase the temperature of the internal electrodes. This, in turn, produces increased outgassing, which may cause a runaway condition. At the same time the voltage may drop, which reduces the pumping speed, making the starting condition even more difficult. If the pump is not clean, more gas may be driven from it than is pumped. Sometimes it is necessary to start and stop the pump a few times before it finally begins to reduce the pressure in a stable manner. Triode and diode ion pumps exhibit different behavior during starting. These differences depend on the design of the power supply and its polarity as well as the cathode shape and material.

At very low pressures the initiation of the pump may take a long time. If a system had been evacuated to 10^{-9} torr by some other means, the ion pump will start in a few seconds; at 10^{-12} torr it may take half an hour. As mentioned before, the performance of the pump at the higher pressures depends to a great

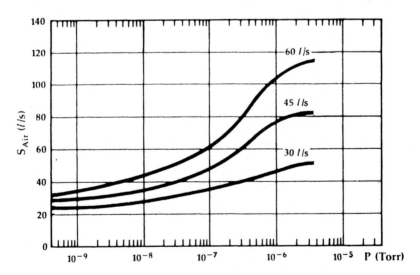

Figure 9.14 Pumping speed of ion pumps before saturation.

degree on the power supply characteristics. One of the important aspects is the requirement for maintaining a current sufficient to obtain higher throughput without overheating. The best performance is obtained when a pump is properly matched to a power unit. A power supply that is too large for a given pump will cause overheating and may damage the pump at high starting pressures. A power unit that is too small will reduce the maximum operating pressure of the pump or, in other words, reduce its maximum-throughput capability. The power unit must supply voltage and current required under different working conditions, automatically limiting the delivered power at high pressures to facilitate starting of the pump under different conditions. The power consumption of the pump strongly depends on pressure and becomes essentially negligible at pressures below 10^{-6} torr (Figure 9.15).

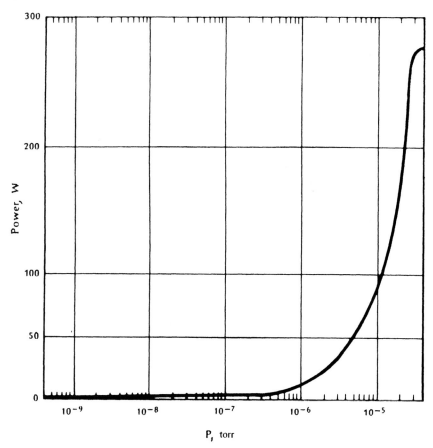

Figure 9.15 Power consumption of a StarCell pump, depending on pressure.

One insteresting and useful characteristic of sputter-ion pumps is the approximately linear relationship between the electrical power used by the pump and the pressure. The pump automatically adjusts the intensity of the pumping activity to the number of molecules present in the system. This should not be surprising because the basic pumping mechanisms of the pump depend on the number of available molecules. the electron "cloud" rotating in the cell of a sputter-ion pump under the influence of a magnetic field (somewhat analogous to a rotor of an electric motor) is relatively stable and is not strongly dependent on pressure. Therefore, the current of the pump, similar to cold cathode gauges (see Chapter 12), can be used as a measure of pressure, as shown, for example, in Figure 9.25. It should be understood that this is only an approximate indication and should not be thought of as a substitute for an ultrahigh-vacuum gauge of good quality. The measurement becomes more uncertain as the pressure is reduced below 10^{-8} or 10^{-9} torr. This is due to leakage currents and other effects that are independent of pressure.

9.6 NONEVAPORATED GETTER (NEG) PUMPS

There is a class of gettering pumps in which the active material is not evaporated or sputtered on a pumping surface but remains in a specially prepared, very porous intermetallic compound. Such pumps were pioneered the by SAES Getters company in Milan, Italy and are often called NEG pumps. The pumping action initially involves an adsorption event followed by diffusion inside the material. (It is ironic that these pumps can be called "diffusion pumps" more than the vapor jet pumps, for which the common name is a misnomer.) The most comprehensive treatment (in high-vacuum textbooks) of the physical basis of the pumping action and properties of various materials was presented by G. Saksaganskii, who devotes 47 pages to this subject in his book.

Usually the pumping material is in the form of thick films or layers deposited on a substrate strip or ribbon, typically a nickel-iron alloy such as constantan, nichrome, or stainless steel and others. The compound has a porosity of 50 to 70%, may be as thick as a few millimeters, and can have a real internal physical area 100 times higher than the projected area (100 to 1000 m^2/kg). The active metallic compound is usually composed of finely dispersed powders of transition metals, which may be sintered, plasma-sprayed, or rolled into the substrate. A common composition has 84% zirconium and 16% aluminum. Other compounds include titanium, vanadium, iron, nickel, and also tantalum, molybdenum, tungsten, and rhenium.

Because of the progressive saturation of the exposed surfaces, the getter must be reactivated by heating to enhance the diffusion process and clean the surface. This activation is achieved by raising the temperature to about 750°C for several minutes in vacuum. Subsequently, the getter is held usually at about

400°C. The performance of the pump is normally judged by its pumping speed for hydrogen. Most common gases present in high-vacuum systems are pumped permanently, but hydrogen can be re-emitted if the temperature is raised above 1000°C. As in other gettering pumps, the total capacity for pumped gas is limited and after a number of activations, the pumping speed declines and the pump saturates. Pump specifications provide information about total accumulation capacity before the getter strip or ribbon becomes brittle and loses its mechanical integrity.

NEG pumps are available as nude cartridges, as glass-encased appendage capsules, or as autonomous flanged pumps that can be attached to any vacuum system. To provide a higher pumping speed in a smaller package, the getter ribbon is often pleated or corrugated (shown in Figure 9.16) and provided with a heater. Water cooling or even air cooling is not necessary for small pumps. A small pump (2.5 cm. diameter, 7.5 cm long), if properly thermally shielded, can be heated in steady-state operation with a 10 watt heating source. Also, for pure hydrogen, the pump can be operated at a reduced speed at room tempera-

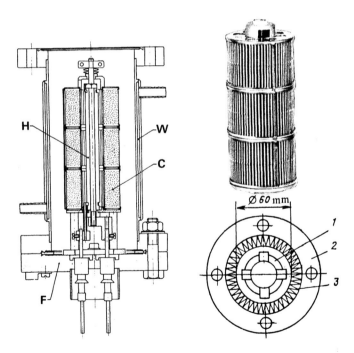

Figure 9.16 The view of a nonevaporated getter pump. Left, cross section schematic; right, corrugated strip design. H and 1, electric heater; W, water cooling (for large pumps); C and 3, the getter ribbon; 2, flange.

ture. Pumps are made in many sizes and a nude cartridge may produce 2000 L/s pumping speed for hydrogen. Typical pumping speeds are 65% for oxygen, 50% for carbon monoxide, and 17% for nitrogen as compared to hydrogen. The hydrocarbons are pumped with a speed that is one hundredth of the speed for carbon monoxide. Such ratios, however, can be misleading because they depend on the quantity of previously pumped gas. For example, Figures 9.17, 9.18, and 9.19 show the pumping speeds as a function of the amount of previously pumped gas (G.L.Saksaganskii, Ref. 9).

Inert gases and methane are not pumped in significant amounts. Therefore, usually a small ion-getter pump is required to provide the pumping speed for these gases. In this manner, ultrahigh-vacuum can be obtained in ultrahigh-vacuum systems with the primary pumping done by NEG pumps, especially, when, after baking, the main residual gas is hydrogen. NEG pump strips are sometimes included for miles, along the entire length of tubing in particle accelerators, and as auxiliary pumping modules for enhance hydrogen pumping speed in autonomous ion-getter (sputter-ion) pumps, which may be called combination pumps.

9.7 COMBINATION PUMPS

Nonevaporated getter pumps can be incorporated into ion-getter pumps to provide a higher pumping speed for hydrogen. For this purpose, it is convenient

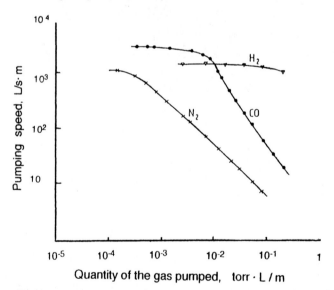

Figure 9.17 Pumping speed per meter of long St101 ribbon, depending on the previously quantity of the pumped gas.(From Ref. 9.)

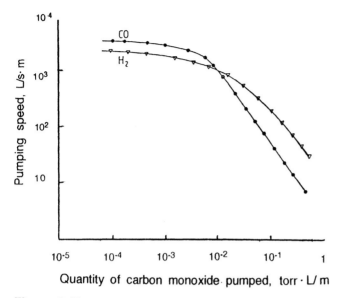

Figure 9.18 Pumping speed for CO and H_2, depending on previously pumped quantity of CO (meter long St101 ribbon). (From Ref. 9.)

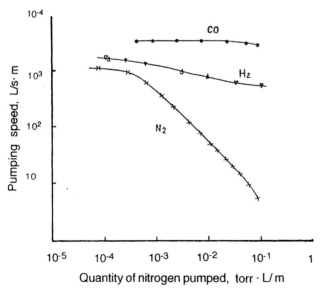

Figure 9.19 Pumping speed for CO, H_2, and N_2, depending on pumped quantity of nitrogen (meter long St101 ribbon). (From Ref. 9.)

to select a NEG composition that can be activated at baking temperatures normally used in the process of obtaining ultrahigh-vacuum (300–400°C). An example of such an integrated pump, using the NEG composition St707, is shown in Figure 9.20. The normal range of the activation temperature for this compound is specified as 600–800°C, but after 12 h bake at 350°C, sufficient pumping action is obtained (after cooling to room temperature) as shown in Figure 9.21. The combination pump performance, as expected, provides higher pumping speed for hydrogen, which is usually the main gas in an ion-getter pump system. The pumping speeds for hydrogen, for three different sized pumps, are shown in figure 9.22, where dashed curves represent pumps without NEG addition and the solid curves represent combination pumps. The corresponding residual gas measurements are shown in Figure 9.23 and demonstrate a sharp reduction of the partial pressure of hydrogen.

9.8 CLEAN ROUGHING SYSTEMS

The initial evacuation of high- and ultrahigh-vacuum systems that are subsequently pumped by ion pumps has two basic requirements: the ultimate pressure of the roughing pumps should be low (near 1×10^{-3} torr), and the rough pumps should not contaminate the system and then the ion pump which can cause starting and operating difficulties. An additional requirement is to use an ion pump with high-enough pumping speed or a system with low-enough out-

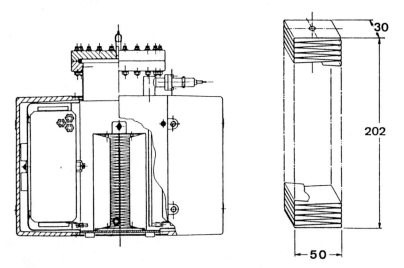

Figure 9.20 Schematic of a sputter-ion combination pump with an integral NEG pump (corrugated strip at right with dimensions in millimeters).

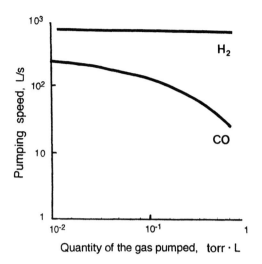

Figure 9.21 Pumping speed of a combination sputter-ion pump, depending on previously pumped quantity of the pumped gas.

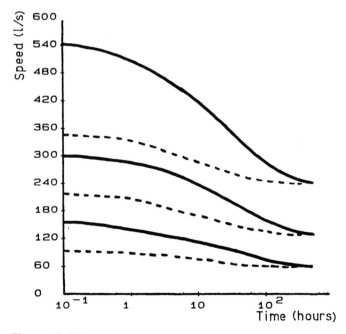

Figure 9.22 Hydrogen pumping speed for three different sputter-ion pumps with (solid curves) and without (dashed curves) nonevaporated getters.

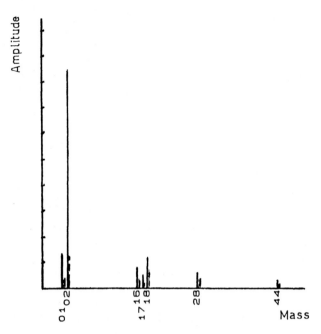

Figure 9.23 Residual gas measurements corresponding to Figure 9.22 showing the effect of greatly enhanced hydrogen pumping efficiency.

gassing rate to be able to reduce the pressure quickly after the high-vacuum valve is opened. It should be recalled that ion pumps have approximately one hundredth of the maximum-throughput capability of other high-vacuum pumps. System design and operation must reflect this characteristic.

With regard to the maximum practically permissible steady-state gas flow into the pump, the ion-getter pumps must be considered as a different class of equipment. Compared to throughput-type pumps of the same nominal pumping speed, they have nearly one thousand times lower throughput capability. This observation immediately suggests that the starting conditions, after the initial evacuation, must be different. It is common in textbooks and operating instructions to specify a pressure level where the pump can be started. This is an unfortunate choice that leads to errors in system design such as the incorrect selection of rough pump type or size, or the size of the high-vacuum pump. There are three items to consider and are illustrated in Figure 9.24. The amount of gas initially entering the pump (after the high-vacuum valve is opened), depends on the volume of the chamber and the pressure (at the end of the roughing cycle), not the pressure alone. The bypass line shown in Figure 9.24 can

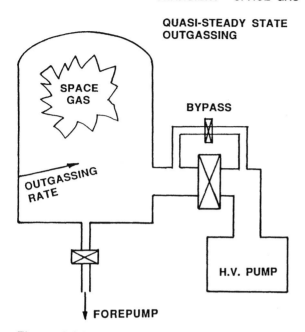

Figure 9.24 Various gas sources after opening the high-vacuum valve.

help in softening the sudden rush of the pumped gas into the pump, but it does not change the two components of the gas contained and emanating from it: the gas in space and the gas emitted from the surface. The selection of the high-vacuum pump size will also depend on the frequency of evacuation events. With all this in mind, a simple gas accumulation estimate should be made before deciding on a particular roughing system or the size of the high-vacuum pump.

Even if some ion pumps can be started at pressures as high as 10 to 20 millitorr, it is not a good idea to do this too often because of the accelerated exhaustion of the cathodes. The cathodes of an ion pump, which are eroded by sputtering, cannot be regenerated but must be replaced. This replacement is not a simple matter because the pump wall has to be cut open and then rewelded. The cathode life depends on the amount of gas pumped, or the integral of pressure and time. Typical values for diode pumps are 200 h at a constant pressure of 10^{-4} torr, 2000 h at 10^{-5} torr, 20,000 h at 10^{-6} torr, and so on. Note that at 10^{-6} torr the life is over 2 years and at 10^{-3} torr only 20 h! For pumping nitrogen only the life of a diode pump at 10^{-6} torr will be about 50,000 h

(5.7 years). Corresponding values for a triode pump are 35,000 h, and for a StarCell pump, 80,000 h.

Ideally, ion pumps should be started at pressures near 1×10^{-5} torr, although in practice starting pressures near 10^{-3} torr are more common. There are several ways to accomplish this, It is possible to use oil-sealed mechanical pumps with well-designed and carefully operated trapping arrangement. Such pumps can produce pressures below 10^{-4} torr with a liquid nitrogen trap or a carefully degassed zeolite trap. However, oil-sealed mechanical pumps are rarely used for the initial evacuation of ion pump systems. Fear of accidental contamination of the system and of the ion pump is sufficient to discourage their use.

Sorption pumps are commonly used as roughing systems for ion pumps (Figure 9.26). This was discussed in part in Chapters 5 and 8. To prolong the saturation period, air ejectors (Venturi pumps) or coarse oil-free mechanical pumps can be used at the beginning. They can remove about 90% of the air before sorption pumps are engaged. Recently developed oil-free high-vacuum mechanical pumps can be very useful. They can produce pressures below 20 mtorr and they can be left pumping on the system for longer periods of time without any possibility of contamination, allowing the system to degas. If the ion pump is large and the evacuation object small (e.g., a microwave tube), the oil-free mechanical pump can be used alone. In general, they should be used

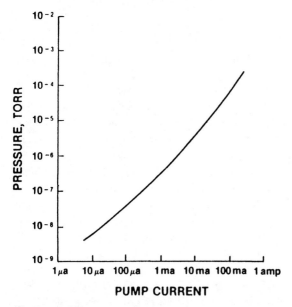

Figure 9.25 Correspondence of pump current and pressure.

GETTERING AND ION PUMPING

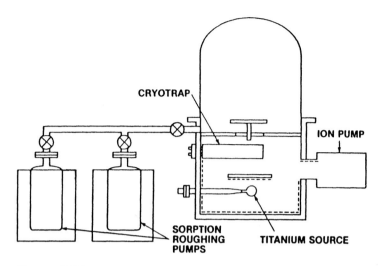

Figure 9.26 Sorption pumps for initial evacuation.

with a single stage of sorption pumping and/or with a titanium getter pump before the ion pump power is started.

In large, expensive installations it may be useful to evacuate the vacuum system to 10^{-5} torr or below using oil-free mechanical pumps followed by cryopumps or turbodrag, or turbomolecular pumps before the system is switched to ion-getter pumps. One example of successful employment of these pumps in large vacuum systems is in the field of high-energy physics, where many of the modern particle accelerators are constructed using ultrahigh-vacuum techniques and are pumped by ion pumps.

REFERENCES

1. Y. Z. Hu and J. H. Leck, *Vacuum*, 37(10), 757 (1987).
2. D. J. Harra, *J. Vac. Sci. Technol.* (Jan./Feb. 1976).
3. G. F. Weston, *Ultrahigh Vacuum Practice*, Butterworth & Company (Publishers) Ltd., London, 1985.
4. P. A. Redhead, J. P. Hobson, and E. V. Kornelson, *The Phyical Basis of Ultrahigh Vacuum*, Chapman & Hall, Ltd., London, 1968; reprinted by AIP, New York, 1993.
5. G. F. Weston, *Vacuum*, 34(6), 619 (1984).
6. M. Audi and M. deSimon, *Vacuum*, 37(8/9), 629 (1987).
7. T. Snouse, *J. Vac. Sci. Technol.*, 8(1), 283 (1971).
8. M. Audi and M. DeSimon, *J. Vac. Sci. Technol.*, A6(3), 1205 (May/June, 1988).
9. G. L. Saksaganskii, *Getter and Getter-ion Vacuum Pumps*, Harwood Academic Publications, Switzerland, 1994.

10. K. M. Welch, *Capture Pumping Technology*, Pergamon Press Ltd., Oxford, 1991.
11. R. W. Roberts and T. A. Vanderslice, *Ultrahigh Vacuum and its Applications*, Prentice Hall, Englewood Cliffs, N.J., 1964.
12. M. Audi et al., *J.Vac. Sci. Technol.,* (A5),4, 2587, (1987).
13. P. Della Porta and B. Ferrario, *Proc. 4th Int. Vac. Congr.*, AIP, 369, (1968).
14. P. Della Porta et al., *Trans. 8th Natl. Vac. Symp., 1961*, Pergamon Press Ltd., Oxford, 229, 1962.

10
Overloading of Vacuum Pumps

10.1 INTRODUCTION

This chapter is primarily concerned with the prevention of overloading high-vacuum pumps and the resulting adverse effects or malfunctions. Although overloading conditions have been discussed in chapters on various pumps, the general subject has been neglected in the past and deserves special attention.

High-vacuum pumps, like other compressors, have basic limitations in regard to the maximum pressure difference (and pressure ratio) and maximum mass (and volume) flow rates that they can produce. Because high-vacuum pumps are usually made to discharge into another pump, their tolerable discharge pressure must be associated with the characteristics of the backing pump. It is also necessary, however, to coordinate the performance of the high-vacuum pump with the performance of the pumping system used for pre-evacuation of the vacuum chamber.

The traditional concept of a single absolute value for the cross-over pressure used for the initiation of high-vacuum pumping is fundamentally incorrect because it is not based on a clear mass flow limitation. To prevent overloading high-vacuum pumps during and immediately after switching from pre-evacuation to high-vacuum pumping, a simple rule must be observed: the cross-over must be performed when the gas mass flow from the vacuum chamber is less than the maximum throughput capacity of the high-vacuum pump.

Typically, at the end of the pre-evacuation period, there are two somewhat distinct gas quantities associated with the vacuum chamber: the gas in the space of the chamber and the quasi-steady outgassing rate. There are distinct pressure decays associated with those two gas quantities. Overloading the high-vacuum pump due to space gas can be prevented by opening the high-vacuum valve slowly or by using a parallel low-conductance by-pass. The overload due to the outgassing rate, however, can only be prevented by following the golden rule of mass flow limitation.

In capture pumps, the maximum throughput value for the crossover condition must be correlated with the period of regeneration (for cryopumps) or cathode replacement (for ion pumps).

There are several reasons for the misconceptions in this area (as expressed by vacuum system designers and users): the reliance on volumetric rather than mass flow in displaying pump performance; the reversal of dependent and independent variables in pumping speed graphs; the use of two or more pumps, which are at times operated in parallel and then in series; the tendency of displaying the pumping speed of the rough- *and* the high-vacuum pump as a single curve; and the fact that high-vacuum pumps are usually multistage devices, so that one stage may be overloaded while others are not.

The tradition of displaying pumping speed versus inlet pressure (instead of inlet pressure versus gas load) and the reversal of dependent and independent variables, appears to have produced a conceptual confusion of cause and effect in several generations of technicians and engineers (and perhaps even an occasional scientist) dealing with the design and operation of high-vacuum systems.

A clear understanding of the limitations of various pumps, and their interrelationships, is necessary to specify the size of the roughing and high-vacuum pumps needed to achieve a desired system performance, to determine the correct cross-over point, or to predict the required evacuation time.

For proper selection of pump size, it is necessary to consider the different system requirements: the evacuation time to a given pressure; maintaining a certain pressure during a process; process cycle time; and the maximum potential gas flow rate. In addition, for complex systems involving vapor jet pumps, Roots blowers, and mechanical pumps, it is important to appreciate the transient events following the abrupt change of system pumping speed. The traditional display of vacuum pump performance in the form of pumping speed versus inlet pressure does not clearly distinguish between the steady-state and transient sections of the performance and does not indicate overload conditions for the high-vacuum pump.

For a full appreciation of high gas load vacuum system design and operation, it is necessary to restate the performance data in the form of the achieved, or maintained, degree of vacuum versus the gas load evolving from the vacuum chamber. Such considerations are required for any vacuum pump, but they are particularly important for capture pumps and vapor jet pumps because improper operation is often the cause of serious malfunctions such as the loss of pumping fluid or back-streaming.

Significant advances have been made in vacuum technology during the last 40 years. With the new knowledge, manufacturers of pumps and systems should relate the pump performance to the specific system requirements, and the users should clearly state the process needs throughout the entire range of process conditions, to specify not only process pressure but also the process gas flow-rate requirement, to appreciate the transient and steady-state situations, and to accommodate the limitations of the pumps to the process operation.

Engineers designing vacuum systems often use certain rules that work well for certain conditions (for example, the changeover (cross-over) from a roughing pump to a high-vacuum pump at 0.1 torr, or to use half of the evacuation time for roughing and half for high-vacuum pumping, or to first close the roughing valve and then open the high-vacuum valve). Depending on system design and operation requirements, most rules can be questioned. Crossover pressure depends on matching the mass flow conditions, not a pressure level. Also, every system should have a different cross-over pressure, depending on the ratio of the pumping speed of the pumps and the permissible maximum mass flow (throughput) for the high-vacuum pump. Even in the operation of the same system, cross-over pressure can be changed, depending on the particular gas being pumped or a change in the performance of the pumps involved. Some of these considerations may be thought of as secondary and, as such, they are rarely discussed in textbooks on vacuum technology. However, it is often the detail of issues that create the difference between success and failure.

10.2 VOLUME FLOW AND MASS FLOW

When the performance of a vacuum pump (or compressor) is presented in the form of a plot of volumetric flow (pumping speed) versus inlet pressure, the basic limitations of the performance may be represented as shown in Figure 3.15. If this graph is simply tilted 90 degrees to show pressure versus pumping speed, it should be immediately suspicious that this double-valued function does not resemble the usual representation of pump or compressor performance normally found in the textbooks on mechanical engineering, fluid mechanics, or pneumatic machinery. The limited power section of the curve is normally omitted because it is not really a part of the intended performance, in other words, the steady-state performance range has reached its limit.

To show this section of the curve (and to try to use the pump in this region) may be analogous to an attempt to deal with the performance of a stalled engine or an electric motor. Yet high-vacuum system operators often attempt to operate in the overload condition, perhaps because they do not see sparks or smoke or hear grinding noises.

For a conceptual understanding of cause and effect relationships, it is much better to plot the performance of the pump as shown in Figure 10.1. Here, we can clearly see the mass flow limit, the separate performance of roughing and high-vacuum pumps, and the reason for wide pressure fluctuations, which often occur when operating near the overload condition. This graph will also show how meaningless it is, for example, to specify the maximum throughput of a vapor jet pump at 40 mtorr, when Q_{max} occurs at 1 mtorr.

In poorly-designed pumps, there is no built-in safety factor in the mass flow limit for each successive stage. In such pumps the collapse of pumping action (upon approaching the overload condition) is sudden and total. In other words, the constant throughput section of the curve is nearly vertical. In such a case, it is nearly impossible to control the gas flow in order to maintain the inlet pressure between, typically, 5 and 50 mtorr. An attempt to do so results in rapid and wide pressure fluctuations because any small variation in gas flow will switch the pumping action from the high-vacuum pump to the rough pump and back again.

In well-designed pumps each successive (downstream) stage is designed with a safety factor in regard to the maximum mass flow capability. Thus, it is possible (upon approach of the excessive gas load condition) to have the inlet stage(s) to be overloaded while the downstream stage(s) may still be function-

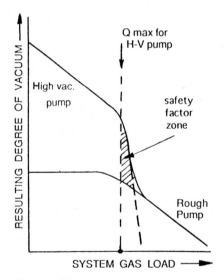

Figure 10.1 Inlet pressure versus imposed gas flow (throughput) for a vapor jet pump.

ing. This useful quality, however, should not be used as an excuse to abuse the pump. First, an early (incorrect) cross-over does not help to shorten the evacuation process and, secondly, no vacuum system should be designed with such a marginal mass flow capacity as to force operation at or above the capability of the pump.

Due of the uncertainties of vacuum gauges and outgassing conditions of a vacuum process, a safety factor of 1.2, at the very least, should be used. In high gas load applications, such as the metallizing of plastic materials or high-vacuum furnaces, a plant that has twice the pumping capacity of the calculated design, will produce satisfactory performance.

In engineering practice, one is often faced with the misconception of treating units such as "sccm" as volumetric flow. This occurs while establishing system design specifications for determining the required pump size and in leak detection practice. The units of mass flow units are "standard cubic centimeters per minute". The word "standard" means that the gas density is atmospheric. If we assume that the process occurs near room temperature, and use the usual units of throughput (Q), the units of sccm can be converted directly to units of torr·liters per second.

$$1 \text{ sccm} = (1 \times 760) / (1000 \times 60) = 0.0127 \text{ torr·L/s} \quad (10.1)$$

To obtain the required pumping speed to maintain a pressure of, say, 1×10^{-3} torr, with a process gas flow of 1 sccm, the relation $S \approx Q/S$ can be used to obtain

$$S = 0.0127 \text{ torr L/s} / 1 \times 10^{-3} \text{ torr} = 12.7 \text{ L/s} \quad (10.2)$$

If the required process pressure is ten times lower, the pumping speed must be ten times higher, etc.

When the flow rate is known, therefore, it is relatively easy to specify the net pumping speed required to maintain a given process pressure. It is more difficult to specify a pumping speed for a required evacuation time because of uncertainties associated with outgassing. When baffles and traps are used at the inlet of the pump, the net pumping speed and the associated throughput are reduced but the maximum throughput value is unchanged.

It should be noted that the ratio of the net pumping speed to the speed without baffles is not the same for different gases. Pump specifications usually provide data for air only. Because the conductances are higher for lighter gases, however, the pumping speed for lighter gases will be reduced less than for air. As an example, if the conductance of a baffle for air is of the same magnitude as the pumping speed for air, the net speed for air will be 50% of the pump speed. But, in the case of helium, the pump speed may be 20% higher than air and the conductance 2.7 times higher. The net pumping speed for helium will be 83% of the pump speed (using the usual reciprocal addition method). For heavier gases, the effect is the opposite.

Regarding the maximum allowed throughput for a particular pump, it is good to remember that the numerical value usually given for air is not applicable for other gases. For the same true mass flow (grams per second), the throughput value for the lighter gases (helium, water vapor) is higher, and for heavier gases (argon), it is lower. To convert the maximum throughput from air to another gas, the throughput must be adjusted according to the ratio of molecular weights and molecular velocities. If Q_1 is the maximum throughput of air, Q_2 is the maximum throughput for another gas, and M_1 and M_2 are the molecular weights of the two gases, then (assuming roughly equal pumping speed)

$$Q_2 = Q_1 (M_1 / M_2) (M_2 / M_1)^{1/2} \quad \text{or} \quad Q_2 = Q_1 (M_1 / M_2)^{1/2} \quad (10.3)$$

This formula may not apply in all cases, but it provides a good estimate for modern high-power vapor jet pumps (for example, the Varian HS-35 for which the helium pumping speed is about 20% higher than air). Actual measurements confirm three times higher maximum throughput for helium and hydrogen as compared to air (Figure 6.11).

For proper steady-state operation of a vapor jet pump, it is important to not exceed the maximum throughput of the top pumping stage, which is often 70% of the value given in commercial bulletins (referred to an inlet pressure of 10 mtorr). There is no standard for measuring the maximum throughput. The old AVS standard recommends plotting a curve for throughput and stating the size of the backing pump used during testing. No reference is made, however, to the simple fact that the throughput values (given at the inlet pressure where the pumping speed is substantially reduced), are measured with a nonfunctioning top vapor jet.

Also, at high gas loads, the size of the backing pump begins to influence the pumping speed. To state the throughput value at 0.1 torr makes a comment about the backing pump rather than the vapor jet pump. At this pressure, only the last stage of the foreline has minimal pumping action but all of the other jets are severely overloaded; the vapor condenses in space without reaching the walls and may pass into the backing pump to produce a rapid pumping fluid loss.

10.3 CROSS-OVER PRESSURE

In the traditional representation of pumping speed curves, it is better to separate the performance of the roughing pump and the high-vacuum pump as shown in Figure 10.2. Note that the mechanical pump curve begins at a maximum mass flow at atmospheric pressure and ends at zero flow under maximum compression (or the achieved limiting degree of vacuum). By analogy, the high-

OVERLOADING OF VACUUM PUMPS

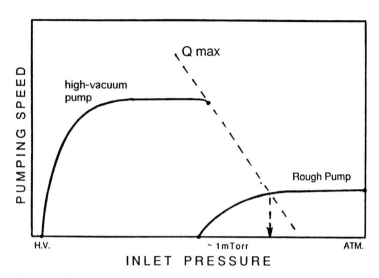

Figure 10.2 Pumping speed versus inlet pressure for a vapor jet pump and a mechanical pump connected by constant gas load line.

vacuum pump curve should also begin at maximum mass flow (maximum throughput) and end at zero flow due to the compression ratio limit and residual gas loads. Connecting these two curves with a smooth solid line implies that operation is possible in the transitional, and transient, region of interaction between the two pumps.

The appropriate time or condition to switch from the rough pump to the high-vacuum pump is during the initial evacuation when the gas load emitted from the chamber is less than the mass flow capacity of the high-vacuum pump. There are simple methods to derive the required mass flow from the measured pressure. This is indicated in Figure 10.2 by a dotted diagonal line that is a line of constant throughput values. The appropriate cross-over pressure is easily obtained from this as shown by the arrow. Then it is only necessary to know the net speed of the roughing pump at the connection to the chamber. The pressure in the chamber can be multiplied by the net speed of the rough pump to obtain the throughput of gas emanating from the chamber.

Both the space gas (volume gas) and the outgassing flow can be expressed mathematically by decay functions, but with markedly different time constants. The pump itself may be considered as an operator, as shown in Figure 10.3. When, in the process of evacuation, the roughing valve is closed and the high-vacuum valve is opened, we may assume that the amount of gas evolved in the chamber remains approximately the same. This is indicated in Figure 10.4 by two arrows. There is, of course, a transient change in pressure due to the sud-

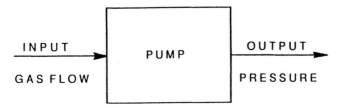

Figure 10.3 A box diagram showing the function of a pump.

den increase of pumping speed. This transition occurs with the time constant $\tau = V/S$ where V is the volume of the chamber and S is the pumping speed of the high-vacuum pump. In typical vacuum chambers, this time constant is less than one second. The decay function associated with the outgassing rate, however, has a time constant of minutes or hours. For purposes of this discussion, the outgassing rate can be assumed to have a quasi-steady state constant value. Thus, we may say that

$$Q = P S_{net} = Q' = P' S'_{net} = P_2 S_2 \qquad (10.4)$$

where P is the pressure before crossover, S_{net} is the net rough pumping speed near the chamber, P' is the pressure after crossover, S'_{net} is the net high-vacuum

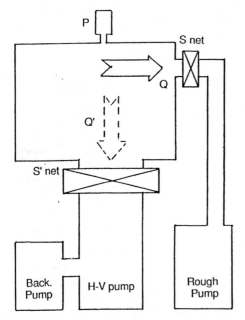

Figure 10.4 Illustrating flow relationships immediately before and after crossover.

pump speed, P_2 is the discharge pressure of the high-vacuum pump (which must never rise above the tolerable forepressure), and S_2 is the backing speed at the outlet of the high-vacuum pump. The cross-over pressure, then, is

$$P = Q_{max}/S_{net} \tag{10.5}$$

where Q_{max} is the maximum mass flow capacity of the high-vacuum pump (maximum throughput) and S_{net} is the net pumping speed of the rough pump at the chamber.

An immediate corollary of matching mass flows is that the larger the roughing pump, then the lower the crossover pressure must be, as illustrated in Figure 10.5.

The suggestion made in Figure 10.4 is not a conservative assumption because it may be expected that the outgassing rate will increase when the pressure drops after crossover. Usually, this does not present a problem, perhaps because the inter-diffusion of gases at the pressures considered here is very high, so that the pressure does not greatly influence the outgassing rate.

Unfortunately, a standard does not exist that can be used to determine the maximum mass flow rate for a high-vacuum pump. Some pumps can tolerate a brief period of operation under overload conditions better than others. In addition, the net pumping speed of the rough pump may not be accurately

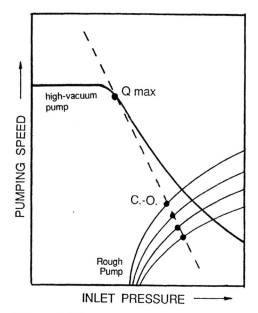

Figure 10.5 Change of crossover pressure depending on the size of the roughing pump.

known, and often the inexpensive pressure gauges used to measure rough vacuum levels are not accurate. The use of safety factors is advised. One rule of thumb may be noted: if, after cross-over, the pressure does not drop quickly, then the high-vacuum pump is overloaded. The pressure reduction should occur almost immediately and it should be according to the ratio of actual pumping speeds of the high-vacuum pump and the rough pump.

The mass flow emanating from the chamber can sometimes be easily determined by measuring the rate of rise of pressure in an isolated chamber. This is illustrated in Figure 10.6. The chamber should be fully loaded with process fixtures and materials. It should be evacuated to the temporary cross-over pressure and the roughing valve closed. The measured mass flow

$$Q = V \Delta P / \Delta t \tag{10.6}$$

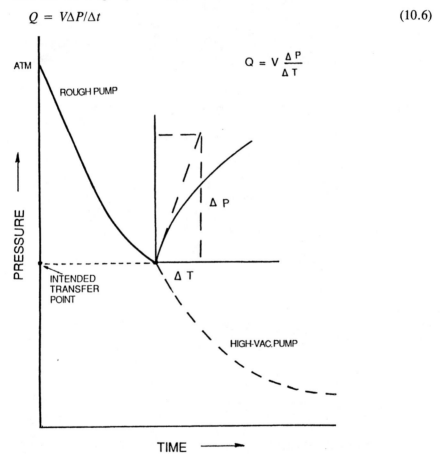

Figure 10.6 Finding the proper crossover mass flow (and pressure) by rate-of-pressure rise measurement.

must be less than Q_{max} of the high-vacuum pump. If it is higher, then a lower cross-over pressure must be used and the experiment repeated.

In processes that use heat (or other energy) after the cross-over, (such as furnaces), the outgassing rate caused by the addition of heat may again exceed the allowable maximum. In this case, the heating rate must be adjusted such that the chamber pressure does not exceed the pressure associated with the location of the Q_{max} point. This is particularly important for vapor jet pumps because each vapor jet has its own mass flow limit. It is possible for the inlet stage to be completely overloaded and, therefore, to produce an excessive amount of backstreaming while the rest of the pump is still functioning. This situation is better in turbomolecular pumps because all stages rotate together on a common shaft. For capture pumps, the limiting mass flow must be tied to the desired period of regeneration (cryopumps) or cathode replacement (ion pumps).

The amount of water vapor in the vacuum system can also affect the crossover condition. The important issue is the base pressure of the roughing pump for water vapor at that time. As noted before, during evacuation, the switch from the roughing pump to the high-vacuum pump should be made at a gas load (throughput) from the vacuum chamber that is lower than the maximum permissible throughput of the fully functioning high-vacuum pump. To find the appropriate crossover pressure, estimate the gas load at the end of the roughing period. One way to do this is to measure the rate-of-rise of pressure of the fully loaded system after a certain period of pumping (Figure 10.6). Another method is to obtain the expected throughput (after crossover) by multiplying the net pumping speed of the roughing pump by the chamber pressure. The difficulty arises in that the net pumping speed depends on the base pressure for water vapor at that particular time.

Figure 10.7 demonstrates the possible changes in the net pumping speed of the roughing pump, which can affect the choice of the crossover pressure when the gas content of the system shifts from mainly air to mainly water vapor. As the base pressure shifts to higher values, the effect is to decrease the net speed at pressures normally encountered. It should be remembered that the slope of the evacuation curve represents the pumping speed and, whenever the evacuation curve becomes nearly horizontal, the pumping speed is nearly zero.

When, after the initial evacuation, a high-vacuum pump is engaged, not only the pumping speed increases but also the overall compression ratio. This can explain why the inlet pressure (after crossover), often drops by 1,000 to 10,000 times (in a matter of seconds) rather than in accordance with the increase of the pumping speed. An example of this behavior is shown in Figure 7.4 where the pressure drop is nearly four orders of magnitude while the ratio of pumping speeds is only about four. Figure 7.4 and Figure 7.13 also clearly demonstrate that in nonsteady state situations, evacuation time does not have a simple inverse relationship with time. The two figures demonstrate that despite the

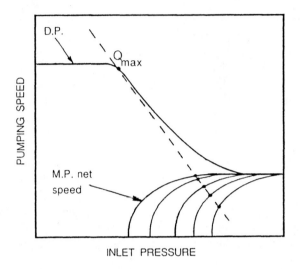

Figure 10.7 Change of crossover pressure depending on the condition of the roughing pump.

large difference between the two high-vacuum pumps, the initial evacuation times are not very different. This is because in this phase, the process of evacuation is influenced more by the transient outgassing rate rather than by the density decay from a constant volume and with a constant pumping speed. In addition, the transient gas flow velocity, after opening the high-vacuum valve, is usually much higher than the velocity that corresponds to a given pumping speed under steady-state conditions.

10.4 THROUGHPUT LIMITS OF VAPOR JET PUMPS

Figure 6.6 is an example of the traditional display where no clear distinction is made between the steady-state plateau section (constant speed) and the transient constant throughput section, where one or more of the inlet stages stop functioning. In other words, in regard to the inlet stage, the pump should be considered overloaded at inlet pressures near the end of the constant speed section. If the backing pump is not oversized, the constant throughput section is clearly defined as shown in the throughput curves of Figure 6.6. A constant throughput means that the amount of gas (number of molecules) removed by the pump at 2×10^{-3} torr is nearly the same as at 2×10^{-2} torr.

The designer of a vapor jet pump should provide a safety factor for the maximum throughput capacity (Q_{max}) in each successive downstream stage to prevent a sudden collapse of the pumping action of the entire pump, when the

OVERLOADING OF VACUUM PUMPS

inlet stage is overloaded. This, however, does not avoid the partial malfunction of the overloaded inlet stage that results, for example, in excessive backstreaming. The location of the Q_{max} point (the end of the constant speed section) is the important design parameter for any high-vacuum pump and it is directly associated with the power used in the pump (see Figure 3.9).

Another important design parameter is associated with linking the high-vacuum pump performance with the backing pump. For economical reasons, the backing pumps often have more than 100 times lower pumping speed than high-vacuum pumps. Thus, it is necessary to have a pumping capacity at the *discharge* stages of a vapor jet pump even at pressures between 0.1 and 1.0 torr in order to prevent the need of very large backing pumps. This does not mean, however, that the vapor jet pump is intended to operate at such pressures at the *inlet* in the high-vacuum stage. It is misleading to specify the throughput capacity at 10 or 100 mtorr without making it clear that this is a measure of a transient characteristic (useful during crossover from roughing to high-vacuum pumps) rather than an indication of a steady-state process pumping performance. If it is necessary to maintain process pressure above the Q_{max} point, then a deliberate impedance must be provided to the flow between the pump and the system (throttling) to keep the maximum gas flow through the pump below the Q_{max} value.

10.4.1 Pressure Fluctuations

Often, the inlet pressure above Q_{max} point becomes unstable, exhibiting large fluctuations which then subside above approximately 25 mtorr. This is usually caused by the interruption of the top pumping jets due to overloading. In general, whenever a curve representing a dependent variable (on a linear rectangular plot) becomes vertical, it cannot be easily controlled. As a result, a small variation of the independent variable will cause large variations of the dependent variable. In the operation of the vapor jet pump, one cannot control the pumping speed or the pressure; what can be controlled is the gas flow (mass flow) by doing something in the chamber or, perhaps, by throttling the inlet valve.

Between inlet pressures of about 1 to 20 mtorr (depending on pump size and design), the mass flow capacity tends to become constant, producing a horizontal throughput versus pressure curve (or, more correctly, a vertical pressure versus throughput curve), so that a small variation of the gas load produces large fluctuations of the inlet pressure. To prevent this, the pump designer has little choice. To produce a less steep slope, it is necessary for each successive pumping stage to have a substantially higher throughput capacity than the preceding stage. In such a case, the total available power must be redistributed such that the first (inlet) stage is the weakest. This is hardly a desirable design direction. Typically, for safety, it is desirable for each successive stage to have about 10% more throughput capacity than the previous stage. In large pumps, which sometimes have five or six stages, this already produces 20 to 30% of "wasted" power. Figure 10.8 shows the separate break-down points of five pump stages. Also note the safety factor zone shown in Figure 10.1.

The other method of increasing the slope of the throughput versus inlet pressure curve is to use large backing pumps, but this is also not a good design direction. An overloaded vapor jet pump with a large backing pump will tend to lose pumping fluid because the denser gas passing through the vapor jet pump will tend to sweep a part of the pumping fluid vapor into the backing pump. Above approximately 30 mtorr, the pressure becomes stable, but then the major portion of the pumping action is due to the backing pump and serious pumping fluid loss may be the result. Often, during prolonged operation at inlet pressures (or rather, gas flows) above the Q_{max} point, it is noticed that the cooling water temperature exiting the pump becomes significantly colder. This is a clear indication that the vapor jets do not reach the pump wall, do not deliver heat energy to the wall, and do not condense inside the pump.

For proper operation, at chamber process pressures above approximately 1 mtorr, it is necessary to limit the gas flow to the Q_{max} (of the first stage) and throttle the inlet valve. Otherwise, one can use smaller valves, baffles, or traps and save a considerable expense, or one can entirely by-pass the vapor jet pump, after the initial evacuation, and continue pumping with the backing or roughing pump. In systems that are exposed to air every half hour, it is not necessary to use enlarged valves, traps, etc., in order to increase the net pump-

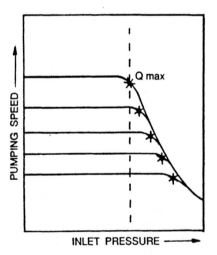

Figure 10.8 Indicating breakdown pressures for each stage of a vapor jet pump.

ing speed. The 20% higher speed obtained by the use of enlarged (high-conductance) components can hardly improve the pump-down time, as can be seen in comparing the two curves of Figure 7.4 and 7.13, but the higher impedance of smaller components can produce the desired throttling effect.

10.5 MASS FLOW LIMITS OF TURBINE-TYPE PUMPS

Turbomolecular pumps display mass flow limitations similar to vapor jet pumps. In Figure 10.9a, the throughput part of the performance display is shown for a 1600 L/s pump (N_2, Varian V-1800). The similarity of performance to that shown in Figure 10.1 is evident. In this particular case, the maximum throughput is reached at about 1×10^{-2} torr, as indicated by the horizontal dotted line. The major difference in response to overloading, compared to vapor jet pumps, is due to the fact that in turbopumps all stages are tied together so that even if the inlet stages are stalled or idled, there is no clear specific malfunction until the motor begins to slow down. There may be subtle effects on the bearing loads and a rise in temperature (due to increased friction in a denser pumped medium), which may shorten bearing life but is normally not immediately perceptible.

It is instructive to note that in Figure 10.9a, the pressure and throughput axes can be reversed, thus plotting the pressure resulting from the pumping action versus the gas flow emanating from the vacuum system. In such a 90° rotated plot, the mass flow limitation (the maximum throughput) can be seen more clearly because it appears as a nearly vertical line that separates the performance of the high-vacuum pump from the performance of the backing pump.

10.5.1. Reduction of Pumping Speed at Higher Pressures

The traditional plot of pumping speed versus inlet pressure often produces conceptual confusion with regard to the cause of pumping speed reduction at the end of the constant speed section. Popular opinion ascribes the reduction to the arrival of viscous flow at the inlet of the high-vacuum pump. This may be associated with the original introduction of the vapor jet pump when, for a number of years, it was held that it was necessary to have the inlet slits of the same dimensions as the mean free path of the pumped molecules. This idea can be questioned immediately by noting that both jet pumps (steam ejectors) and turbines (aircraft compressors) work very well near atmospheric pressures. The shift of the performance shown in Figure 10.9 indicates that the drop in pumping speed at the higher inlet pressure has essentially nothing to do with the type of flow, but is basically due to the effect of power and its association with the maximum steady-state throughput capacity of the pump (while the inlet stage is still fully functioning).

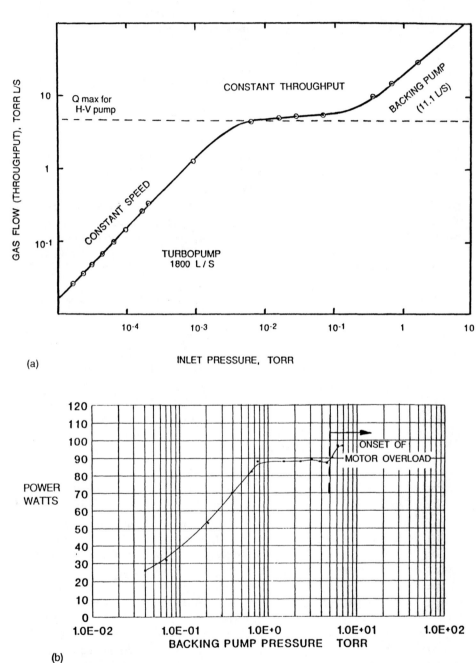

Figure 10.9 (a) Flow versus inlet pressure for a turbomolecular pump. (b) Power consumption of a 250 L/s hybrid turbopump (Varian $V - 250$ M) for nitrogen.

The overloading of the turbopump motor can be clearly seen in Table 10.1 where it is shown that, as the overload condition is approached, the power remains constant while the rotational speed (rpm) is gradually reduced with higher discharge pressure. This is also reflected in Figure 10.9b.

When viscous flow conditions are approached at the inlet of the high-vacuum pump, the limiting compression ratios are reduced but not the volumetric pumping efficiency. This is a well-known effect associated with the absence of genuine pressure gradients in molecular flow. As a result of the gradual reduction of the compression ratios, the base pressure of the pump increases. This is illustrated in Figure 10.10, which shows that the speed of reduction originates from the inability of the pump to cross the line of maximum throughput. If the backing pump were larger (counteracting the effect of the low compression ratio), and the power adequate, the pumping speed near viscous flow should become higher rather than lower that at the molecular regime (roughly 20 L/s/cm^2 for an orifice with air at room temperature instead of 11.6 L / s / cm^2).

The dimensions of throughput are equivalent to power, which, in an isothermal process can written as

$$W = Q \ln P_2/P_1 \tag{10.7}$$

In a high-vacuum pump, this power is a very minor part of the total power. In a vapor jet pump, most of the power is consumed in reboiling the motive fluid after condensation (which is a necessary step for the separation of the pumped gas and the pumping fluid). In turbopumps, power is consumed in bearings and gas friction at the discharge part of the pump. Also, in most compressors, particularly valveless compressors such as turbopumps and jet pumps, some power is required for the maintenance of the pressure difference across the pump, even at net zero flow.

Table 10.1 Compression Ratios for Nitrogen for a Macrotorr-type Turbopump (Varian's V-250) Demonstrating the Onset of Motor Overload

Outlet Pressure (torr)	Compression Ratio	RPM	Power (Watts)
0.01	2.0×10^8	56,000	13
0.38	2.0×10^8	56,000	65
0.76	2.0×10^8	56,000	85
1.52	1.9×10^8	52,000	90
2.74	8.7×10^7	47,000	90
3.88	3.9×10^7	43,000	90
5.00	6.0×10^6	38,000	90

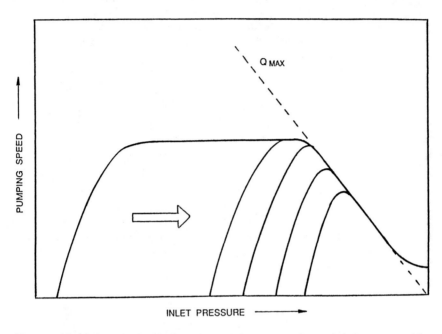

Figure 10.10 Illustrating deterioration of the compression ratio when approaching viscous flow conditions.

The total power may be expressed as

$$W = Q \ln (P_2/P_1) + S (P_2 - P_1) + W_f \qquad (10.8)$$

where Q is throughput (mass flow), P_1 and P_2 are the inlet and exhaust pressures, S is the pumping speed, and W_f may be called wasted power. In turbopumps and mechanical rough vacuum pumps, this is mainly friction. For multistage devices, power consumed in each stage can be added to give the total power.

The compression ratio and pressure difference are somewhat independent characteristics of a pump (or compressor), depending on the design and operating conditions. Their relationship, demonstrated in Figure 10.11, can be obtained by deliberately increasing the exit pressure of a high-vacuum pump at different preset gas loads and observing the effect on the inlet pressure. The values indicated can be associated with the performance of a hybrid turbopump with nitrogen. It may be appreciated how difficult it is to specify a single numerical value for the maximal compression ratio; it can even become unmeasurable when inlet pressures reach 10^{-10} torr range.

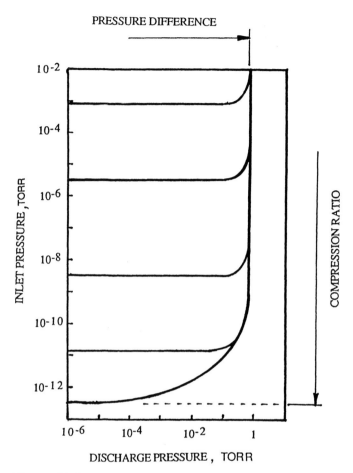

Figure 10.11 General relationship between inlet and discharge pressures at different gas flow conditions illustrating the distinction between compression ratio and the maximum sustainable pressure difference.

10.6 SIMILARITIES BETWEEN PUMPS

Various vacuum pumps have been developed by different people at different times. This has resulted in different terminology and different ways to display the performance. For example, compression ratios are associated with turbopumps, but not with vapor jet pumps or the conventional rotary vane pumps. Even the units of measure are often different. For example, the pumping speed for vane and piston pumps is expressed in m^3/h in Europe and CFM in the United States, while for high-vacuum pumps it is in L/s.

Mechanical rough-vacuum pumps also have overload conditions and limiting compression ratios and it is important to know what these limitations are. For example, in leak detection work with helium as a tracer gas, it is necessary to know that the pressure ratio for helium in double-staged oil-sealed vane pumps is near 10^7. This enables the mass spectrometer leak detectors of the ContraFlow™ type to pump out helium (introduced through a leak) to a helium background level of about 10^{-10} torr when the helium content in the atmosphere at the discharge side of the pump nears 4×10^{-3} torr.

Modern, high-powered vapor jet pumps usually have compression ratios for helium in the order of 10^6. The actual value depends on the density of the vapor jets and the number of stages, which is very similar to turbomolecular pumps. Modern compound turbopumps can also produce similar compression ratios. The similarity of these basic relationships can be seen in Figure 10.12, where pressure ratios are shown related to the bulk velocity of the gas entering the inlet passage of the pump.

For basic design purposes, it may be instructive to represent the performance of any displacement-type vacuum pump in the form of a plot of compression ratio versus mass flow or throughput. Note that this does not require any more data than is normally obtained. This form of plotting will take into account the fact that the compression ratio really does not stay constant under all conditions. Also, it demonstrates the influence of discharge pressure, which was neglected in Figure 10.1. The various accommodations necessary to match transfers between parallel and connected pumps in-series can then be done, based on mass flow considerations rather than pressure.

It may be said that the speed-versus-pressure curve is perhaps of more interest to a designer or test engineer than the user, who should be interested in the pressure obtained in the vacuum chamber, depending on the amount of gas emanating from it. After all, when we measure the performance of the pump, we measure the gas flow and pressure at the inlet of the pump. The pumping speed values are then derived from these basic measurements.

10.7 CRYOPUMPS AND ION-GETTER PUMPS

The overload situation in pumps that do not exhaust pumped gases are somewhat different than compressor-type devices like turbopumps or vapor jet pumps. In addition to a maximum gas flow capacity limit (maximum inlet pressure in steady-state operation), they have a limit of a maximum total gas accumulation. Figure 8.12 reflects the typical periods between regeneration of cryopumps and the cathode replacement of ion-getter pumps, depending on the amount of previously ingested gases. The cryopump response to a momentary overload will depend on its cryogeneration capacity as well as the heat capac-

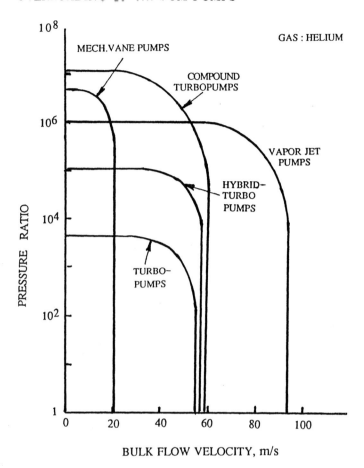

Figure 10.12 Typical relationship between pressure ratio and the bulk flow velocity at the pump inlet.

ity of cryoarrays. If the amount of gas causing the overload is sufficiently high, the temperature of the pumping surfaces may rise, resulting in the release of previously pumped gases. Typically, compared to vapor jet pumps, cryopumps can absorb somewhat higher momentary pressure bursts but the value for the tolerable continuous throughput for a typical 20 cm diameter pump is lower (as shown in Figure 10.13). The broader bands for the cryopumps and ion-getter pumps in Figure 10.13 indicate that the actual value for Q_{max} depends on the desirable (or tolerable) period between regenerations. This situation can also can

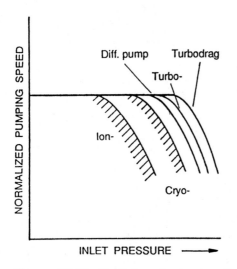

Figure 10.13 Typical overload regions for various pumps.

be reflected in terms of cost per unit of throughput (or power consumption) per unit of throughput (Figure 8.13). The overload associated with the total amount of previously pumped gases can be established by monitoring the pumping performance for hydrogen.

For ion-getter pumps, the need for replacing the cathodes can be associated with a general degradation of performance, particularly the difficulty in starting the pump, or a sudden failure due to electrical shorts produced by the mechanical fractures of cathode structures.

REFERENCES

1. M. H. Hablanian, *J. Vac. Sci. & Technol.*, *A6*, 1177, (May/June 1988).
2. M. H. Hablanian and A. A. Landfors, *J. Vac. Sci. & Technol.*, *1*, 1, (1974).
3. M. H. Hablanian and P. R. Forant, *Soc. Vac. Coaters, 19th Ann. Conf.*, (1976).
4. M. H. Hablanian and K. Caldwell, *Soc. Vac. Coaters, 34th Ann. Conf.*, (1991).
5. M. H. Hablanian, *J. Vac. Sci. & Technol.*, *A10*(4), (1992).

11
Ultrahigh Vacuum

11.1 INTRODUCTION

The region between 10^{-9} and 10^{-12} torr is somewhat arbitrarily defined as ultrahigh vacuum. In dealing with such low levels of pressure, the main interest of course is not in pressure as such, as a force per unit area, but in corresponding number density or molecules per unit volume. This, in turn, is associated with the cleanliness of surfaces, or the amount of gas adsorbed on the surface. Many important applications of ultrahigh vacuum are associated with the requirement of keeping surfaces clean. If we introduce a gas-free surface into a vacuum chamber at 10^{-6} torr at room temperature, it will typically have a monolayer of adsorbed gas formed on it within a few seconds. At 10^{-10} torr the time of monolayer formation may be several hours, which permits surface measurements and surface process work (such as thin-film deposition) without the interference from adsorbing gases. During the last 30 to 35 years ultrahigh-vacuum techniques moved from the laboratory environment into important industrial applications; for example, processing of certain microwave and x-ray tubes and thin-film deposition in molecular beam epitaxy systems.

There are two basic requirements for the achievement of ultrahigh vacuum. The pumping speed must be very high and the gas evolution inside the chamber very low. The number of gas molecules in a chamber is always directly proportional to the total gas evolution rate (including leaks) and inversely proportional to the pumping speed. In engineering practice, it is not only too expensive but essentially impossible to provide the required pumping speed. Even if the process were to be conducted at satellite (space shuttle) conditions, the pumping speed would be limited to the speed of an orifice discharging into a "perfect" vacuum (about 11 L/s · cm^2 for air at room temperature). The more practical and less expensive route is to reduce leakage (including permeation through gaskets) and reduce outgassing rates by some means of surface treatment.

11.2 DEGASSING BY BAKING

One of the most important ingredients of ultrahigh-vacuum technique is the use of metal gaskets, which not only eliminates the concern about permeability but also reduces the outgassing rate compared to elastomer gaskets by 10 to 1000 times. The other simple but basic requirement is to elevate the temperature of the entire vacuum chamber (including pressure gauges) at some point during the evacuation to increase the rate of gas evolution and therefore accelerate the removal of gas from the chamber. This process is, rather prosaically, called "baking." There are other means of imparting energy to adsorbed molecules which usually are mostly water vapor, such as electron or ion bombardment, high-intensity light, and perhaps others. The most common method, however, is baking, usually between 200 and 400°C. The temperature limit is given by practical considerations such as external oxidation of stainless steel chamber walls.

In systems that cannot be baked, surface pretreatment is significant. There are a variety of chemical cleaning and reverse plating processes that can be helpful. In addition, it is possible for some alloys to produce a large crystal grain structure on the surface to reduce the amount of gas contained in the grain boundaries. Such "glassification" may reduce the outgassing rate 10 to 100 times. When baking is impossible or inconvenient, reduction of the intrinsic outgassing rate is very important. However, when baking is a part of each evacuation cycle it is possible that a rougher surface may have an advantage because, after cooling, it can become an adsorbing medium.

A qualitative representation of the baking process is shown in Figure 11.1. Baking is initiated so that the higher temperatures are reached after most of the air is pumped out. This prevents oxidation of internal surfaces. The length of the baking period depends on applications. Usually at least several hours are required; often, vacuum chambers are baked overnight (15 h) or longer. To appreciate the baking process in more quantitative terms, see Chapter 2.

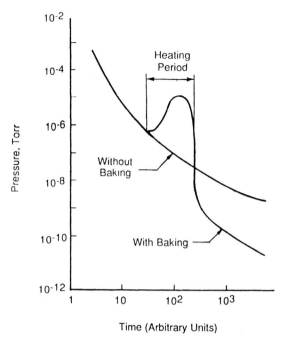

Figure 11.1 Baking process.

11.3 ALL-METAL SYSTEMS AND COMPONENTS

A typical material for ultrahigh-vacuum chamber construction is stainless steel (type 304), although for small systems glass can be used. In special cases aluminum is usable but is limited in the temperature range. The choice of stainless steel is fairly obvious. It does not rust and therefore has a smoother surface and lower outgassing rate compared to structural steels. It can be welded and brazed, it can be machined (even if with some difficulty), it is adequately strong and hard for making demountable joints, and it is nonmagnetic, which can be useful for some applications. For making demountable joints stainless steel can be conveniently allied with soft copper gaskets because the thermal expansion coefficients of both metals are similar. There are two basic design requirements for the construction of all-metal seals. To produce a tight seal suitable for high and ultrahigh vacuum, the gasket has to be subjected to plastic deformation (i.e., the sealing forces must be high enough to exceed the yield point of the gasket material). The second requirement is that some elastic energy must be stored in some components of the sealing system so that the pressure on the gasket is maintained during heating and after cooling.

There are many possible designs. However, a system introduced under the trademark Con-Flat (Varian Associates) has found almost universal acceptance for flanged joints below 30-cm diameters. A schematic drawing of the Con-Flat seal is shown in Figure 11.2. Both flanges that make the seal are identical. They contain a circular knife edge that penetrates into a flat copper gasket shaped as a simple washer. Due to the particular geometry of the knife edge, the copper gasket is pushed outward in the radial direction as the flange bolts are tightened. The dimensions of the flanges and the gasket are accurate enough to bring the expanding gasket to a stop at the inner periphery of the flange and then be extruded also in the axial direction. After the closure, the elastic energy may be stored in all three elements: the gasket, which has been extruded and cap-

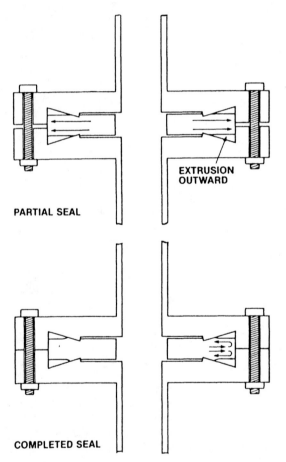

Figure 11.2 Design and action of ConFlat flanges.

tured in the corner; the bolts, which have been slightly stretched; and the flanges, which have been slightly bent. Because one side of the knife edge is rather flat, the deformed annular area is rather large and the copper gasket can support a great force without further extrusion. This, combined with the similarity of the thermal expansion coefficients, assures that the system does not develop leaks after cooling. Such joints can be baked many times without developing leaks, but a new gasket should be used after demounting.

In addition to the need for precise dimensions, the ultrahigh-vacuum demountable joints must have precise control of material properties. Con-Flat flanges of good quality are made from selected stock to assure correct metallurgy and low level of inclusions in the microstructure. Even if the stainless steel is processed to the tightest military and commercial specifications, it may be inadequate for ultrahigh-vacuum applications. For small flanges that are made from bar stock, attention must be paid to the possibility of leaks along the flow lines produced during hot rolling, particularly near the center of the bar. Small leaks can be detected by mass spectrometer leak detectors (Chapter 13) through materials as thick as 2 cm. Leaks are sometimes discovered only after machining, cleaning, and baking. Large flanges should be made from cross-rolled plate to minimize the possibility of leaks.

To ensure adequate repetitive sealing, the metal gaskets also require a high level of quality control. Only electronic grade, oxygen-free, high-conductivity (OFHC) copper should be used for gaskets. Their hardness must be controlled and their surfaces must be free of any debris, which might interfere with proper sealing. The bolts and nuts that are used to assemble the flanges must also be of superior quality, having superhigh tensile strength, to avoid relaxation through a small amount of plastic flow resulting from high forces. When high-quality materials are used, the Con-Flat flanges can withstand temperature cycling at a rate of about 10°C per minute in the range -195 to 500°C.

As mentioned earlier, Con-Flat flanges have become almost a standard in the high-vacuum industry, although it is possible to make adequate seals with other systems: for example, for larger diameters (above 35 cm), the flange system shown in Figure 11.3, where a metal O-ring, placed in a confined corner is compressed directly between asymmetric flanges. The gasket can be copper or gold. To reduce size and weight, the flanges are short and they are held by strong clamps spaced several centimeters apart.

Bakeable ultrahigh-vacuum systems also require all-metal valves and other components, such as electrical and mechanical penetrations (feedthroughs), sight glasses, pressure gauge connections, and so on. Two examples of all-metal valves are shown in Figures 11.4 and 11.5. The angle valve contains a welded stainless steel bellows which isolates the drive mechanism from the vacuum space. The valve seat may or may not have a knife edge. Specifications of such valves should include the information regarding whether or not the valve can

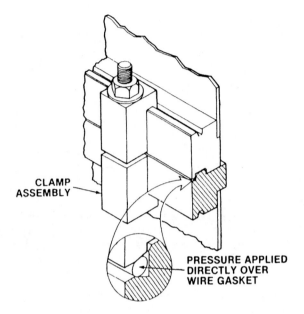

Figure 11.3 Metal wire seal system for large flange joints.

be baked in closed position and how many closures are possible before the seal has to be replaced. The slide valve shown in Figure 11.5 has a cup-shaped valve plate which is sufficiently flexible to spread radially and enter the corner seat. The ridge shown in the circled detail sketch is gold-plated and the seal is pro-

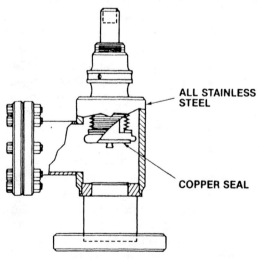

Figure 11.4 Example of a small all-metal valve.

ULTRAHIGH VACUUM

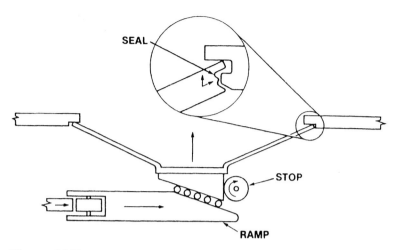

Figure 11.5 Mechanism and seal plate for larger valves.

duced at this gold-plated edge. Needless to say, such valves require precise concentric machining to effect a seal with a minimum of deformation. The details of the drive mechanism are not shown. It contains a beam that carries the valve plate and contains a ramp with needle bearings to provide an upward sealing force after the plate comes to a stop.

In addition to components shown in Figures 11.2–11.5, there are a great variety of commercially available specialized hardware items, valves, high current and high voltage electrical feedthroughs, sight ports, etc. Two examples are shown in Figure 11.6, a sapphire (aluminum oxide) window and a Con-Flat flanged cross. Not all components used in ultrahigh systems can be baked at high temperatures. At intermediate UHV levels, a limited amount of polymer composition (viton) gaskets are permissible. The swing valve shown in Figure 11.7 has two viton O-ring gaskets—one in the sealing seat and the other at the endplate. The sketch in the photograph shows the inner linkage, which produces a sufficient mechanical advantage to provide the sealing force but makes the valve body as short as possible. Bakeable varieties of such valves have heavy flanges to accommodate a metal gasket seal. A highly specialized precise variable leak valve and a typical quick-connect coupling are shown in Figure 11.8.

11.4 THROUGHPUT-TYPE PUMPS AND ULTRAHIGH VACUUM

11.4.1 Introduction

The development of ultrahigh-vacuum techniques had a profound effect on vacuum chamber design. It stimulated equally important improvements in pump

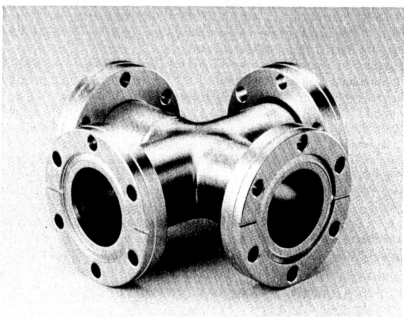

Figure 11.6 Examples of ultrahigh-vacuum hardware. Top, a glass-to-metal seal-type sight port; bottom, ConFlat flanged crosspiece.

design and in the understanding of some basic aspects of pump performance, as well as their limitations. Design improvements also helped to clarify certain elusive concepts, for example, the meaning of the term "ultimate pressure" of a pump. Regarding partial pressures, the ordinary dual-stage sliding-vane mechanical pumps have demonstrated full pumping speed at pressures as low as 10^{-10} torr for gases nearly absent at the discharge. Modern vapor jet pumps appeared in the 1960s with improvements in all performance aspects. Using high-performance pumping fluids, pressures as low as 10^{-9} torr were achieved with only ambient temperature baffles and below 10^{-11} torr with liquid nitrogen cryobaffles. Even the extreme high-vacuum range was obtained with vapor jet pumps assisted by gaseous helium or liquid-helium cooled traps. The fear of accidental backstreaming events, however, shifted the users to turbomolecular pumps which, after nearly 50 years of stagnation, finally became the major pumping preference. Turbopump improvements continue, driven by the need for cleaner pumping methods and rapid achievement of lower pressures. The trend is to optimized staging arrangements to increase the limiting pressure ratios (particularly for hydrogen), the introduction of light-weight, ceramic, grease-lubricated bearings, magnetic bearings, and at least ten times higher permissible discharge pressures. This allows entirely oil-free backing pumps, the use of smaller backing pumps, and the achievement of ultrahigh pressure range even in the presence of light gases such as hydrogen.

The achievement of ultrahigh-vacuum is not likely to be a simple matter for many years to come because it depends not only on the performance of the pump but also on the design and operation of the vacuum system. The interaction between the pump and the system can be complex and it is sometimes said that the ultrahigh-vacuum technique is an art. When a technical discipline is called an art, it often means that the observed occurences depend on the proverbial secondary or even tertiary effects that have been neglected during theoretical modeling. Designers and manufacturers of vacuum pumps are usually concerned with obvious primary performance factors, such as pumping speed, lowest attainable pressure, etc. When the use of these pumps is extended to ultrahigh-vacuum, relatively minor improvements are made to reduce outgassing, provide metal gaskets, and make the pump fully or partly bakeable. The lower the pressure the more important it is to understand the usually unreported characteristics of pumps, their interaction with the system, and, when multiple pumps are used, their interdependence.

In general, the inlet pressure of a high vacuum pump can be expressed as a summation of various gas flow effects and compression ratio limits

$$P_1 = \left(\Sigma \tfrac{Q_i}{S_i}\right)_{ext} + \left(\Sigma \tfrac{Q_i}{S_i}\right)_{int} + \Sigma \tfrac{P_{2i}}{K_i} \qquad \text{(Eq. 4.35)}$$

where P_1 is the inlet pressure, Q_i is the gas flow rate (throughput) for a particular gas, P_{2i} is the partial pressure of the particular gas at the discharge end

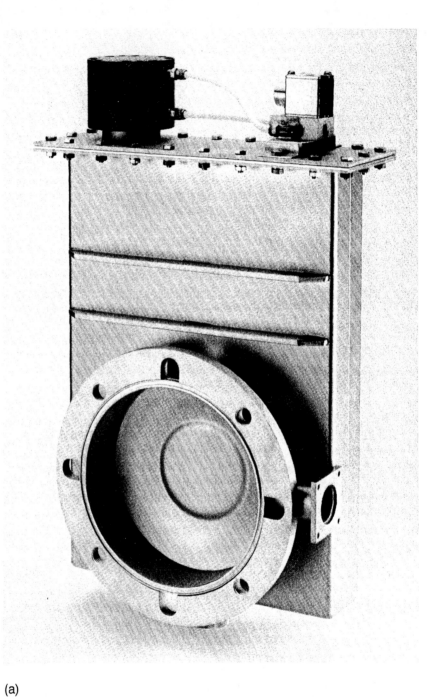

(a)

Figure 11.7 An ultrahigh-vacuum type valve with viton seals and internal swing mechanism.

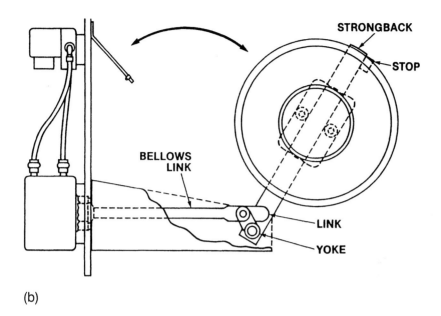

(b)

of the pump, and K_i is the limiting compression ratio for the particular gas. The first term represents external gas loads and the second term gas loads from sources inside the inlet sections of the pump itself. What is generally taken to be the ultimate pressure of the pump (not system!) is given by the last two terms of the summation. It cannot be emphasized too strongly, however, that the manufacturers and users of vacuum pumps traditionally have ignored the last term (for nearly 100 years). This is certainly true for mechanical pumps and vapor jet pumps. The compression ratio effects and limits are only associated with the more recently developed turbomolecular pumps. Without knowing the gas composition and the compression ratio limits for each gas present in the system, it is generally not possible to know what caused the ultimate pressure in a vacuum system.

Because of the enormous range of densities involved in evacuating a vacuum chamber from atmospheric pressure to ultrahigh-vacuum, it is not possible, for practical reasons, to do the required pumping with one integral pump. Typically, at least two pumps are connected in series. To understand the interaction between those pumps and their effect on inlet pressure, it is necessary to know not only the total pressure established in the conduit between the two pumps, but also the partial pressures of various gases. Yet, even if a partial pressure analyzer were used at the discharge side of the pump, it is not a simple matter to make definitive conclusions from the known pump performance parameters.

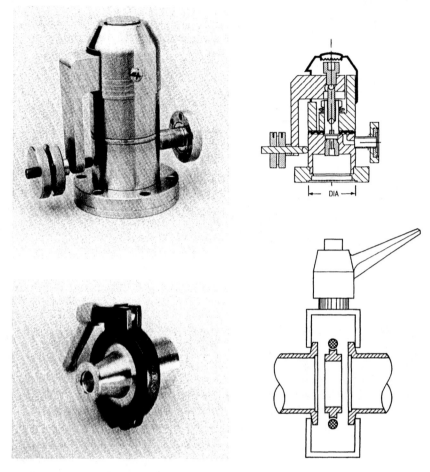

Figure 11.8 A valve for precise, small variable leak control and a clean quick-connect coupling.

Most high-vacuum pumps do not operate entirely in the molecular flow range. Therefore, gas interactions can occur in the downstream parts of the pump, which are not easily predictable from the primary performance data. Even if compression ratios are known for the particular pump and gas, the presence of a second gas can alter the expected compression limit values. Therefore, it is very important for high-vacuum technologists to be aware of the various effects normally not given in the technical bulletins or textbooks on high-vacuum technology and to understand the qualities and peculiarities of a particular pump and its construction details. The traditional display of pump performance does not represent the entire story. It does not give, for example, any information about the influence of discharge pressure on pump performance.

11.4.2 Mechanical Pumps

Most high-vacuum technologists do not associate ultrahigh-vacuum with ordinary mechanical vacuum pumps. The reason for this is, usually, the incorrect interpretation of the performance display graphs (pumping speed versus inlet pressure, which is traditional in high-vacuum technology but not in the field of fluid mechanics) and an incomplete understanding of the term "ultimate pressure." There are two basic important associations: the existence of pumping action in regard to partial pressures and the effect of performance of mechanical pumps on the performance of ultrahigh-vacuum pumps.

As noted before, various modern high-vacuum pumps have been developed by different people in different times and, as a result, their performance is often described using different measures and even different terminology. For example, compression ratio is usually associated with turbomolecular pumps but hardly ever with vapor jet pumps or mechanical pumps; and, for example, what is called "tolerable forepressure" of vapor jet pumps is actually nothing more than the maximum pressure difference that a pump is capable of maitaining as associated with molecular pumps.

High-vacuum pumps, or rarefied gas compressors, of the positive displacement type maintain their full pumping speed in the ultrahigh-vacuum range as long as their construction is adequate to maintain a certain pressure ratio across the pump. The ordinary oil-sealed, doubble-staged, sliding vane mechanical vacuum pumps usually have very high pressure ratios. The pressure ratio for air is often as high as 10^8, i.e., lowest inlet pressure is 1×10^{-5} torr and the discharge pressure atmospheric. Single-stage pumps produce a limiting pressure ratio of nearly 10^5. The various oil-free multi-staged pumps are typically limited to a compression ratio of 10^4, although a ratio as high as 10^5 can also be produced by some special four-stage piston pumps.

In the course of the work with counterflow (Contra-Flow) leak detectors, it was discovered that the dual-stage mechanical pumps had full pumping speed for helium at partial pressures at the inlet as low as 10^{-10} torr. In the Contra-Flow leak detector, helium introduced into the foreline space as a result of a leak is pumped out, at those pressure levels, in a few minutes by the mechanical pump. The typical discharge pressure of helium in the atmosphere is 5×10^{-3}, which indicates a pressure ratio of 10^7 across the pump. The existence of pumping speed at such low partial pressures was directly measured using a mass spectrometer (as a helium gauge) with small standard helium leaks serving as flow meters (see Figure 5.7 and 5.12).

When throughput-type pumps are used in ultrahigh-vacuum systems, it is important to understand the pressure ratio limits of high-vacuum pumps and backing pumps for various gases. The ultimate pressure of dual-stage, oil-sealed mechanical pumps is usually given in commercial bulletins and textbooks as 1×10^{-4} torr, measured by gauges that do not deal simply with condensible gases. The contribution to this pressure comes from a few separate sources: leaks, oil vapor and its fractions, and outgassing. Because the working module of mechanical pumps is usually submerged in oil, designers often omit seals between smooth, flat sections of the pump. This often produces minor leaks that do not interfere with the basic performance specifications but may affect some ultrahigh-vacuum experiments. Mechanical pumps have very low pumping speeds compared to high-vacuum pumps. Their inlet pressure is influenced by outgassing at higher pressures, which are not usually associated with outgassing.

When molecular flow is reached at the inlet of a pump, the possibility of backstreaming is evident. Extensive studies of the amount of backstreaming have been published. Typical prevention methods include the use of traps and the use of continuous gas bleed to keep the conditions in the conduits near viscous flow. Usually the estimates of pressure necessary to suppress backstreaming are associated with the diameter and length of the conduit and the molecular weight of the oil. However, a simple experiment can demonstrate that sometimes the backstreaming consists of rather light fractions of the oil. Figure 11.9 is a schematic of a simple system used to study the efficacy of foreline traps. Here, the residual gas analyzer can be alternatively connected to the mechanical pump using a line with and without a zeolite trap. The results are shown in Figure 5.17. When the valves were opened and closed to channel the flow through the trap or through the open line, the changes in the pattern on the RGA screen (Figure 11.9) were noticed as fast as the valves could be operated. This indi-

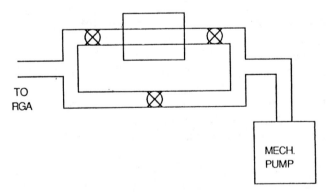

Figure 11.9 Schematic of the system used to study the efficiency of zeolite traps.

cated that the cracking patterns in the RGA were not entirely due to the instrument itself but due to the noncondensible elements of vapor flow. Properly sized zeolite traps can be effective. They do not impede the evacuation time and remain fully active for at least one month (see Sections 5.4 and 5.6).

11.4.3 Turbomolecular Pumps

In regard to turbomolecular pumps, the achievement of ultrahigh-vacuum is, at least in the conceptual sense, very simple. It is only necessary to obtain a sufficiently high compression ratio and reduce outgassing at the inlet sections of the pump. In principle, regarding the basic pumping mechanism, turbomolecular pumps can function at any pressure encountered in high-vacuum technology (including atmospheric) (see the recently published general articles by J. Henning). There are some obvious limitations in commercially available pumps. They cannot be baked at the usual temperatures employed in ultrahigh systems and they require some bearing lubrication, which makes them not entirely free of the possibility of hydrocarbon contamination. Even magnetic levitation bearings do not eliminate the baking limit. Thus, the quoted turbomolecular pump ultimate pressure is near 1×10^{-10} torr. No doubt, with extensive baking and with the assistance of liquid-nitrogen traps or gettering pumps, the final pressure can be lowered to 1×10^{-11} torr or lower. The same result can be achieved using two turbopumps in series.

Until a few years ago, turbopumps were made with rather low compression ratio for light gases. This poor compression ratio for hydrogen and helium is due to practical engineering considerations—primarily the limited number of stages and the issue of size and cost. In older pumps, the residual gas at the inlet to the pump consisted mainly of hydrogen (over 99%, reported by J. Henning). Most of the hydrogen originated from the oil-sealed mechanical backing pumps. Because of bearing lubrication and backing by oil-sealed mechanical pumps, the operation of turbopumps requires certain precautions. To prevent contamination of the vacuum system the pump must be vented during deceleration. A special, automatically acting, small valve is often provided for this purpose at the midsection of the pump. Also, the vacuum system should not be exposed to an operating mechanical pump environment when the turbomolecular pump is stopped. One of the advantages of turbopumps is the possibility of omitting inlet valves (involving expense and additional outgassing), because, unlike operating vapor jet pumps, turbopumps can be exposed to atmosphere without subsequent malfunction or damage.

In recent years, hybrid and compound pumps tolerate much higher discharge pressures, up to 20 torr or higher. There is no clear dividing line between hybrid and compound pumps. The term hybrid may be reserved for pumps that have more or less the same number of stages or impellers, same motor, same size and cost as the conventional pumps, but have at least ten times higher discharge pressure and compression ratios than ordinary pumps. As the name implies, hybrid pumps have different types of impellers held on the same shaft thereby optimizing the pumping speed and compression ratio requirements. An example of such a pump (Varian V-250) is shown in Figure 11.10, where the right side represents the new design and the left side the conventional arrangement of axial-bladed stages. The new design has 8 axial-flow bladed rotors (and stators) and 3 peripheral molecular drag disks. This pump produces a compression ratio for helium over 10^5 torr and for hydrogen over 10^4 torr. These values do not represent an intrinsic limit. By careful design, with a small increase in cost, at least another factor of 10 can be obtained.

Compound pumps are somewhat larger. The longer shaft makes it more difficult to mount both bearings in the forevacuum side and the bearings are often magnetic. Such pumps produce compression ratios for helium of 2.6×10^8 and for hydrogen 3×10^6. They often have 10 axial stages and 5 cylindrical drag sections. The air or nitrogen compression ratios are much higher but, in typical high-and ultrahigh-vacuum systems, there is usually no air present either at the inlet or the discharge end of the pump. However, the compression ratios of the last pump are sufficient to maintain pressures near 10 torr at the discharge while producing ultrahigh-vacuum in the 10^{-10} torr range at the inlet. This permits the use of oil-free roughing and backing pumps to eliminate the possibility of mechanical pump backstreaming. Because of higher compression ratios and lower conductances in the downstream sections of the new pumps, backstreaming is substantially reduced even if oil-sealed pumps are used for backing. This trend is demonstrated in Figures 11.11 a & b, which show RGA spectra for conventional and new pumps. In this set of experiments the turbopump was not vented but connected to an operating backing pump, which is a condition that should be avoided in practice. Despite the improved performance, and especially when pumps are frequently vented to air, it is important to observe a proper venting procedure. It is best to vent the pump from the inlet side (or through the specially provided venting port and venting valve), while the backing pump is operating.

Generally, the ultimate pressures of turbomolecular pumps are limited only by the maximum values of the compression ratios for light gases (mainly hydrogen) and the outgassing rates of the materials used at the inlet side of the pump. If the compression ratio is insufficient, two pumps can be used in series. If higher pumping speed is required, the performance of the turbopump can be enhanced by the addition of water-vapor pumping cryo-cooled surfaces or titanium sublimation pumps. Turbopumps usually have aluminum rotors and have internal electric motors, so they cannot be baked at temperatures often used for ultrahigh-vacuum degassing. Pressures in the 10^{-10} torr range can be obtained in typical ultrahigh-vacuum metal gasket inlet flanges after relatively mild

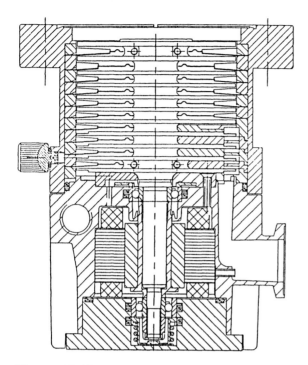

Figure 11.10 Left side, cross section of a conventional turbomolecular pump and right side, a hybrid pump.

baking. Even 10^{-12} torr pressures have been reported after some heroic measures, including the machining of parts by sharp diamond tools in inert gas atmosphere and coating the rotor with ceramic films (quartz) to increase thermal emissivity and lower the temperature.

11.4.4 Vapor Jet Pumps

Vapor jet pumps, popularly known as diffusion pumps, are momentum transfer pumps. Like turbopumps, as any other compressor, they also are characterized by a finite compression ratio, a maximum sustainable pressure difference, and a mass flow limit. Their internal gas generation possibilities, however, are complex and, as a result, the achievement of ultrahigh-vacuum requires a more careful technique. Vapor jet pumps, both mercury and oil driven, were developed long before ultrahigh technology was realized and before residual gas instruments became available. Therefore, early designs (which are commercially available now) had extremely poor compression ratios for light gases. Such pumps sometimes have zero pumping speed for hydrogen and helium, for ex-

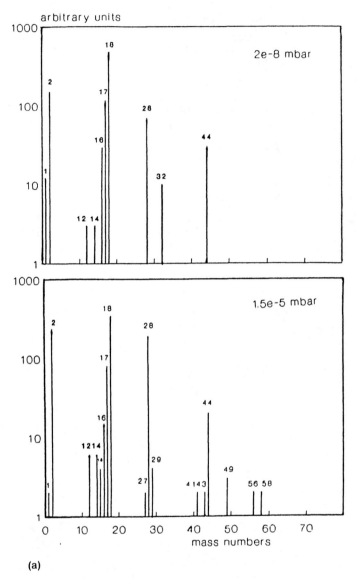

Figure 11.11 Residual gas content for turbomolecular pumps. (a) Conventional pump with top, running and bottom, stopped. (b) The same for a hybrid pump (Varian V 60/10).

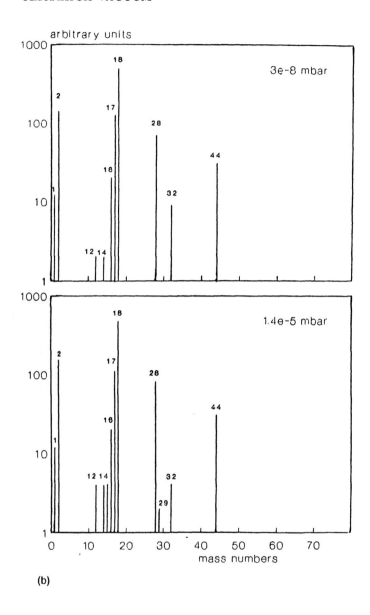

(b)

ample when the heater voltage is 10% low. Great variations of the compression ratio for light gases also exist in turbopumps but, in vapor jet pumps, the variation in compression ratio for helium can range from 10 to 10^8. It is obvious that to achieve ultrahigh-vacuum a pump with certain qualities must be selected.

In addition to such a basic issue as the compression ratio, great variation exists in other performance aspects such as the influence of the type of motive fluid used, accidental exposure to air while heating, and overload conditions. Also, vapor jet pumps are often associated with the performance as it was obtained in nonultrahigh-vacuum systems, and the limitations of the system and the pump are confused. Many textbooks quote pumping speed and "ultimate pressure" limitations in the high-vacuum range.

The existence of pumping speed at extremely low partial pressures should be obvious. After all, even if only a few molecules of a new gas are released in a vacuum chamber they have an excellent chance to flow into the pump and be removed. However, due to the association of the term ultimate pressure with the traditional speed versus pressure graphs, this point is very often overlooked by vacuum pump manufacturers and users.

The fact that there is life below the usually perceived ultimate pressure was directly observed in 1962 in the extreme high-vacuum system (shown in Figure 11.12), at the National Research Corporation (later part of Varian Associates). After obtaining ultrahigh pressure, helium was introduced into the cham-

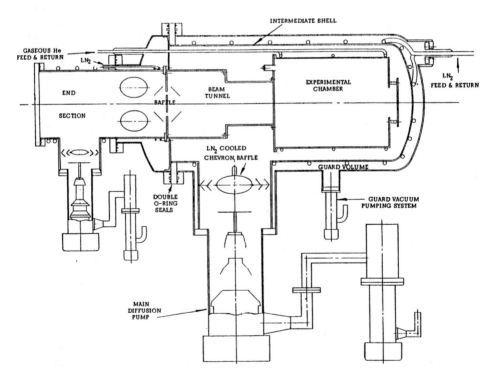

Figure 11.12 Schematic of an ultrahigh-vacuum system with vapor jet pumps.

ber through a leak valve to pressure in the 10^{-4} torr range. When the leak was stopped, it was quickly re-pumped to extremely low pressure, following the expected exponential decay associated with a constant pumping speed.

This extreme high-vacuum system deserves closer examination to demonstrate the general ultrahigh-vacuum capabilities of commercially available vapor jet pumps. It is constructed of 304 stainless steel and can be baked to a few hundred °C by radiant heaters and then cooled to 10 K by gaseous helium circulated from an external helium refrigerator. The first experimental chamber was 45 cm in diameter and 60 cm long. Ultrahigh-vacuum gauges were located on the internal wall of the experimental chamber (a B-A gauge and a magnetron, Redhead-type, cold cathode gauge). Pressures in the 10^{-10} torr range were routinely obtained using only liquid nitrogen traps. When helium refrigeration was used, pressure quickly dropped below the minimum signal detection level of the gauges. The pressure was estimated to be near 1×10^{-13} torr. Later experience demonstrated that the chamber design was unnecessarily elaborate. Guard vacuum is not required if all the joints are made with bakeable metal gaskets and, when vapor jet pumps with ejector stages are used in the forelines, the secondary pumps in series are not required. A much simpler system (using liquid helium cooled traps) achieved similar pressures (see Figure 11.13).

Without the auxiliary devices, such as cryotraps or sorption traps, the lowest achievable pressures of vapor jet pumps are usually limited to the 10^{-10} torr range, provided that an appropriate pump is used, with fluids of lowest vapor pressure (DC-705 and Santovac-5), and shielding from direct backstreaming. Examples of shielding include water-cooled or ambient-cooled baffles or cold-cap-baffle combinations placed inside the pump (usually in an expanded body section at the inlet of the pump).

When vapor jet pumps are used in high-vacuum systems, certain precautions must be observed otherwise backstreaming of the motive fluid is possible. The most important items are noted below.

1. When the pump is operating, the maximum sustainable pressure difference must *not* be exceeded! Because foreline pressure gauges are often not accurate, exceeding 50% of the permissible discharge pressure should be avoided.
2. When the pump is operating, the inlet pressure at the pump must be kept below the point where the constant pumping speed begins to decline, usually near 1×10^{-3} torr.
3. During evacuation of a system, containing a high-vacuum valve, the correct cross-over pressure must be observed. Never switch from the rough pump to the high-vacuum pump at any point above the Q_{max} limit (see Chapter 10).

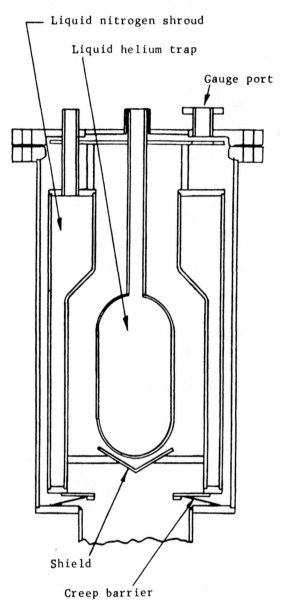

Figure 11.13 A simple, ultrahigh-vacuum system using a vapor jet pump.

4. The procedures for evacuation, starting the pump, degassing (baking), and the introduction of liquid nitrogen trap cooling (particularly in valveless systems), must be carefully considered and observed to prevent any migra-

11.4.5 Summary of Ultrahigh Vacuum

To achieve ultrahigh-vacuum using throughput-type pumps (in addition to the necessary conditioning of the system), all that is necessary is to select a pump with an appropriate pressure and pumping speed performance. This may sound like a truism, but it is the primary requirement and is often not clearly defined in textbooks associated with the high-vacuum technology. One source of confusion is the traditional display of pump performance in the form of speed versus inlet pressure, which does not clearly distinguish between steady-state and transient performance regions, and does not indicate the existence of limiting pressure ratios. If the three types of pumps discussed in this section are considered as rarefied gas compressors (discarding the historical burden of mystification), their performance can be shown in the achieved pressure versus flow rate. Depending on the particular interest, the pressure term can be chosen to be the compression ratio or the pressure difference and the flow may be mass flow or volume flow.

In the molecular flow regime there is, of course, no escape from the possibility of migration of internally generated gases from the pump to the system. Techniques for the prevention of this migration must be carefully considered and maintained for all three pumps types. Perhaps the cleanest pumping system would be a turbomolecular compound pump with magnetic bearings in series with an oil-free backing pump. But even in this case some measure of skepticism may be in order.

Generally speaking, turbomolecular pumps have an advantage compared to vapor jet pumps in regard to their performance during overload conditions. The designer of a multistage pump has little choice. Each following stage should have a slightly higher mass flow capacity than the preceding stage. In turbopumps, the overloading of upper stages results in lower pumping speed but does not produce other adverse effects. In vapor jet pumps, the overloading of the inlet stage can produce serious backstreaming of the motive fluid while the downstream stages may still function normally. Vapor jet pumps may provide the less-expensive system, but they require more careful operational procedures. A clear understanding of such "secondary" issues should be very helpful in the process of designing and operating ultrahigh-vacuum systems.

11.5 CAPTURE PUMPS

Capture pumps (ion-getter pumps, cryopumps, titanium sublimation pumps, nonevaporative getter pumps, and sorption pumps) have the advantage of functioning in an isolated system (i.e., without exhausting the pumped gas to outside atmosphere). The obvious associated disadvantage is that the gas is accumulated inside the pump and, therefore, the amount of pumped gas is limited. After a certain period of pumping, the pump must be regenerated, or the worn parts of the pump must be replaced. Sensitivity to the amount of previously pumped gases must be considered in the process of pre-evacuation and throughout the continuous operation of the pump. The concern is particularly significant to achieve ultrahigh-vacuum conditions. The avoidance of operation at pressures that are associated with the designated overload region (see Chapter 10) is very important.

In general, the constancy of pumping performance of capture pumps may be less assured. For example, the pumping speed of ion-getter pumps tends to decline at lower pressures. This is particularly true when the ultrahigh-vacuum region is approached. In pumps that rely on freshly deposited, chemically active films for their function, the pumping surfaces tend to become saturated, and require periodic re-establishment of an active surface. Cryopumps tend to develop lower pumping speed for helium and hydrogen (after a certain period of pumping). The pumping action of the cold surface for some gases, however, may be enhanced by the presence of layers of other gases previously pumped.

During the initial evacuation, both ion-getter and cryopumps can be sensitive to the presence of inert gases, which can saturate the inner cryosorption panels of the cryopumps and cause subsequent re-emission in ion-getter pumps. In addition to care during the initial evacuation and cross-over conditions, capture pumps can re-emit previously pumped gases into the vacuum system. In ion-getter pumps, the high energy exchanges on the surface or the gradual erosion of the pumping surface can release previously captured gas. For cryopumps, it is important to keep the temperature of the pumping surfaces constant. If, for example, a significant amount of thermal radiation or a heated gas is produced in the vacuum system, the cryopump should be shielded.

Despite these cautionary notes, all capture pumps, alone or in combination, can be used to achieve ultrahigh-vacuum. The isolation from the outside atmosphere actually makes capture pumps more suitable for ultrahigh-vacuum work, provided that system or process gas evolution is not very high. Ion-getter pumps are predominantly used in systems that are kept under vacuum for a long time, for example, high energy particle accelerators and similar installations, and many scientific instruments in which creation and maintenance of clean surfaces is required.

Ion-getter pumps were, from the beginning, associated with ultrahigh-vacuum technology, so much that even the earliest paper by L. Hall notes 2×10^{-10} torr performance. It may be said that the development of ion-getter pumps actually precipitated the entire technology of ultrahigh-vacuum component. One obvious advantage of such pumps is that they can be baked, together with the main UHV system, at temperatures as high as 400°C for thorough degassing. In recent years, the development of the oil-free roughing pump and oil-free turbopumps has simplified the initial evacuation procedures. The system can be pumped during baking by clean turbomolecular or turbodrag pumps to the high-vacuum pressure range. After that, the 10^{-12} torr range of vacuum can be achieved routinely.

REFERENCES

1. P. A. Redhead et al., *The Physical Basis of Ultrahigh Vacuum*, American Institute of Physics, New York, 1993.
2. G. F. Weston, *Ultrahigh Vacuum Practice*, Butterworths, London, 1985.
3. L. D. Hall, *Rev. Sci.I nst.*, 29,5, (1958).
4. P. A. Redhead, *J. Vac. Sci. Technol.*, A12, 4, 904, 1994
5. M. H. Hablanian, *J. Vac. Sci. Technol.*, A12, 4, 897, 1994
6. K. M. Welch, *J. Vac. Sci. Technol.*, A12, 4, 915, 1994
7. J. F. O'Hanlon, *J. Vac. Sci. Technol.*, A12, 4, 921, 1994
8. P. W. Palmberg, *J. Vac. Sci Technol.*, A12, 4, 946, 1994
9. H. F. Dylla, *J. Vac. Sci. Technol.*, A12, 4, 962, 1994
10. Th. Lindblad et al., *Vacuum 37*, (3/4), 293, (1987)
11. B. Cho et al., *J. Vac. Sci. Technol.*, A13, 4, 2228, (1995).
12. H. Ishimaru and H. Hisamatsu, *J. Vac. Sci. Technol. A 12*, 4, 1695, (1994).
13. M. Miki et al., *J. Vac. Sci. Technol*, A12, 1760, (1994).
14. M. Audi & L. Dolcino, *J. Vac. Sci. Technol.*, (Varian VR-3) (1986/1987).
15. M. H. Hablanian, *Vacuum 41*, 7-9, 1814, (1990).
16. M. H. Hablanian, *J. Vac. Sci. Technol.*, 19(2), 250, (1981).
17. M. H. Hablanian, *J. Vac. Sci. Technol.*, 18(3) 1156, (1981).
18. N. S. Harris, *Vacuum*, 28, 6-7, 261, (1978).
19. D. M. Hoffman, *J. Vac. Sci. Technol.* 16(1) 71, (1979).
20. G. Lewin, *J. Vac. Sci. Tecnol.*, A3(6) 2212, (1985).
21. M. H. Hablanian, *J. Vac. Sci. Technol.*, A4(3) 286, (1986).
22. J. Henning, *Vakuum-Technik*, 37, 5, 134, (1988).
23. M. H. Hablanian, *J. Vac. Sci. Technol.*, A11(4) 1614, (1993).
24. G. Levi, *J. Vac. Sci. Technol.*, A10(4), 2619, (1992).
25. D. P. Sheldon and M. H. Hablanian, *Trans. Amer. Vac. Soc.*, New York, 159, (1967).
26. M. H. Hablanian and P. Vitkus, *Trans. Amer. Vac. Soc.*, New York, 140, (1963).
27. M. H. Hablanian and J. M. Maliakal, *J. Vac. Sci. Technol.*, 10,1, 58, (1973).
28. D. J. Santeler, *J. Vac. Sci. Technol.*, 8,1, 299, (1971).
29. M. H. Hablanian, *J. Vac. Sci. Technol.*, A10(4) 2629, (1992).
30. L. H. Dubois, *J. Vac. Sci. Technol.*, A6(1) 162, (1988).

12
Vacuum Gauges and Gas Analyzers

12.1 INTRODUCTION

Vacuum technology deals with 15 orders of magnitude of change in pressure, from 760 to 10^{-12} torr. It should not be surprising that a single transducer cannot span the entire range of interest. At least two different types of instruments must be used, one at the high-pressure range (760 to 10^{-3} torr) and the other in the high- and ultrahigh-vacuum region. The better the vacuum, the more difficult it becomes to measure the pressure accurately. The amount of substance available for making the measurement becomes so small that direct pressure measurement becomes impossible. The force per unit area (i.e., the pressure) is much too small for the direct mechanical methods normally used at higher pressures.

In high vacuum, rather than measuring the pressure, what is actually measured is the gas density or the number of particles per unit volume. This can then be related to pressure assuming that the temperature and gas species are known. What is often of interest for vacuum technologists is the condition of the surface (i.e., the quantity and type of the gas in an adsorbed phase of the surface rather than the density in the space). Those two cannot easily be related, so the surface preparation and analysis for some applications become important techniques depending on interactions of temperature, time, pressure, adsorption phenomena, surface structure, and so on.

The gases used to measure pressure in vacuum systems can be divided into two categories: those measuring total pressure, and gas analyzers that measure partial pressures of different gases in a mixture. The many types of gauges commonly used to measure total pressure can be categorized according to four operating principles: force measuring, momentum transfer, heat transfer, and electrical charge transfer (ionization). Partial pressure gauges are specially adapted mass spectrometers using gaseous ionization sources. The most common are magnetic sector instruments and radio-frequency "mass filters" called quadrupoles.

Units used to measure pressure are listed in Table 12.1. The various types of total pressure gauges and the normal pressure ranges for which they are employed are shown in Figure 12.1.

12.2 FORCE-MEASURING GAUGES

Pressure-responsive elements depending on the elastic deformation of a sensitive solid member with changes in gas pressure are (apart from liquid manometers) among the earliest devices for measuring gas pressures. Mechanical manometers measure total gas pressure in a more fundamental way than do gauges operating on the gas kinetic or ionization principles. Figure 12.2 shows typical forms of pressure-sensitive elements used in mechanical manometers.

Various forms of transducers are available to convert the mechanical movement by optical or electrical means, either to get a remote indication or control of pressure or to extend the range of the instrument. Practically any method suitable for measuring small displacements can be adapted, bearing in mind that the forces available are rather small. Mechanical manometers have been made using strain gauges, inductive methods including differential transformers, and capacitance methods.

For higher pressures, Bourdon tubes are widely used. These tubes are made in a large variety of elastic materials, depending on the particular application.

Table 12.1 Units of Pressure

1 atmospheric pressure	=	1.0133×10^5 N/m^2 or Pa
	=	1.0133×10^6 dyn/cm^2
	=	1.0133 bar
	=	14.696 psi
	=	760 torr
1 millitorr (10^{-3} torr)	=	1 micrometer (Hg)
1 microtorr	=	10^{-6} torr
1 nanotorr	=	10^{-9} torr
1 picotorr	=	10^{-12} torr

VACUUM GAUGES AND GAS ANALYZERS

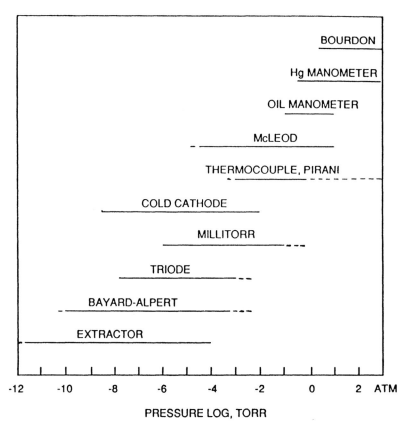

Figure 12.1 Pressure ranges of major vacuum gauges.

Glass or quartz tubes are also used for corrosive vapors and for bakeable conditions.

12.2.1 Bourdon Gauges

In the simplest types of Bourdon or capsule gauges, the element is enclosed in a glass-fronted leaktight case and the movement transmitted mechanically to a pointer visible through the glass. The inside of the case and the mechanism is exposed to the vacuum system and consequently, subject to damage by process conditions. Therefore, for many applications, a preferred arrangement is to use two separate capsule stacks, one of whcih is evacuated and sealed and acts as a reference, the other connected to the vacuum system, so that the linkage mechanism can be open to the atmosphere.

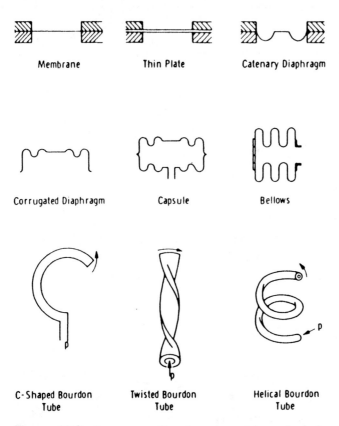

Figure 12.2 Pressure-sensitive elements used in mechanical manometers.

It should be noted that not all Bourdon gauges are alike. The small (3 to 4 cm diameter) gauges showing the range 0 to 15 psi typically do not produce meaningful readings below 2 psi (100 torr). Large instruments having a dial of 20 cm diameter with a pointer going twice around the periphery are usually marked 0 to 760 torr, but even they cannot be relied upon to be accurate below 10 torr.

10.2.2 Diaphragm Gauges

The status of the art of now widely used, commercially available diaphragm gauges has dramatically improved in recent years with the advent of electronic or, as they are sometimes known, capacitance manometers (Figure 12.3). Diaphragm deflections as small as 10^{-8} in. can be resolved, which in the most sensitive, thinnest membrane models, corresponds to an applied pressure of 1×10^{-5} torr. The useful measuring range of such an instrument extends from

VACUUM GAUGES AND GAS ANALYZERS

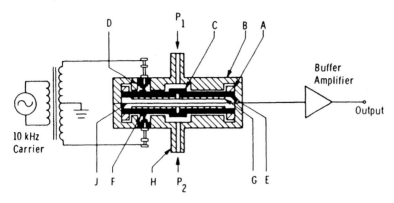

Figure 12.3 Capacitance manometer; (a) substrate support, (b) sealed housing, (c) screen, (d) jumper wire, (e) ceramic substrate, (f) film termination, (g) gold film, (h) pressure part, (j) diaphragm.

about 10^{-4} to 1 torr. In thicker membrane units, the measuring range extends to above 1 atm. Systems are available where analog and digital readouts are provided for continously monitoring or controlling pressure. In the range of 10^{-4} to 100 torr, these are the most accurate continuous-reading, total pressure-measuring instruments available. Typical accuracies range from 0.05% at 1 torr and above to 3% at 10^{-3} torr. Input pressure to the sensor is applied to a thin, radially tensioned metal membrane, welded to massive supporting rings, which deflect in approximate proportion to the applied pressure.

Stationary capacitor plates consisting of a metal film deposited on a ceramic insulating disk are mounted adjacent to the membrane, one on either side so that the diaphragm becomes the variable element of a three-terminal capacitive potentiometer. By arranging the element in a bridge circuit excited by a voltage of 10 kHz, the capacitance changes caused by the diaphragm motion unbalances the bridge and produces an output of 10 kHz of amplitude precisely proportional to pressure. The sensor output signal feeds into a metering console which accurately scales the pressure signal with a three- or four-decade range switch, provides a meter display of pressure value, and changes the 10 kHz into a dc voltage analog of pressure using phase-sensitive, synchronous, demodulation techniques.

Materials used in the sensors are metals and ceramics, fully compatible with the most critical vacuum applications. Some models can be baked and actually measure pressure at temperatures up to 300 to 400°C. Certain models require (and some users prefer) the provision of an external vacuum reference provided by a high-vacuum pump. Usually, valves are used to enable the connection of the inlet and the reference port to the vacuum reference, while at the same time blanking the inlet port so that the sensor can be zero-set. Other models have their own sealed-in vacuum reference. It is a good idea to use clean isolation valves with sensitive capacitance manometers so that the transducer can be kept under vacuum when the system is exposed to air. This will present unnecessary stress on the diaphragm, which may lead to small but significant zero shifts.

Diaphragm gauges, as other force measuring gauges, have the advantage of measuring the actual pressure independent of the gas composition. Also, a single-sided transducer can be made so that only the material of the membrane is exposed to the vacuum environment. This will minimize the possibility of contamination. Capacitance manometers made for high-vacuum applications usually have excellent accuracy. The most sensitive modules are often operated at a somewhat elevated temperature in order to make the transducer less dependent on ambient temperature changes. When molecular flow conditions are approached, the temperature difference between the transducer and the system may necessitate small corrections due to the errors associated with the thermal transpiration effect. When the diaphragm is operated in an ac mode, difficulties will be encountered with condensable vapors. A monomolecular film of low-vapor-pressure oil on the membrane will not cause serious trouble, but droplets of liquid will disturb the measurements.

In recent years, several new pressure transducers have been introduced that utilize microelectronic and microfabrication techniques developed by the semiconductor industry. Most of the available devices, however, do not have the wide range usually required in high-vacuum applications, and their internal and external construction may not be suitable for a high-vacuum environment.

12.2.3 McLeod Gauge

By combining a liquid manometer with a means of compressing a sample of gas, as is done in the McLeod gauge, the range over which pressure can be measured can be extended considerably below the practical limit of 10^{-3} torr for such manometers. The essential elements of a McLeod gauge consist of a glass bulb with a capillary tube extension on the top, a sidearm connecting to the vacuum system, and some means of raising and lowering the liquid level within the gauge (see Figure 12.4).When the mercury level in the gauge is lowered below branch point A, the bulb of volume V is connected to the system through sidearm B. The gas in the bulb is then at the same pressure as that in the system. When the mercury level is raised, the bulb is cut off from the sidearm and the sample of gas compressed into capillary C_1. Capillary C_2 is in parallel with a section of sidearm B and has the same bore as C_1, so that the surface tension or capillary effect is the same. The difference in level of the mercury in C_1 and C_2 is therefore due to the pressure difference resulting from

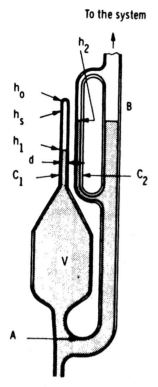

Figure 12.4 McLeod gauge (from C. M. Van Atta; Ref. 11).

compression of the sample from the large volume V into the small volume of C_1 above the mercury level.

The pressure of the compressed gas in the closed capillary is proportional to $(h_2 - h_1) + P_0$, in which h_1 and h_2 are the heights in millimeters of the mercury in capillaries C_1 and C_2, and P_0 is the pressure in the system still present in C_2. Since the compression ratio is typically very large, P_0 is negligible compared with $h_2 - h_1$. The pressure of the compressed sample of gas is thus just equal to $h_2 - h_1$ torr within the limit of reading error when h_1 and h_2 are measured in millimeters. If the system contains permanent gas only during the compression cycle, according to the general gas law,

$$PV = P'V'$$

in which P and P' are the pressures before and after compression, respectively, V is the volume of the bulb in liters (i.e., the volume of the closed portion of the gauge above the cutoff point A), and V', the volume of the closed capillary above the mercury level h_1, is given by

$$V' = \frac{(h_0 - h_1)a}{1000}$$

where h_0 is the effective height of the closed end of capillary C_1 and a is its cross-sectional area in square millimeters. Then

$$PV = \frac{(h_2 - h_1)(h_0 - h_1)a}{1000}$$

which holds for all values of h_2 and h_1 as the mercury is raised in the system. It is also evident that

$$(h_2 - h_1)(h_0 - h_1) = \frac{1000 PV}{a} = \text{constant}$$

From the above, the pressure P is given by

$$P = \frac{(h_2 - h_1)(h_0 - h_1)a}{1000 V}$$

Pressure readings determined by a McLeod gauge are valid only if the gases in system obey the general gas laws. If the system contains condensable vapors, a refrigerated vapor trap has to be used in the vacuum line connecting to the gauge. The trap, however, will also condense mercury vapor, so that a steady stream of mercury vapor will be set up flowing from the gauge to the trap. Because mercury vapor pressure at room temperature is relatively high (1 to 1.5 mtorr) compared to the level of desired measurements, collisions between the flowing mercury vapor and gas molecules from the vacuum chamber will set up a pressure difference and, consequently, an error.

McLeod gauges played an important role in vacuum measurements until recent times because they can be considered to be primary gauges, because the measurement is based on basic measurements of volumes and heights. Their popularity is much decreased lately because the measurement is not continuous, mercury is toxic, and the instrument in general if often fragile, large, and requires careful attention. Also, other vacuum gauge calibration methods have been developed based on volume expansion and orifice flow techniques.

12.3 HEAT TRANSFER GAUGES

Thermocouple and Pirani gauges use the variation of the thermal conudctivity of the gas with pressure as a means for measuring the pressure. In the past, these gauges were used to measure pressure in the range for which the mean free path was comparable to or greater than the dimension across which the flow of heat occurred—in other words, in the free molecular regime, but this range can be extended to atmospheric pressure by utilizing heat transfer by convection.

12.3.1 Thermocouple Gauge

Thermocouple gauges are generally useful in the range of pressure from 5000 to 1 mtorr, but this can be extended to atmospheric pressure. Some unreliability in the readings below 10^{-2} torr is caused by varying response due to varying gas composition, aging due to contamination or corrosion, and changes in sensitivity due to temperature variations. A typical thermocouple gauge is shown in Figure 12.5, and the gas species dependence on calibration of a typical thermocouple gauge is shown in Figure 12.6.

A thermocouple gauge that provides temperature compensation for the sensitivity loss with temperature is shown in Figure 12.5. A thermistor is connected in series with a measuring meter, and the sensitivity loss with temperature is compensated by the negative-temperature-coefficient thermistor, which increases the metering circuit sensitivity with temperature.

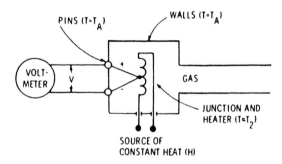

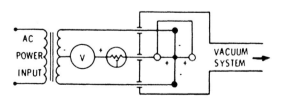

Figure 12.5 Thermocouple gauge.

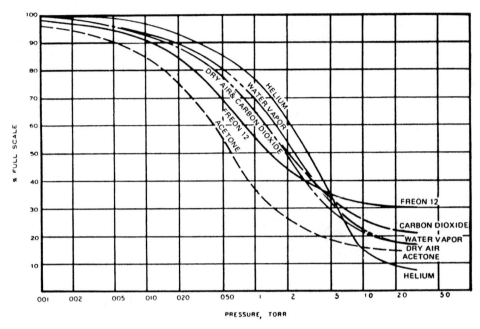

Figure 12.6 Sensitivity of thermocouple gauge with various gases.

A thermistor instead of a wire is used as the heated element in the thermistor gauge, which is also available commercially.

12.3.2 Pirani Gauge

In the Pirani gauge, the heat loss from the wire exposed in the vacuum system is measured electrically with a Wheatstone bridge network, which serves both to heat the wire and to measure its resistance. A schematic of the Pirani gauge and its electrical control circuit is shown in Figure 12.7. The gauge element and the compensating element are as nearly identical as possible. The gauge element is mounted in an envelope open to the system under measurement, and the compensating element is sealed off in a glass envelope at a pressure of less than 1 mtorr.

The bridge voltage is held constant. A change in the pressure in the open envelope causes a change in the temperature of the exposed filament which, in turn, changes the filament resistance and unbalances the bridge. The amount of unbalance is indicated on a microammeter calibrated in units of pressure. Conversion factors are used for various gases. The operating range is usually from 1 to 1×10^{-3} torr. Special gauge heads are also available to increase sensitivity in the low-pressure range to 10^{-4} torr.

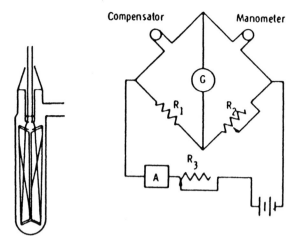

Figure 12.7 Pirani gauge.

The principal advantage of Pirani gauge is that the response curve of the bridge current as a function of the pressure may be essentially linear over a very wide pressure range by the proper choice of circuit constants.

12.3.3 Convection Effects in Thermal Gauges

Both Pirani and thermocouple gauges can be modified to detect convection currents that permit the extension of the measurement range to atmospheric pressure. These gauges essentially measure the rate of heat transfer from the heated element (such as wire) to the gas and subsequently to the gauge envelope, which is usually at room temperature. When the gas density is high and gas molecules can travel freely from the hot to the cold surface, the heat transfer rate is proportional to the number of molecules (hence to pressure). At higher pressures (above a few torr), the conductivity of the gas is not proportional to pressure because even though the number of molecules increases, they have to collide many times before they can cover the distance between the hot and cold elements.

A wire or a ribbon element can be heated either at constant voltage, or constant current, or constant power. All of these have some advantages and disadvantages, all will keep the wire at higher temperature when the pressure is low, and all will have a rather slow time response to pressure changes because of the heat capacity of the heated element, which must reach a new temperature. Perhaps the best method is to operate the wire at a constant temperature and measure the heat energy required to keep the temperature constant. This provides two basic advantages. When the wire is kept hot, at higher pressures convection currents become measurable and the range is easily extended from 1 torr to atmosphere and the time response is greatly improved (below 1 s for a change in pressure of over five orders of magnitude). Often there is no need to change the design of the sensor, only heating and read-out circuitry.

A word of caution is necessary regarding the use of thermal vacuum gauges with different gases. Such gauges must be calibrated for the gas for which they are used (see, for example, Figure 12.6). Apart from issues of accuracy, there exists the possibility of a serious hazard when gases of higher molecular weight (compared to air) are used with uncorrected gauges. Heavier gases have poorer heat transfer rates, so when the gauge (which normally is calibrated for air) is used with argon, the pressure may rise to several atmospheres while the gauge dial displays vacuum values. This can result in an explosion at any weak part of the vacuum system, including the gauge envelope.

In the convection mode, some thermal gauges may change their pressure indication, depending on whether they are mounted horizontally or vertically. This is because convection currents are influenced by the geometric configuration and may shift when the gauge body is rotated into a different position. Therefore, the heated wire should be placed as often as possible, into a symmetrical position relative to all gauge body surfaces.

Also, some gauges that utilize gas convection effects can be significantly affected by gas velocities in the vicinity of the internally located compensating resistors (Figure 12.7). This occurs during rapid evacuation of relatively small volumes by relatively large pumps. The convection heat transfer rate at higher pressure is proportional not only to pressure, but also to the velocity of the gas because during a rapid evacuation the condition shifts from natural convection to forced convection. For example, some gauges indicate a pressure increase (up to 1000 torr) instead of a decrease during rapid evacuation and it may take as long as 3 seconds to regain the correct pressure values. Generally, in transient conditions, the behavior of both pumps and gauges should be understood in order to obtain the optimum system design.

12.4 SPINNING ROTOR GAUGES

In recent years a version of a friction or viscosity gauge has been perfected to the point that it has become useful for gauge calibration work. The principle of operation consists of detecting the degree of deceleration of a magnetically levitated spinning metal sphere when it is placed in a rarefied gas environment. The arrangement is shown in Figure 12.8. A small (4 to 5 mm diameter) sphere, usually an ordinary steel ball bearing, is magnetically suspended in a small stainless steel tube that is welded to a usual high-vacuum flange at its open end. This short small tube and the metal sphere are the only objects exposed

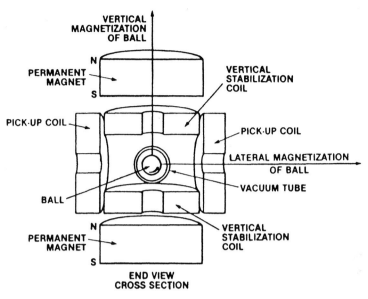

Figure 12.8 Spinning rotor gauge.

to vacuum. The ball is rotated by external electromagnetic coils at a frequency of 400 Hz; then the driving force is removed and the ball allowed to decelerate. The rate of deceleration is proportional to the number of gas molecules (or atoms) that come into contact with the ball. The change in angular velocity is proportional to gas pressure and inversely proportional to the mean molecular velocity.

The rotation of the ball induces a synchronous ac voltage in a pair of coils currounding the ball due to its magnetic moment. This signal is processed by high-speed digital counting circuitry to produce a pressure indication. The above indicates that the proportionality factor of this gauge depends on gas species and the length of the measurement depends on the pressure range. The behavior of the instrument regarding effects of temperature, loss of power, initial calibration, and long-term effects of surface changes of the ball, including contamination, have been studied extensively at the National Institute for Science and Technology (NIST) and a number of papers published in the vacuum journals. The accuracy of the gauge in the range 10^{-2} to 10^{-6} torr is claimed (by manufacturers) to be close to 1% of reading with satisfactory long-term repeatability. The gauge can be mounted on a metal gasket flange and baked in an ultrahigh-vacuum system (after removing the driving and detector coils).

12.5 IONIZATION GAUGES

For a high- and ultrahigh-vacuum region, a variety of ionization gauges are used. Ionization gauges are divided into two categories, cold cathode and hot cathode.

12.5.1 Cold Cathode Gauges

Ever since Penning's original publications in 1936 and 1937, cold cathode gauges have been very popular for use in industrial vacuum installations. This popularity stems from the fact that the simpler designs conveniently cover a pressure range extending from 10^{-2} to 10^{-7} torr while retaining relative simplicity and robustness compared with a thermionic ionization gauge of similar range. Among the advantages of cold cathode gauges over hot cathode gauges are the absence of a hot filament to cause thermal decomposition of gases and the greater overall sensitivity.

A number of disadvantages of the cold cathode gauge are well known, particularly the tendency for calibration discontinuities, which limits the accuracy, possible errors due to pumping action, and delay in starting the gauge at low pressure. With a better understanding of cold cathode gauge operation and a study of the characteristics of discharges in magnetic fields, improved designs have been produced that have given more reliable calibration. At the same time, pressure ranges of cold cathode gauges have been extended to much lower values.

Figure 12.9 shows a schematic of a commerically available (magnetron type) discharge gauge that covers a range of 10^{-2} to 10^{-9} torr. In the ultrahigh-vacuum range, a cold cathode gauge may not start after it is turned on (i.e., when the high voltage is applied) until some event initiates an electrical discharge. The triggering event may be due to an energetic particle or radiation. The period of waiting is roughly inversely proportional to the pressure and at 10^{-10} torr it can be several hours. The triggering of the gauge can be achieved by a momentary pressure pulse, by a flash of intense light, or by the presence of a radioactive material. The starting period can be reduced to less than 1 second in the 10^{-10} torr range by a small radioactive source, and to less than one minute in the 10^{-11} torr range by using, for example, an americium source of 0.9 microcurie at 4.10^{-11} torr (references 13 and 14).

12.5.2 Hot Cathode Ionization Gauges

Conventional hot cathode ionization gauges of the simple triode design have a limit of 10^{-8} torr as the lowest measurable pressure. For a high-vacuum range of 10^{-2} to 10^{-8} torr, this is adequate. Until 1950, this was the only type of hot cathode ionization gauge available (Figure 12.10).

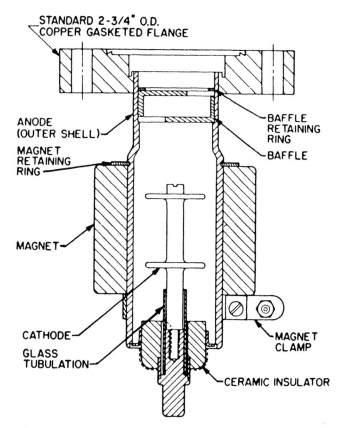

Figure 12.9 Cold cathode gauge (approximately actual size).

Bayard-Alpert Gauge

To extend the low-pressure limit below 10^{-8} torr, Bayard and Alpert replaced the cylindrical positive ion collector with a fine wire located inside the grid and the filament was placed outside the grid structure. The differences between a conventional ionization gauge and a Bayard-Alpert ionization gauge are shown in Figure 12.11. The conventional ionization gauge, cannot measure pressures below about 10^{-8} torr. When electrons from the hot-wire filament strike the grid, they generate x-rays that impinge on the ion collector and eject photoelectrons. This effect shows up as a contribution to the current flow in the ion collector's measuring circuit, since the x-ray generated current is about the same as the ion current at 10^{-8} torr. In the Bayard-Alpert ionization gauge the ion collector consists of a small wire which, because of its favorable position, collects most of the ions formed inside the grid. Due to the small size of the

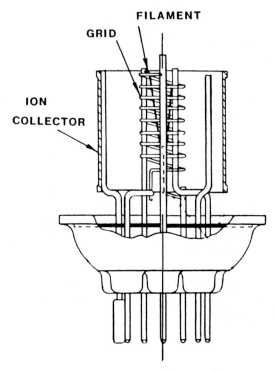

Figure 12.10 Ionization gauge of triode type.

wire, only a small fraction of the x-rays produced at the grid impinge on the collector.

The inverted ionization gauge geometry has three major advantages over a standard ion gauge. First, the small surface area of the ion collector minimizes the x-ray effect; second, the potential distribution between the grid and the ion collector is such that almost all the volume within the grid is available for efficient ionization; and third, the efficiency of ion collection is increased due to the central location of the collector.

The gauge shown in Figure 12.11 uses a helical grid that can be outgassed by resistive heating or by electron bombardment. In addition, a conductive coating of platinum is applied on the inside walls of the gauge tube, which is maintained at the filament potential. This prevents the surface charge buildup usually associated with glass surfaces, and provides repeatability and uniform sensitivity at the ultrahigh-vacuum range.

A Bayard-Alpert gauge of the nude type with a closed grid structure and ultrafine collector wire is shown in Figure 12.12. This gauge has an x-ray

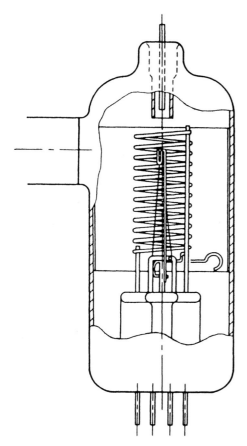

Figure 12.11 Bayard-Alpert ionization gauge (approximately actual size).

background of 2×10^{-11} torr. Outgassing of this gauge is done by electron bombardment. The sensitivity of this gauge, S, defined by

collector current = $S \times$ electron current $\times$ pressure

is 24 A/A · torr for nitrogen.

Millitorr Gauge

A hot cathode ionization gauge designed especially for high-pressure ranges is shown in Figure 12.13. The "Millitorr" gauge provides accurate pressure measurement from 1 to 10^{-6} torr. The Millitorr gauge operates similarly to the other hot cathode triode ion gauges, except that it produces electron paths of very short length and is operated at very low emission current. The thorium-coated

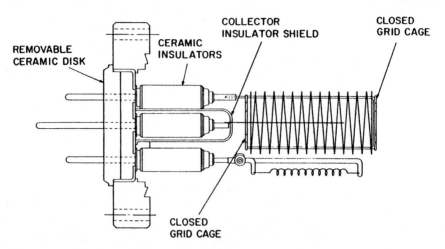

Figure 12.12 Nude ionization gauge

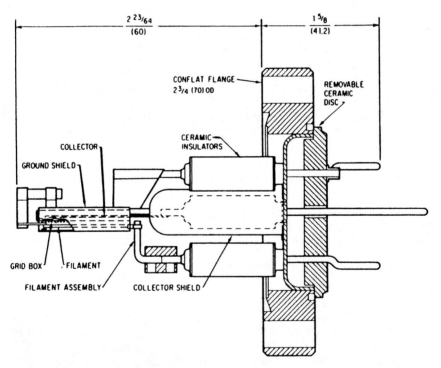

Figure 12.13 Millitorr ionization gauge.

iridium filament permits operation at higher pressures. Electrons from the filament travel through the grid lattice and across a short space in the grid box, and are collected immediately on the solid back of the grid box. The ions produced by the electron collision with the gas molecules are attracted to the collector, which indicates the pressure. The electron arrangement ensures linear measurement at high pressure by providing electron paths of uniform length throughout the operating range.

Extractor Gauge

Figure 12.14 shows the electrode arrangement in an extractor gauge, which is capable of measuring pressures of 10^{-12} torr. Ions formed within the grid are attracted toward the shield, which is at filament potential. Most of the ions pass through the hole in the shield and are focused onto the grounded ion collector by the action of the hemispherical ion reflector at grid potential. The x-ray limit of this gauge is reduced because the collector wire subtends a very small solid angle at the grid. The extractor gauge is normally operated with filament, bulb, and shield at +200 V, grid-to-filament voltage of 105 V, ion collector at ground, and an electron current of 2 mA or less. The sensitivity factor under these conditions is similar to that of a Bayard-Alpert gauge, and for the gauge shown, is 13 per torr for nitrogen.

The x-ray limit of the extractor gauge has been estimated to be about 3×10^{-13} torr, and the lowest pressure measured with the gauge is near $\times 10^{-12}$ torr. The extractor gauge is much less affected by electronically desorbable gases on the grid than is the Bayard-Alpert gauge. Since this is frequently a more serious limitation to the measurement of very low pressures than the x-ray limit, this represents a very distinct advantage of the extractor gauge.

Sensitivity of Ionization Gauges for Various Gases

The relative sensitivity of ionization gauges for different gases is determined primarily by gas ionization cross sections under electron impact. However, it is difficult to calculate absolute gauge sensitivities directly from the cross sections for two reasons:

1. The electron paths are not known in detail in most gauges.
2. The fraction of the ions produced that are collected is not known.

Measuring System as Sinks and Sources of Gas

Almost all parts of an ultrahigh-vacuum system pump and/or release gas to some extent. It is often essential to consider the measuring system as a dynamic combination of sinks and sources of gas. At very low pressures, it is important to assess the effect of these sinks and sources on the pressure being measured. The essential information required, if the observations are to be interpreted correctly, is the magnitude and location of the sinks and sources in each

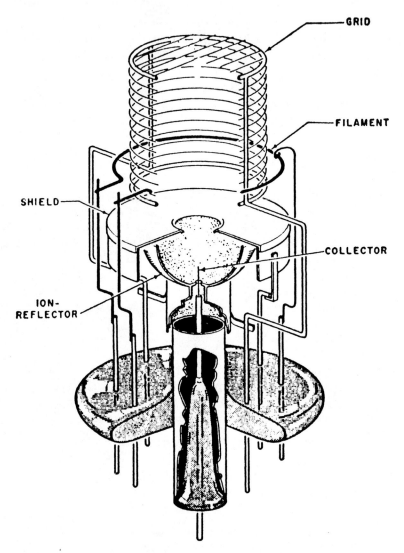

Figure 12.14 Extractor-type ionization gauge. (From Ref. 1.)

gauge and connecting tubes. This problem can be divided into three categories: pumping and reemission of gas in gauges; gas interactions at hot surfaces; and sinks and sources in connecting tubing and gauge electrodes.

The pumping speed of a gauge is dependent on the number of gas molecules present, the electron current, the grid accelerating potential, and the temperature of the gauge walls. Reemission of previously pumped gases, especially inert gases, constitutes a source of gas that increases the pressure reading. Reemission probabilities are of such a magnitude that a system cannot be pumped more than about four orders of magnitude in pressure without becoming limited by reemission unless the pumping surfaces are regenerated in some way. A corollary to this is that a gauge operated for 15 min at 10^{-5} torr cannot subsequently be used below 10^{-10} torr without additional outgassing.

Some chemically active gases such as H_2, H_2O, CO_2, and hydrocarbons can dissociate at the surface of a hot cathode of an ionization gauge. The dissociation products are usually highly reactive and may interact with other gases or surfaces to produce new gaseous products. Thus hot cathodes may produce considerable changes in the gas composition of an ultrahigh-vacuum system, causing errors in pressure measurement.

It has been observed that an ionization gauge inserted directly into a large chamber have a pressure reading higher by a factor of 10 than an identical gauge enclosed in an envelope and attached to the chamber through a tube. The first gauge is called "nude" and the second "tubulated." The readings of the gauge arose from the vapor from the vapor jet pumps, and the discrepancies in the two gauge readings arose primarily because of vapor adsorption in the tubulation. Subsequent experiments have shown that this effect extended to ultrahigh-vacuum range and also in systems free of oil vapors and water vapors. For example, a tubulated hot filament ionization gauge separated from a nude gauge of similar design by a length of glass tubing may read lower by a factor of 10 for CO. Smaller factors were found for H_2 and N_2, and both gauges indicate the same pressure for He. In general, the divergence depended on the pressure and on the temperature of the tubulation. This phenomenon may be attributed to adsorption and desorption of gas on the walls of the tubulation.

The choice of the most suitable type of gauge to be used for a particular experiment or system depends mainly on the following gauge characteristics:

1. Lowest measurable pressure. Can the gauge measure to a sufficiently low pressure for the particular experiment?
2. Gauge pumping speed. Is the gauge pumping speed sufficiently low to prevent serious errors in pressure measurement? This is particularly important in experiments where the system pumping speed is very low.
3. Effect of chemically active gases. When measuring pressure in systems with chemically active gases present it is important to reduce chemical effects at hot filaments and to minimize the effects of electronic desorption.
4. Sensitivity.
5. Mechanical ruggedness.
6. Interference by stray magnetic fields. In some experiments the stray fields from magnetic gauges cannot be tolerated.
7. Reemission of previously pumped gas.

Two factors that make total pressure gauges useful even when they approach a mass spectrometer in cost and complexity are:

1. The total pressure gauges are nearly always easier to outgas than a mass spectrometer.
2. The higher sensitivity of most total pressure gauges.

12.5.3 Gauge Calibration and Operation

Thirty years ago, vacuum gauges were generally calibrated against the McLeod gauge, which measures pressure in an absolute way, with the pressure in a vacuum system defined as the force exerted by the gas per unit area. Up to a pressure range of 10^{-4} torr, this was a reliable method, provided that necessary precautions were taken to read the pressure on the McLeod gauge.

Ionization gauges used in high- and ultrahigh-vacuum ranges can be calibrated indirectly using the following techniques:

1. The sensitivity variations of the gauge with pressure is measured over the range from UHV to about 10^{-3} torr. The gauge is then compared with a McLeod gauge at about 10^{-4} torr to obtain the absolute sensitivity.
2. Gas is allowed to flow from a reservoir at a measured high pressure through two or more constrictions of accurately measured conductance to high- or ultrahigh-vacuum pumps of known speed, creating calculable pressures in the intermediate volumes.
3. Gas at a known higher pressure and smaller volume is expanded into a larger measured volume. The new pressure is determined by the ratio of the volumes.

All three techniques rely on simple predictions of gas kinetic theory; namely, that in a static expansion, the pressure and the volume are inversely related and that the equilibrium rate of flow of gas through a conductance C is given in the molecular flow range by $Q = C \Delta P$, where ΔP is the pressure difference and C is a constant.

Calibration of high-vacuum gauges is usually made with air or nitrogen. Fortunately, the gauge sensitivity with water vapor is not very different from that with air so that for general evacuation of vacuum chamber, the "nitrogen equivalent" pressure readings produce sufficient accuracy. For other gases, corrections according to the sensitivity ratio can be used (Table 12.2 and reference 10).

When measurements are made using a different gas with an ionization gauge calibrated for nitrogen, the following procedure should be followed. First, the residual reading (the "ultimate pressure") before the experiments began should be subtracted, then the value obtained should be corrected according to the

Table 12.2 Sensitivity of Ionization Gauges for Various Gases Relative to Argon

	Triode and B-A	Cold cathode
Argon	1	1
Helium	0.13–0.17	0.14
Neon	0.22–0.26	
Krypton	1.33–1.53	
Xenon	2.0–2.22	
Air	0.81	
Nitrogen	0.53–0.75	0.59
Hydrogen	0.28–0.30	0.31
CO	0.48–0.82	
Oxygen	0.61–0.94	0.58
CO_2	0.9–1.16	0.76

sensitivity ratio. For example, for helium, if the system pressure before helium was introduced was 1×10^{-7} torr and then increased to 1.1×10^{-6} torr due to introduction of helium, the actual helium pressure will be

$$6(1.1 \times 10^{-6} - 1 \times 10^{-7}) = 6 \times 10^{-6} \text{ torr}$$

Sensitivity for helium is approximately six times lower than for nitrogen. The justification for subtracting the residual pressure, which exists prior to the experiment, can be ascertained by noting the meaning of the ultimate pressure in Eq. (4.17), Section 4.4.

The calibration of high- and ultrahigh-vacuum gauges is not done by a simple and straightforward procedure because the amount of matter available for measurement is extremely small. Normally, commercially available gauges are not calibrated.

Typically, the high-vacuum ionization gauges with external collectors (triodes) will have a variation of about 5% in sensitivity even if purchased from the same manufacturer. The ultrahigh-vacuum gauges with internal collectors may have a total variation of 10 to 20%. It is only in the last 15 years that well-established calibration procedures have been developed in standards laboratories in several countries that have a substantial vacuum industry (including the U.S. National Institute of Science and Technology). The calibration service for a gauge tube (transducer) and the associated controller will more than double the price of the instrument.

In industrial practice, one encounters two types of vacuum technologists. Some insist on recording pressure readings with three-place decimal accuracy. The digital meters are particularly seductive in this respect. Others insist that high-vacuum gauges are only "an order of magnitude" accurate. For industrial measurements the absolute accuracy is often less important than precision (i.e., the degree of repeatability from gauge to gauge and day to day).

It is advisable to maintain a few secondary standard gauges for intercomparisons that can be traced to a calibrated instrument. With careful work and maintenance, overall precision uncertainty of about 10% is achievable in the high-vacuum region. In the ultrahigh-vacuum region both accuracy and precision become more and more difficult. One of the reasons is that the measurements approach the ultimate intrinsic residual readings of the instrument.

It is, of course, most important that the gauge be used in the same manner in which it was calibrated. It is in this area where most engineering problems occur. The calibration is made in order to establish the basic sensitivity of the transducer (gauge tube). During calibration measurements, the residual base pressure is either subtracted or the measurements are made far removed from the base pressure, so as to make it negligible. However, often the instrument is used in practice to record the base pressure of the system, which is typically a measurement of the worst possible accuracy. To illustrate, consider Figure 12.15, in which several "identical" ionization gauges are compared to a better outgassed sixth gauge. The point is that the calibration is made where a definite sensitivity slope has been established.

If all these gauges were to be calibrated, their corrected readings would form a single line at the right side of the figure. However, many technicians and

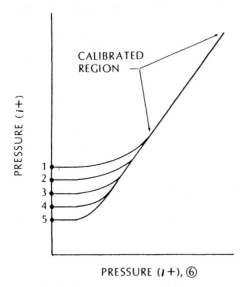

Figure 12.15 Comparison of six ionization gauges.

engineers use the gauge readings obtained at the residual condition (left side of the figure), where large deviations are common. There are many reasons for the existence of these large deviations. They depend on the degree of outgassing and contamination of the tube.

Consider several additional items that may help to develop an attitude of moderation in using high-vacuum gauges. Most high-vacuum gauges do not measure pressure. Rather, they measure particle density. The conversion is not certain when high-density gradients exist in the apparatus.

The gauge can affect the measurement system. In a typical ionization gauge at 10^{-6} torr there is 10^{-10}g of gas; the force exerted on the walls is about 10^{-6} g/cm^2; the number of molecules in the volume of the tube is about 10^{12}, but as many as 10^{17} may reside in a monolayer on the surface of the bulb, about 10^{13} on the surface of the filament, and 10^{15} on the grid. Obviously, adsorption and desorption possibilities exist. The gauge temperature can produce uncertainty of 5% due to absolute temperature effect on density alone, even disregarding outgassing effects.

Some ionization gauges have serious glass charging effects near 10^{-4} to 10^{-3} torr. The signal can shift by a factor of 2 or 3 unless a conductive layer is used to stabilize the surface potential. Even nude gauges are not free of environmental effects, and a second grid is sometimes used to provide a stable electric field. Figure 12.16 illustrates this effect, including the hysteresis effect, which may be a source of an additional puzzlement.

Figure 12.17 shows a bell-jar vacuum system with a typical arrangement of pumps, valve, and gauges. The expected deviation between gauges 1 and 2 should be near a factor of 2 because of some conductance limitations along the flow path between the chamber and the underside of the valve. However, un-

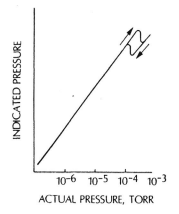

Figure 12.16 Shift in pressure indication due to charge effects on glass envelope.

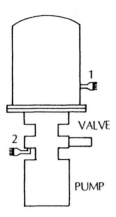

Figure 12.17 System with alternative gauge locations.

der transient conditions, during evacuation, the upper gauge may read 10 or 100 times higher pressure after the valve is opened. This is because the gauge has been exposed to atmosphere and has excessive quantities of gas on its surfaces, subsurface pores, and oxide layers. A period of time (few minutes to a few hours, depending on the pressure level) and degassing of the gauge are required before the expected factor of 2 or 3 ratio between the two gauges is established.

Another important consideration is the general location of the gauge in a system. The distribution of pressure (or density) of the gas in a system at high vacuum is rarely uniform. In the system shown in Figure 12.18, gauge 1 should read a higher pressure than gauge 2, because gauge 2 is downstream toward the pump, gauge 1 is affected by the additional outgassing (and possibly permeation) from the large O-ring in the door and because there is a large object

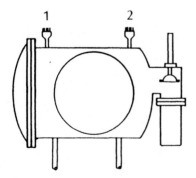

Figure 12.18 System with two gauges.

VACUUM GAUGES AND GAS ANALYZERS

placed in the chamber which creates an additional impedance between the two gauges.

Ionization gauge tubes in clean systems may last a long time (years). They may be turned off when not in use, but usually they remain lit as long as high vacuum exists in the system so that the pressure is monitored at all times. The effect of time on the sensitivity of the gauge is shown in Figure 12.19. It is seen to be reasonably good provided that the gauge is not abused or contaminated.

When measuring pressures below about 10^{-4} torr, it is important to establish an intimate connection between the gauge and the system and to reduce any unwarranted outgassing at the joint. Normally, unless there is some compelling reason to do so, extended tubes, valves, flexible connectors, and so on, must be eliminated.

For measurement above approximately 5×10^{-8} torr, a single O-ring in a specially designed compression seal is permissible. Below that pressure, a ConFlat flange with a metal gasket is mandatory. Otherwise, the gauge will read a pressure due to the outgassing in the connecting tube rather than the system pressure (see, e.g., Figure 6.18).

12.5.4 Degassing

All hot-wire ionization gauge instruments have a degassing feature. The degassing process is not an exact technique because the degree of degassing required to reduce the "parasitic" gas content of the gauge depends on its condition or,

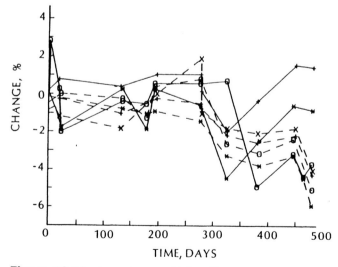

Figure 12.19 Long-time sensitivity effects. (From Ref. 9.)

in other words, its recent history of operation and exposure to higher pressures. Typically, it is not necessary to degas the gauge at pressures above 10^{-5} torr. In the 10^{-6} torr range, a few minutes of degassing are sufficient. Below 1×10^{-9} torr, an hour or more may be required.

Figure 12.20 is a schematic representation of the basic degassing process. Because the gauge tubulation represents a barrier of communication between the gauge and the system, very large differences can exist between them unless the gauge gas content is reduced. In a system that can reach high vacuum (10^{-7} torr) in a few minutes, starting from atmosphere, if the gauge has been exposed to the atmosphere, especially on a humid day, the gauge reading will lag the system pressure by 10 or more times. The gauge tubulation has, typically, a conductance of about 10 L/s. When the filament is lit, the desorbing gases may take a long time to leave the gauge. Even nude gauges do not have an infinite conductance, which may be an equivalent of 500 L/s, at best.

After degassing is completed, there are basically three possibilities. The pressure indication may proceed as in Figure 12.20. This is the idealized situation even though a doubt may remain whether the degassing was adequate. If the degassing step is repeated and pressure returns to the previous value, the reading is likely to be correct, provided that the measurement is not near the final residual reading of the instrument. The second possibility is that the pressure indication will be reduced after degassing but increase (after the gauge is cooled) to the level of the upper dashed curve in the graph. This implies that degassing was not necessary, especially if a second more prolonged attempt does not change the values. The third possibility, sometimes encountered at ultrahigh vacuum, is a slow gradual increase of pressure after a thorough de-

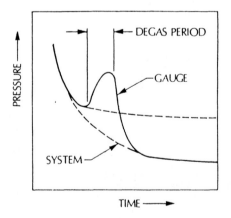

Figure 12.20 Degassing process.

gassing period. When the gauge is heated and then cooled, it will act as a pump until an equilibrium is established. At ultrahigh vacuum, equilibrium may be reached in only a few hours. If partially condensable vapors are present in the vacuum system, the equilibrium time may be days.

The use of ionization gauges requires a certain skill and experience in operation and interpretation of results. At extreme ultrahigh vacuum, under some conditions the gauge may be the principal source of gas in the system. But even at a higher pressure, the gauge can be a source of gas when the pumping speed is low. A simple example is shown in Table 12.3 for a system that has a speed of 3.5 to 4 L/s (provided by a trapped mechanical pump). The RGA values are total pressure measurements provided in the instrument. Both instruments are uncalibrated (i.e., as purchased and generally used in the laboratory for several months). Degassing at these pressures does not change the values (133 Pa = 1 torr). The question may be asked: What is the real pressure? It would appear that nearly half the pressure indication is due to the gauge itself.

Similar situations occur even in the presence of very high pumping speeds when the pressures are in the ultrahigh-vacuum region. It is a good educational (if not sobering) experience to observe the indicated pressures of two identical gauges in the same vacuum system. They may show not only large differences but also contribute a significant quantity of gas to the system. In one such example, in a system with a 15,000 L/s nominal pumping speed, the pressure of 4×10^{-10} torr dropped to 2×10^{-10} torr when either of the gauges was turned off. In this case the B-A gauges were tubulated and unfortunately the partial pressure changes were not recorded. Therefore, even after the proper operation and degassing treatment, ordinary gauges (which do not have means of subtracting their own residual contribution to the indicated pressure) cannot be expected to provide meaningful readings near their ultimate pressure measuring range.

Table 12.3 Interaction of Two Pressure Measuring Instruments

	Pressure indication	
Condition	Ion gauge (Pa)	RGA (Pa)
Ion gauge off	—	1.5×10^{-3}
Both on	1.6×10^{-3}	2.3×10^{-3}
RGA off	1.5×10^{-3}	—
Both on	2.3×10^{-3}	1.7×10^{-3}
Ion gauge off	—	1.3×10^{-3}

12.6 NEW TRENDS OF GAUGES AND SENSORS

One primary requirement of any instrument is to not disturb the environment that the instrument is intended to measure. At very low pressures, in the high- and ultrahigh-vacuum region, this basic requirement is very difficult to achieve. For example, it has been said that both hot-wire and cold-cathode gauges are "chemical factories" because of various reactions and effects of filaments at high temperature and high voltages present in those gauges. At very high- and ultrahigh-vacuum conditions, a gauge can be a significant (and sometime even a predominant) source of gas not only in its own envelope but in the entire vacuum chamber. It can also provide significant pumping action due to occurrences similar to those associated with the performance of ion-getter pumps.

Unfortunately, despite the great advances in vacuum technology during the last 40 years, the transducers used for pressure measurements have not changed very much. In the same period, the development of microelectronic devices have permitted enormous improvement in electric and electronic controllers; in low-level signal detection; stable, reliable breadth of function signal displays; and computer interfaces. It is generally desirable to reduce the physical size of the transducer and thereby limit the degree of interaction with the medium being measured. The smaller size usually decreases the sensitivity, however, and it may also contaminate more rapidly. Although much can be done in this regard, miniature transducers are presently being used only at higher pressures. Some examples of recently-introduced instruments are discussed in the next section.

12.6.1 Universal Controllers

Vacuum gauge controllers of relatively small size have been introduced that can operate many transducers simultaneously such as capacitance manometers, thermocouple gauges, pirani gauges, and cold- and hot-cathode ionization gauges. Ten independent sensors can be operated simultaneously by a controller of the same size that, a few years ago, could operate only one.

12.6.2 Extended Range Thermal Gauges

Both pirani- and thermocouple-type gauges have been extended to handle atmospheric pressure. An example of these gauges is shown in Figure 12.21. It can be seen that the scale of the instrument is not uniform (i.e., the length of the band scale used for each decade of pressure has large variations, with the section between 10 and 100 torr being the shortest). The reason for this can be appreciated by reviewing the heat transfer relationship shown in Figure 2.1 and realizing that the curve in the region between conductance and convention tends to be nearly horizontal. Therefore, even with the most accurate electronic circuitry and carefully selected compensating resistors in the sensors, it is dif-

Figure 12.21 An example of a vacuum gauge based on gas conduction/convection effects.

ficult to maintain absolute accuracy of better than ±20% across the entire range of the instrument. The references to accuracy and precision in technical literature tend to be comments about the quality of the electronics rather than the intrinsic properties of the sensor itself.

12.6.3 Combined Sensors and Controls

In many cases, utilizing modern electronic devices, it is possible to attach the signal conditioning and read-out circuitry directly to the sensor. An example of such an instrument is shown in Figure 12.22. (Note: the connector on the left is for remote operation.) For more complex instruments, only the initial signal conditioning and preamplifier circuitry may be attached to the sensor, which greatly reduces the problems with connecting cables.

12.6.4 Combined Sensors

It is often convenience to place two different sensing devices in one envelope in order to widen the sensing range of the instrument. For example, one gauge of the same size as shown in Figure 12.22 includes a diaphragm sensor and a pirani gauge, thus giving a pressure range from 1 mtorr to 1,500 torr. This

Figure 12.22 An example of a vacuum gauge with the controller attached to the sensor.

combination eliminates the uncertainty of gas composition, associated with convection gauges at the higher pressures. The diaphragm gauge measures a force at higher pressures that is independent of the type of gas (see Figure 12.23).

12.7 MASS SPECTROMETERS OR PARTIAL PRESSURE GAUGES

The basic objective in partial pressure measurements is to provide a signal that can be accurately related to the number density of a particular species of molecule in that region of a vacuum system where an experiment is being performed. Operation of mass spectrometers can be divided into four functional steps:

1. Ions are created from the gas, usually by electron impact.
2. The ions are accelerated to certain kinetic energies in a chosen direction and focused onto the entrance aperture of an analyzer.
3. The ions entering the analyzer are subject to an arrangement of electric and/ or magnetic fields which separate them on the basis of their charge-to-mass ratio.

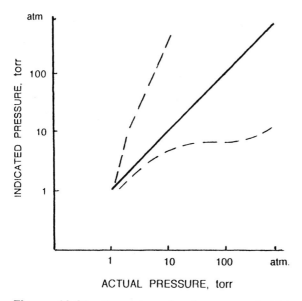

Figure 12.23 Comparison of performance at the higher pressure scales. Solid line, diaphragm gauge; dashed lines, convection gauge calibrated for air but used with helium (upper curve) and argon (lower curve).

4. The separated ions are detected upon arrival at a collector. Ions of a selected charge-to-mass ratio are brought to the collector by adjustment of the ion-accelerating potential and/or the analyzer fields.

General characteristics of mass spectrometer operation are:

1. *Sensitivity.* The sensitivity in amperes of ion current per unit of pressure should be high, as high or higher than in ionization gauges.
2. *Resolving power.* The resolving power must be adequate to separate clearly the gases that are likely to occur in the system either as residual gases or those deliberately introduced.
3. *Outgassing rate.* The outgassing rate of the mass spectrometer when operating must be very low.
4. *Peak shape.* The peak shape, that is, the profile of the collected ion current versus the independent variable used to scan the spectrum, should be as nearly flat-topped as possible, if quantitative monitoring of individual mass peaks is required.
5. *Magnetic requirements.* The presence of a magnet or its size, weight, and stray magnetic field can in some cases be a decisive factor in the choice of an instrument for a specific experiment.
6. *Background currents.* Because collected ion currents in mass spectrometers

are small, it is important that spurious currents appearing at the collector be minimal.
7. *Scanning speed.* If the experiment requires the rapid comparison of different signals, the scanning speed may be important. The scanning speed affects items 1, 2, and 4 above.

12.7.1 90-Degree Magnetic Sector Spectrometer

This instrument can be used either with a permanent magnet and voltage scanning or with a constant ion energy and a scanning electromagnet. Both the ion source and the electron multiplier collector are mechanically separate from the analyzer region (Figure 12.24).

The magnetic mass spectrometer utilizes the fact that charged particles follow a curved path in an electromagnetic field. This dispersion is spatial and the radius of curvature is given by

$$r = \frac{mv}{Be}$$

where B is the magnetic field intensity, m is mass, e is the charge, and v is the ion velocity. The ion velocity is given by

$$v = \sqrt{\frac{2eV}{m}}$$

where V is the ion accelerating potential. Substitution of this expression for velocity gives

$$r = \frac{1}{B}\sqrt{\frac{2Vm}{e}}$$

The radius of curvature is thus dependent on the initial ion energy.

Magnetic spectrometers vary from low cost, low resolution to high cost, high resolution units. Significant characteristics of the low-resolution magnetic mass spectrometer include easily understandable operation, slow scan speed, and a basically nonlinear mass display. For high resolutions, above several hundred amu (atomic mass units), magnetic instruments are often used.

12.7.2 Quadrupole Spectrometer

The absence of an analyzer magnet is one of the most important characteristics of spectrometers of this type, Figures 12.25 and 12.26). Ion source and collector are easily separated from the analyzer, and thus addition of an electron multiplier is not difficult. Transmission probability of the ions can be varied

VACUUM GAUGES AND GAS ANALYZERS 465

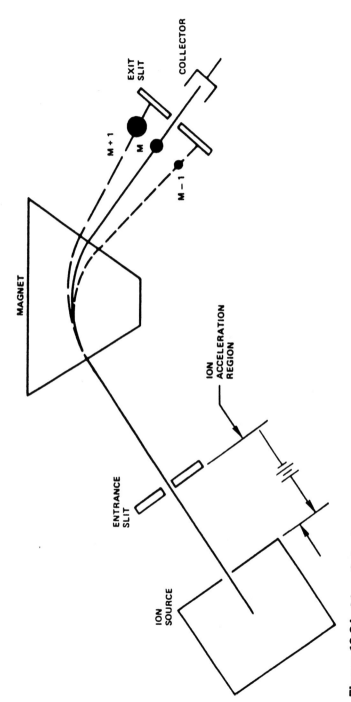

Figure 12.24 Magnetic bend mass spectrometer.

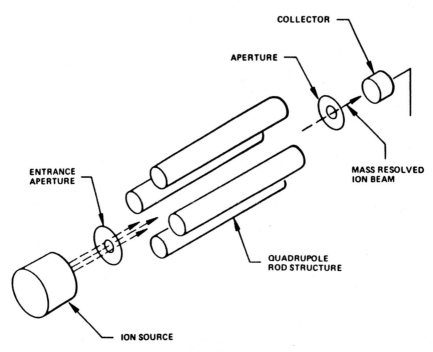

Figure 12.25 Quadrupole mass spectrometer.

and increased at the expense of resolving power. The ion source and collector are connected in a straight-line path. The transmission of x-rays and light from source to collector can thus cause spurious (pressure-independent) currents. Ions entering the analyzer portion from the ionizer are subject to time-dependent

Figure 12.26 An RGA instrument with attached preamplifier section (Stanford Research Systems).

fields which drive these ions into oscillatory motion perpendicular to the quadrupole axis. If the oscillatory motion becomes too large, the ions strike the metal rods and are neutralized.

Equal but opposite radio-frequency (RF) + dc voltages are applied to the rod pairs. The (+) pair has a positive dc voltage with a superimposed *RF* voltage applied. The (-) pair has a negative dc voltage equal in magnitude with the (+) pair and a superimposed RF voltage 180° out of phase with the *RF* of the (+) pair.

The positive pair of rods tends to neutralize all ions above a certain mass and the negative rods tend to neutralize all ions below a certain mass. The positive pair acts as a low-mass pass filter, and the negative pair acts as a high-mass pass filter. By allowing a slight overlap in the passbands, the analyzer can be made to pass ions only in a narrow mass range. The conditions for critical tuning are

$$V_{ac} = \frac{0.703 m r_0^2 w^2}{4e}$$

and

$$V_{dc} = \frac{0.237 m r_0^2 w^2}{8e}$$

where r_0 is the inscribed radius of the rod structure, w the *RF* frequency, V_{ac} the peak *RF* voltage, and V_{dc} the dc voltage.

The advantages of the quadrupole are that it has a high scan speed, a linear scan, and a mass range on the order of 1000. On the other hand, the broad variations in tuning capability, from uniform peak width to uniform resolution, lead many users to classify it as a nonquantitative instrument. This tunability, however, is frequently beneficial to the researcher, whose needs can be more fully satisfied with variable tunings. The quadrupole can also be used to advantage in specific process applications to optimize sensitivity. The quadrupole ionizer is the simplest type of ionizer. The only energy requirement for the ions is that the energy must be low enough so that the ions spend several cycles in the *RF* field.

Potentially, quadrupole mass spectrometers have advantages for the high mass region of the spectrum. To a first approximation, the peak width in mass units, rather than the resolving power, is independent of mass number.

As an example of a scan, Figure 12.27 shows a mass spectrum obtained in a liquid-nitrogen trapped vapor jet pump system. Peak 19 was unusually high in this spectrometer due to emission from ceramic material used in the tube. Peak 16 was also unusually high. Normally, peak 18 (water vapor) is higher. Peaks 50, 51, 52, 77 and 78 indicate the phenyl group associated with DC-704

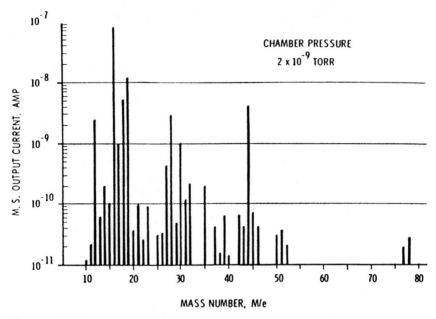

Figure 12.27 Mass spectrum in a vapor jet pump system (with an LN trap).

and DC-705 pumping fluids. Figure 12.28 shows a mass spectrum of a baked ultrahigh-vacuum system with a sputter-ion pump. The most prominent peak is hydrogen.

The interpretation of m/e peak distribution obtained with an RGA instrument is not always simple. It is necessary to become familiar with typical peak patterns produced by certain gases and gas mixtures. RGA instruments are more complex than total pressure gauges in mechanical construction and in the information they display. Molecules can be doubly ionized (for example, peak 20 together with peak 40 for argon). They can sometime's combine with only one electron missing (for example, peak 36 together with the typical group of peaks at 18 for water). Molecules can fractionate in the process of ionization, producing well-known but not always easily recognizable patterns when mixed with others. It is useful to monitor the background pattern of a given vacuum system and then look for the differences in peak height produced after a certain time of operation (after gases are introduced into the vacuum chamber in the course of the operation). Despite these complications, partial pressure measurements provide a wealth of information about the state and performance of the vacuum system. They are becoming indispensable for monitoring background gases and controlling gas mixtures deliberately introduced into the vacuum system as required in some operation processes.

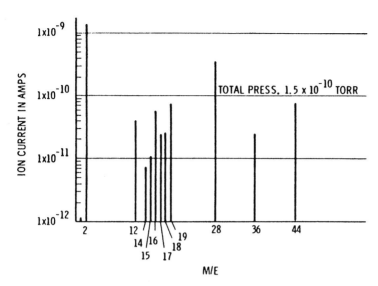

Figure 12.28 Mass spectrum in an ion pump system.

Compared to total pressure ionization gauges, the size and mechanical complexity of mass spectrometers make it rather difficult to fulfil item 3 of the desirable characteristics listed in the beginning of this section. A recent addition to the art of quadrupole mass spectrometer design has been introduced (reference 19), which shows promise due to a substantial reduction of size and the amount of material exposed to the vacuum system. The design consists of 16 small parallel rods forming 9 parallel quadrupole beam tunnels, thus providing adequate sensitivity. The small size (no larger than a typical ionization gauge) permits operation at higher pressures (up to 10 mtorr), which is useful when measuring parts per million content in a mixture.

REFERENCES

1. P. A. Redhead, J. P. Hobson, and E. V. Kornelsen, *The Physical Basis of Ultrahigh Vacuum*, Chapman & Hall, Ltd., London, 1968. Also AIP/AVS reprint, 1993.
2. A. Berman, *Total Pressure Measurement in Vacuum Technology*, Academic Press, INc., Orlando, Fla., 1985.
3. R. K. Fitch, *Vacuum*, *37*(8/9), 637–641 (1987).
4. P. Nash, *Vacuum*, *37*(8/9), 643–649 (1987).
5. W. Steckelmacher, *Vacuum*, *37*(8/9), 651–657 (1987).
6. J. H. Batey, *Vacuum*, *37*(8/9), 659–668 (1987).
7. F. M. Mao and J. H. Leck, *Vacuum*, *37*(8/9), 669–675 (1987).

8. M. J. Drinkwine and D. Lichtman, *Partial Pressure Analyzers and Analysis*, American Vacuum Society Monograph, 1980.
9. S. D. Wood and C. R. Tilford, *J. Vac. Sci. Technol.*, *A3*(3), 542 (1985).
10. R. L. Summers, *NASA Tech. Note*, *TND-5285*, June, NASA, Lewis Res. Center, Washington, DC (1969).
11. C. M. Van Atta, *Vacuum Science and Engineering*, McGraw-Hill, New York, 1965.
12. G. L. Weissler and R. W. Carlson, *Vacuum Physics and Technology*, Academic Press, New York, 1979.
13. B. R. F. Kendall and E. Drubetsky, *J. Vac. Sci. Technol.*, *A14*(3), 1292 (1996).
14. K. M. Welch et al., *J. Vac. Sci. Technol.*, *A14*(3), 128 (1996).
15. C. L. Owens, *J. Vac. Sci. Technol.*, 2, 5, 104 (1966).
16. P. D. Levine and J. R. Sweda, *J. Vac. Sci. Tecnol.*, *A14*(3), 1297 (1996).
17. C. R. Tilford et al., *J. Vac. Sci. Technol.*, *A13*(2), 485 (1995).
18. A. R. Filipelli and P. J. Abbot, *J. Vac. Sci. Technol.*, *A13*(5) 2582 (1995).
19. R. J. Ferran and S. Boumsellek, *J. Vac. Sci. Technol.*, *A14*(3), 1258 (1996).
20. P. H. Dawson, *Quadrupole Mass Spectrometry and Its Applications*, Elsevier, Amsterdam, 1976.

13
Leak Detection

13.1 INTRODUCTION

Like many technical disciplines, leak detection has experienced a rapid growth since World War II. The development of the electronics industry and increased activities in space exploration stimulated the development of new equipment and processes requiring controlled environments that had to be free of significant leakage. Today, many areas of science and manufacturing enjoy the increased capability for locating leaks in both evacuated and gas-filled containers. Detection of leaks using helium as a trace gas has become a useful tool in a wide range of industries. Analysis of gas with a mass spectrometer has been known since 1920, but it was not until World War II that the concept was adapted as a means of locating leaks.

Development of nuclear devices required utmost integrity of gaseous diffusion processing systems, a tightness far beyond all existing means of measurement or detection. As in other phases of the project, the uniqueness of the requirement demanded a new and different technique for identifying and measuring leakage paths smaller than previously known to exist.

The technique and tracer gas selected were tailor-made for this stringent requirement. Helium was selected as the tracer primarily because, as the lightest inert gas, it penetrates small leaks readily. Helium is also nontoxic, nonhazardous, nondestructive, plentiful, relatively inexpensive, and is present in the atmosphere only in minute quantities (5 ppm). These features enable it to go through small passages without affecting parts or processes, and may be detected easily, reliably, and economically.

The mass spectrometer leak detection technique depends upon separation of helium from other gases in a vacuum. It is accomplished by imparting an electrical charge to a sample quantity of gas, moving the sample through a magnetic field, and collecting the helium ion current. The current is used to drive a meter, actuate an alarm, or illuminate a display. Ionization, separation, and collection take place within the spectrometer tube, which is the heart of the instrument.

Escalating requirements for electronic tubes and other devices with critical operating conditions increased the need for the sensitive and reliable leak detection provided by the helium mass spectrometer. Growth was accelerated in the late 1950s by the space exploration program. Nearly in everything in space vehicles, from hermetically sealed electronic components to fuel tanks and transfer lines and valves, had to be leak free or nearly so. The commercial market for appliances and other consumer goods, spurred on by the developments of the space age, began to expand in the 1960s. Helium leak detectors appeared on production floors testing beverage can ends, pressure transducers, torque converters, and the like. In some applications, the units were automated to the point where the need for operators was virtually eliminated.

By 1970, a number of new developments had given the helium mass spectrometer leak detector far more sensitivity than was needed for most requirements. Hence, in subsequent designs, sensitivity was traded for speed. Designs were simplified and cycles automated. Helium leak detectors became a vital industrial tool capable of being integrated into computer and microprocessor-controlled production lines. Most of the judgment and guesswork previously required by the operator had been replaced by automation.

Improvements to simplify manual operation, reduce maintenance, and increase productivity of the helium leak detector have made it possible to expand its usage and acceptance to areas not previously considered. Food, pharmaceutical, medical, and diagnostic equipment industries are beginning to utilize helium leak detection techniques as the appliance, automotive, chemical, and similar industries have been doing in the past 25 years.

Before examining the principles and techniques of the helium mass spectrometer method, it is desirable to have an understanding of the fundamentals of leak detection. An introduction to the nature of leaks and the special terminology used in leak measurement are presented in the next section, and various conventional leak detection methods are surveyed.

13.1.1 Leak Detection

A leak may be defined as an unintended crack, hole, or porosity in a containing wall that allows the admission or escape of liquid or gas. The basic functions of leak detection are the location and measurement of leaks in sealed products, which must contain or exclude fluids. For the majority of products, a leak test procedure is a quality control step to assure device or system integrity, and is a one-time, preferably nondestructive test.

Leak detectors, the instruments or systems designed to locate and/or measure leaks, range in complexity from a tank of water (in which bubbles from a leak can be viewed) to highly sophisticated systems using radioactive tracer gases and gas reclaiming systems. The full range of approaches is detailed later. The basic considerations in choosing a leak detector are sensitivity, reliability, and ease of use.

13.1.2 The Need for Leak Detection

Even with today's complex technology, it is, for all practical purposes, impossible to manufacture a sealed enclosure or system that can be guaranteed to be leakproof without first being tested. The fundamental question in leak detection is: What is the maximum acceptable leak rate consistent with reasonable performance life of the product?

Anyone who manufacturers or uses closed vessels needs leak detection. A partial list of typical users includes:

Any industrial pressure vessel manufacturer
Manufacturers using tubular elements, such as refrigeration equipment
Industrial aerosol container manufacturers
Beverage canning industry
Hermetically sealed instrument manufacturers
Any manufacturer using bellows in products
Vacuum chamber manufacturers
General R&D laboratory users
Hermetically sealed electronic component manufacturers producing, typically:
    relays, connectors, quartz crystals, Reed switches, etc.
Vacuum tube manufacturers

What type of leakage should these products avoid? Some examples are given below.

Sealed packed foods	Oxygen, water vapor, bacteria
Semiconductor IC packages	Oxygen, water vapor
Pharmaceuticals	Bacteria
Cardiac pacemakers, other implants	Corrosive body fluids
Chemical process systems	Oxygen, water vapor
Vacuum tubes—all kinds	Air (degradation of vacuum)
Vacuum process equipment	Air, oxygen, other gases
Watches (conventional or digital)	Water, vapor or liquid

Sealed temperature and pressure sensors — Air (degradation of vacuum)
Cryogenic storage and transport units — Air (degradation of vacuum insulation)

Escape of Materials Through Leaks

Refrigeration and air-conditioning systems — Loss of refrigerant
Beer and beverage cans — Loss of carbonation
Shock absorbers — Loss of oil or compressed gas
Hydraulic systems, torque converters — Loss of operating fluid
Electrical power transmission — Loss of insulating gas
Aerosol spray cans — Loss of propellant

13.1.3 The Source of Leaks

Leaks in newly manufactured products are most commonly caused by imperfect joints or seals by which various parts are assembled to form the finished product. Typically, these include welds, brazed joints, soldered joints, glass-to-metal seals, ceramic-to-metal seals, O-rings and other gaskets, compound-sealed joints, and others.

Leaks caused by imperfect or too-thin sections of the containing wall represent another class of leaks. For example, in a plastic bottle, a particular area may leak. Microscopic examination may show many minuscule "cracks" rather than one or two holes, Similarly, in ring pull-tab can tops, if the score mark is too deep, the remaining metal may be too thin to prevent gas diffusion through numerous microscopic "cracks." Still another class of leaks consists not of holes or cracks in the usual sense. Instead, the molecular structure of the containing wall itself is arrayed in such a way as to permit gas diffusion or permeation through the wall.

A different type of leak, known as a virtual leak, is not really a leak but is the semblance of a leak due to the evolution of gas or vapor within a part of a system. Virtual leaks can be caused by improper welds, threaded connections, or press-fit joints. This type of leak often causes problems in vacuum systems.

13.2 SIZES OF LEAKS AND UNITS OF MEASUREMENT

It is important to avoid even extremely small leaks when constructing high- and ultrahigh-vacuum devices or process chambers. Detection, sealing, and avoidance of small leaks is also very important in production of hermetically sealed devices. Two examples are given below are to illustrate the quantitative effects of the presence of small leaks in vacuum systems or devices. The first regards a vacuum system in which a continuous action of a pump is available. To achieve the desired lowest possible residual pressure in the system, three different limitations must be considered: limitation of the pump itself, internal surface outgassing, and leaks. It is convenient to measure the outgassing quantity and the rate of leakage in units of throughput, Pa · m^3/s (or torr · L/s). To recall, when temperature and gas species are known, throughput units can easily be converted into mass flow units (g/s) or number of molecules or moles per second. Consider available total net pumping speed in a vacuum chamber to be 100 L/s and assume the presence of a leak of 10^{-7} atm · cm^3/s. Note that atmospheric or standard cm^3/s is not a volume flow unit, but throughput, because reference is made to a pressure (and a temperature) at which the flow is measured. If the temperature and the gas species are known, the pressure term in this case can be directly converted to density and hence to mass flow. Assuming that the outgassing rate in our system is negligible compared to the leak and the pump is not near its limit, the pressure in the system will be $p = Q/S$.

$$p = \frac{10^{-7} \text{atm} \cdot \text{cm}^3}{s(100 \text{ L/s})} \left(\frac{\text{L}}{1000 \text{ cm}^3}\right)\left(\frac{760 \text{ torr}}{\text{atm}}\right) = 7.6 \times 10^{-10} \text{ torr} \qquad (13.1)$$

Using a different set of units with an equivalent leak of 1.0×10^{-8} Pa · m^3/s, we obtain

$$p = \frac{10^{-8} \text{Pa} \cdot \text{m}^3/\text{s}(1000 \text{ L})}{(100 \text{ L/s})(\text{m}^3)} = 10^{-7} \text{ Pa} \qquad (13.2)$$

This simple illustration can provide a quick assessment of required leak detection sensitivity. It is evident that when the system has a high pumping speed (1000 L/s or more) and the best required vacuum is let us say, 10^{-7} torr, one or two leaks in the range of 10^{-7} atm · cm^3/s will be of little importance.

For the second example, consider a small evacuated and sealed device such as an electrometer tube that has to function at a pressure below 10^{-5} torr. Using the same leak rate as in the first example, we can assume that it remains constant because the pressure difference between the external atmosphere and the vacuum inside remains practically constant. In this case the rate of pressure increase will depend on the volume, assuming no strong adsorbing surfaces are present. If the volume is 2 cm^3, the time to reach 10^{-5} torr will be $t = V(\Delta P)/Q$.

$$t = \frac{2 \text{ cm}^3 \cdot 10^{-5} \text{torr} \cdot \text{L}}{(7.6 \times 10^{-8} \text{torr} \cdot \text{L/s})(1000 \text{ cm}^3)} = 0.26 \text{ s} \qquad (13.3)$$

Obviously, this leak rate is totally unacceptable. Even a leak rate of 10^{-10} atm · cm^3/s will be too high, producing a life of only about 7 h. Thus the art of sealing small devices that must remain at high vacuum for a number of years challenges the art of most sensitive leak detection methods.

There exist many coarser leak detection methods which are described briefly below. However, in high-vacuum industry the most commonly employed method is based on mass spectrometry using helium as a tracer gas. The sensitivity of this method would permit the detection of a leak involving a 10% reduction in the pressure of an automobile tire over 1000 years. Another tangible example of the sensitivity of such instruments is that they can easily detect helium emanating through the skin (near the blood vessels at the wrist) if a person a few seconds before has ingested about 1 cm^3 of helium through the lungs in the process of normal breathing.

13.2.1 Leak Measurement

Leaks are measured by the rate at which a gas or liquid flows into or out of the leaks, under certain conditions of temperature and pressure. Leak rates can be defined in two ways:

1. In terms of the application, such as:
   a. 1 oz of refrigerant R-12 in 2 years at 70 psi (28 g at 575 kPa), or
   b. 65 cm^3 of oil per year at 0 psi and 140°F (101 kPa and 60°C), or
   c. 10 cm^3 of CO_2 in 3 months at 60 psi (at 508 kPa)
2. In terms of the leak detection method used, such as:
   a. 2 bubbles/s (1/8 in. or 0.31 cm diameter) when pressurized to 40 psi or 372 kPa (bubble method)
   b. 1.8×10^{-7} std cm^3/s of helium at 1 atm (helium method)
   c. 2 psi pressure decay in 5 min at 60 psi or 508 kPa (pressure decay method)

While each of these examples is a legitimate description of leak rate for a specific test method, the general accepted unit of leak rate for leak detection is std cm^3/s, because it contains the actual units of flow rate, namely mass/time. The term "std cm^3/s" is an abbreviated form of "cubic centimeters of gas (at standard temperature and pressure) per second." The term "atm cm^3/s" is also used. This term describes the same volume flow rate, but defines it at whatever atmospheric pressure and ambient temperature prevail at the time. In leak detection, it is considered to be virtually interchangeable with std cm^3/s (except at locations substantially above sea level).

In vacuum work, where pumping speeds are measured in liters per second and pressures in torr, the term "torr · L/s" is widely used. This term is also used to express vacuum system leakage, since it defines the volume rate as well as the pressure at which the volume is measured. In rough calculations, this unit is taken to be equivalent to the previous two. More accurately:

$$1 \text{ std cm}^3/\text{s} = 0.76 \text{ torr} \cdot \text{L/s}$$
$$1 \text{ torr} \cdot \text{L/s} = 1.3 \text{ std cm}^3/\text{s} \tag{13.4}$$

Although std cm³/s is the leak rate terminology used for product leak detection, other units of measure may be used in different application areas, such as the torr · L/s used to express vacuum system leakage. A table showing the relation among other units of measure that may be encountered in various leak detection applications is provided in the Appendix.

13.2.2 Sizes of Leaks that Affect Product Life

The maximum acceptable leak rate for a given product depends on the nature of the product. Since the cost of leak detection increases as the specified leak rate decreases, it follows that testing for unnecessarily small leaks can incur unnecessary test expense. Some examples of product leak specifications are given in Table 13.1. Note the wide range of sizes. In the chemical process industry, the product flow rate is typically high. Therefore, a leak in the process equipment may not significantly affect yield until it reaches 0.1 to 1 std cm/s. On the other hand, the body implant, such as a pacemaker, will require exacting standards to ensure guaranteed safe operation in a corrosive environment (body fluid) for a predictable time interval. In general, static sealed systems require more strict specifications than dynamic systems. It can be seen that the terms "free from leaks" or "zero leakage" really have no meaning, since all of these products can have leaks smaller than the specified leak without impairing operating life. Attempts to test for "zero leakage" will lead to needless expense and frustration.

13.2.3 Distribution of Leaks by Size

The curve in Figure 13.1 shows the general distribution by size of commonly encountered leaks based on observations over many years for many products. Knowedge of this distribution is helpful in establishing reasonable leak rate

Table 13.1 Examples of Leak Rate Specifications for Various Products and Industries

Product or system	Leak rate specification (atm · cm³/s)	Comment
Chemical process equipment	10^{-1} to 1	High process flow rates
Torque converter	10^{-3} to 10^{-4}	Retention of fluid
Beverage can end	10^{-5} to 10^{-7}	Smaller leaks, if present have negligible effect
IC package	10^{-7} to 10^{-8}	
Pacemaker	10^{-9} to 10^{-10}	Long life, implanted in body

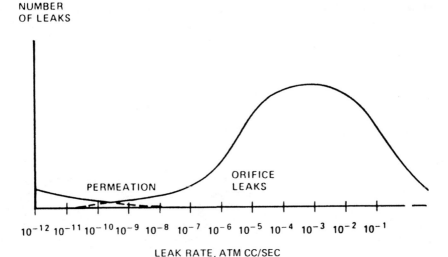

Figure 13.1 Typical distribution of leaks encountered in industrial products.

specifications for a given product. There are two regions because there are two means of gas transfer: through holes and by permeation.

Although industrial leak rate specifications range from 10^{-9} to 1 atm · cm³/s, the majority of products have leak rate specifications lying in a narrower range, from 10^{-6} to 10^{-1} atm · cm³/s. The upper part of this range is covered by bubble testing, down to 10^{-4} std cm³/s. Other methods overlap the bubble method and extend well below its lower limit. The helium method can detect leaks smaller by a factor of 1,000,000. Leaks larger than 10^{-1} atm · cm³/s can usually be spotted visually.

The large class of leaks caused by incomplete welds, brazes, seals, and so on, usually does not extend below 10^{-7} atm · cm³/s. Smaller leaks are usually plugged by water vapor from the atmosphere. Baking, when feasible, can reopen these leaks by evaporating the water vapor.

Some products, which have to be reliable for an extended period—say five years or more—may warrant testing at higher sensitivity (10^{-8} to 10^{-9} atm · cm³/s) for the reassurance of having conservative test results. An example is the pacemaker, which must function for years in a difficult environment.

For most products, however, looking for leaks 100 or 1000 times smaller than the acceptable limit will incur unnecessary additional test expense without improving reliability. For example, it would not make economic sense to test automotive torque converters or ring pull-tab beverage can ends at 10^{-8} or

10^{-9} atm · cm³/s; in fact, it would lead to costly rejection of perfectly serviceable products.

13.3 METHODS OF LEAK DETECTION

13.3.1 Leak Location and Measurement

The basic functions of leak detection are the location and measurement of leaks. These functions are carried out through the use of standard leak test techniques, which are usually selected according to the configuration of the part to be tested, the economics of the test, and the nature of the system. These techniques depend on the use of a tracer gas (or liquid) passing through a leak and being detected on the other side, and are applicable to most of the leak detection methods, including the helium method.

Leak Location

Leak location is the testing approach used to find the precise location of individual leaks. It is usually a qualitative procedure only. The techniques used are very dependent on the skill and alertness of an operator looking for something that is not known to exist until found. Leak location does carry out the very valuable function of identifying the sources of leaks in order to facilitate repair, remanufacture, or, in some cases, even redesign.

Leak location is carried out by means of two techniques, termed the detector-probe mode and the tracer-probe mode. In the detector-probe mode (Figure 13.2), the test piece is filled with tracer gas, and the exterior is scanned with a probe that is attached to the inlet of the leak detector. The probe con-

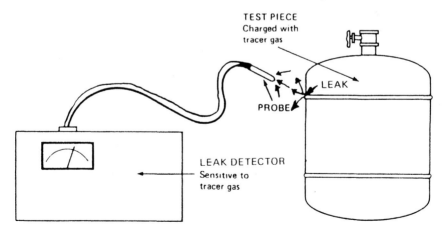

Figure 13.2 Detector-probe leak location technique.

tinuously admits (or "sniffs") some of the air directly in front of the test piece. This air is inducted to the analytical portion of the leak detector, where any of the tracer gas that may be leaking from the test piece is detected.

In the tracer-probe mode (Figure 13.3), the leak detector is used to evacuate the interior of the test piece, and a probe is used to discretely spray test gas on suspected leak sites. Any leaks are evidenced when the tracer gas flows through the evacuated test piece and is detected by the leak detector.

Leak Measurement

Leak measurement in the method actually used to measure the total or partial leakage of a device or system. The most reliable approach is initially to test a system or device for total leakage, and then for individual leaks, which are isolated by using one of the leak location techniques. Use of a leak location technique to measure total leakage usually results in poor reliability and high cost.

The two standard leak measurement techniques are known as the inside-out mode and the outside-in mode. In the inside-out mode, the test part is filled with tracer gas, and is placed in a test fixture that is subsequently evacuated. Any of the tracer gas that flows out through a leak is captured in a volume surrounding the test part. The contents of this volume are analyzed by the leak detector as representative of total leakage. This approach is shown schematically in Figure 13.4.

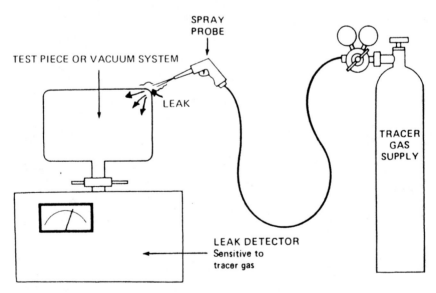

Figure 13.3 Tracer-probe leak location technique.

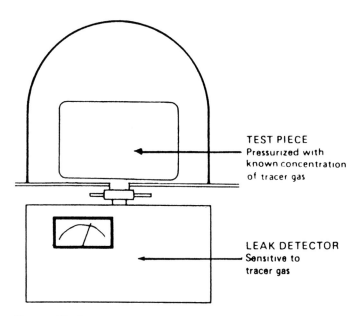

Figure 13.4 Inside-out leak measurement technique.

In the outside-in mode, the test part is evacuated and is placed in a volume containing the tracer gas, which flows through all leaks to the interior of the test part, where it is detected. The outside-in method is shown schematically in Figure 13.5.

With either of these approaches, the tracer gas may be allowed to accumulate before detection, or may be detected continuously. In general, continuous detection yields a faster test with adequate sensitivity; however, circumstances sometimes require the accumulation of tracer gas prior to analysis.

13.3.2 Leak Detection Methods

There is a wide range of methods available for finding and measuring leaks in products or assemblies. These methods include a range of technology from trivial to highly sophisticated, with a corresponding range of equipment costs. It is obviously necessary to choose a leak detection method that fulfills the defined needs of sensitivity and speed.

Acoustical

Acoustical leak detection, a variation of inside-out testing (Figure 13.6), uses the sonic (or ultrasonic) energy generated by a gas as it expands through an orifice. Acoustical leak detection is widely used in testing ductwork for leaks,

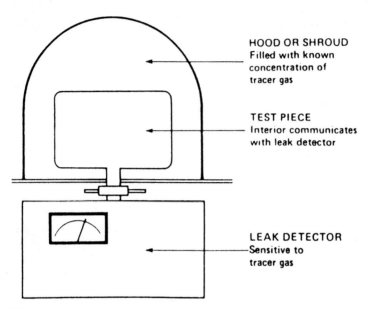

Figure 13.5 Outside-in leak measurement technique.

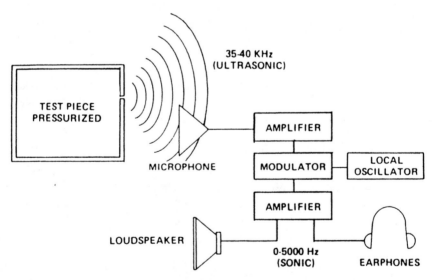

Figure 13.6 Ultrasonic leak test setup.

and in testing high pressure lines. It requires modest instrumentation and is fairly simple and fast. Its sensitivity, however, is limited to about 10^{-3} std cm³/s.

Bubble Testing

Bubble testing, a form of inside-out testing, is the most prevalent method of leak detection in today's industry. In its simplest form, bubble testing consists of pressurizing a part, placing it in a water bath, and looking for a stream of bubbles from any leaks present (Figure 13.7). This can also be accomplished by applying a soap solution to the part.

This method has some benefits. The observer can ignore leaks that are irrelevant; for example, temporary seals, installed for test, can leak and be ignored while the body of the part is carefully examined for leaks. Equipment costs and complexity are minimized; worker skill requirements are modest. As in most inside-out test methods, sensitivity can usually be increased by increasing the charging pressure.

The method has drawbacks, however. It requires wetting, which frequently requires a drying cycle before painting or final assembly. It does not measure total leakage from the part. Its success depends on the continuous attention of the operator. And it is limited, generally, to a sensitivity of about 10^{-4} std cm³/s, although this varies widely with the test arrangement and the design of the part.

The sensitivity of this method can be improved by proper lighting, use of a dark background, and use of a fluid with a low surface tension (to permit smaller bubbles to break away). Ultimately, the sensitivity limit is reached when

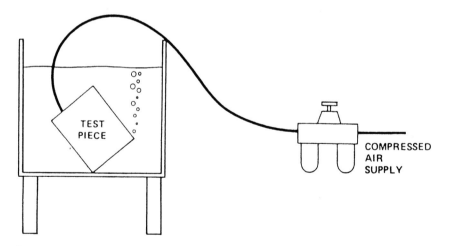

Figure 13.7 Bubble testing.

the streams of bubbles look like such fine hairs (or the bubbles are so infrequent) that the observer no longer sees them.

Pressure Decay/Vacuum Decay

Long a standard practice in the plumbing trade, this technique is a rough test method in many segments of industry. In its simplest form, it requires only a pump or compressor and a pressure gauge (Figure 13.8). Its sensitivity is proportional to the wait time (within limits), and its range is generally limited to approximately 10^{-4} std cm^3/s. One of the major drawbacks to a pressure decay test is its sensitivity to temperature fluctuations, since they affect the pressure being observed. Such fluctuations can lead to considerable inaccuracies, most obvious when the pressure rises during a pressure decay test! A vacuum decay test (the first of the outside-in test methods) is less vulnerable to temperature fluctuations, but it is limited in pressure differential to atmospheric pressure, and may be influenced by virtual leaks.

Thermocouple Gauges

Thermocouple gauge meters respond when acetone or other gases are squirted through a system leak. This fact is used and abused in a tracer-probe form of testing in laboratories throughout the world. The abuse is related to the fact that acetone can seriously degrade O-ring seals and certain insulating materials occasionally used on vacuum systems. Generally, the method may be used on any metal-sealed component or joint in the system as long as great care is taken to prevent the acetone from contacting any organic material used on the system. Acetone must be applied sparingly and with great caution, since it is a powerful solvent, dangerously flammable, and mildly toxic. Indiscriminate squirting of acetone around a vacuum system is ill-advised—and possibly ille-

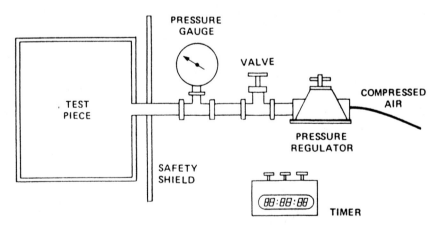

Figure 13.8 Simplified pressure decay test setup.

gal—in industrial installations due to OSHA regulations. The sensitivity range for this method is limited to approximately 10^{-2} std cm^3/s. Response will be upscale if the acetone goes through the leak or downscale if it plugs the leak.

Ion Gauges and Ion Pumps

Ion gauges and ion pumps operate as tracer-probe leak detectors on the same fundamental principle: namely, the pressure signal is electronically nulled and an attempt is made to alter the leak rate either up or down and detect the change in the pressure signal. Since these instruments measure total pressure and have considerable variations in sensitivity for the various gas species, response to a probe gas entering the system through a leak may be much less than anticipated. In addition, most real leaks have a conductance which is not a simple known function of the atomic mass of the leaking gas. This technique works best when the system is at its leak-limited pressure. Under these circumstances, the sensitivity may be sufficient to locate leaks as small as 10^{-7} std cm^3/s. If, however, the system pressure is high and/or limited by other sources of gas—including other small leaks—sensitivity is greatly reduced. (Care must be taken in spraying acetone or other conductive volatiles or gases to avoid electrical paths between gauge pins.)

Halogen Leak Detectors

Halogen leak detectors are typically used in the detector-probe mode, requiring that the system be pressurized with a gas containing an organic halide, like one of the Freons. The exterior of the system is then scanned with a sniffer probe sensitive to traces of the halogen-bearing gas (Figure 13.9). A disadvantage of the halogen leak detector is that it responds to a variety of other gases; including cigar smoke. Halogen detectors are also available for use on vacuum

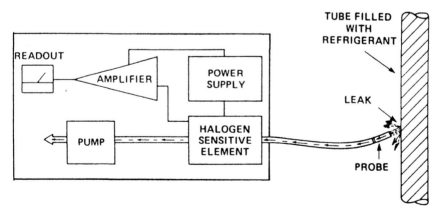

Figure 13.9 Halogen leak detector.

systems. Some halogen detectors can be used to locate leaks as small as 10^{-7} std cm^3/s.

Dye Penetrant

This method is an adaptation of a technique used to find cracks in metals and defects in welds. It uses a low-viscosity fluid that exhibits a high rate of surface migration. This fluid is painted on one side of a suspected leak site and, after a time, is detected on the far side of the part (Figure 13.10). Some fluids are detected when they glow under ultraviolet light. This process is only applicable when there is access to both sides of the part.

The major advantages of this test are its conceptual simplicity, low cost, and inherent "record keeping." The disadvantages are (within its range of sensitivity) its limited applicability, the time response, and the fact that, in many applications, the part cannot be washed clean enough for use even if it does not leak. The sensitivity of this method can be as high as 10^{-6} std cm^3/s.

Thermal Conductivity

This detector probe method relies on the use of a tracer gas with a thermal conductivity different from that of air. In one sensitive arrangement, the sample gas is pumped past a thermistor arranged in a bridge with a reference thermistor. The imbalance of the bridge is displayed on the leak rate meter (Figure 13.11). Leaks of 10^{-6} std cm^3/s (of selected gases) can be located with this method.

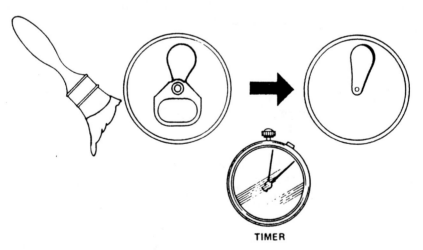

Figure 13.10 Dye penetrant testing.

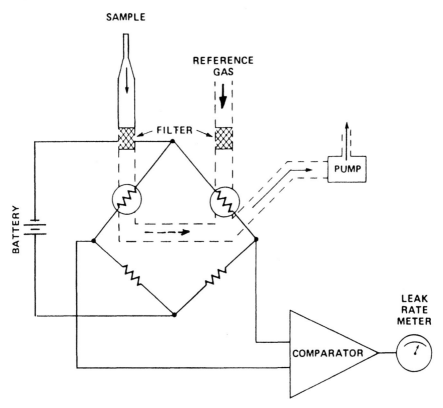

Figure 13.11 Thermal conductivity-sniffer test diagram.

Radioisotope

This method is useful only for testing hermetically sealed components. When the radioisotope method is used, the hermetically sealed components to be tested are placed in a chamber, which is evacuated and filled with radioactive tracer gas (typically, krypton-85). After the tracer diffuses through any leaks present in the components, it is pumped from the chamber and reclaimed. Any residue is flushed from the surfaces of the parts. Radiation from the tracer that has diffused through any leaks in the components is measured by a radiation detector (Figure 13.12). This method has a sensitivity of approximately 10^{-11} std cm^3/s, but it involves an expensive installation, including monitoring and control devices. (The radioisotope is from 4 to 10 times as costly as a helium installation, depending on the degree of isolation required.) It also requires a radiation safety officer.

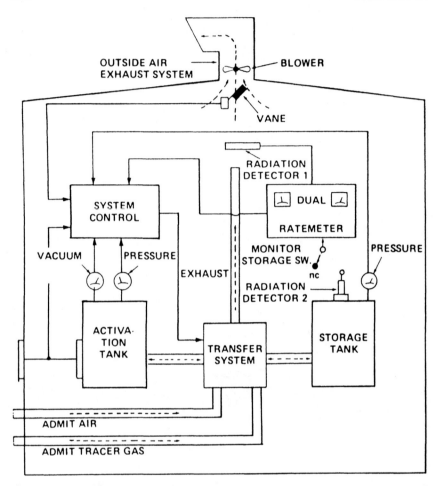

Figure 13.12 Radioisotope testing.

Helium Permeability Detector (Helitest)

The operation of this detector is based on very large differences in the permeability of various gases through glass. A very small quartz capillary is attached directly to a small ion-getter pump and the indication of a permeation leak through the quartz is obtained from the increase of the pump current when excess helium enters the pump. The instrument is portable, battery operated, and used in the sniffer mode. Its sensitivity is ten times better than that of typical thermocouple or halogen-based leak detectors and almost approaches the sensitivity of the mass spectrometer leak detector when the latter is used in the sniffing mode. It can detect changes of 2 parts per million of helium in the gas surrounding the quartz capillary. This is roughly comparable to a leak rate of $5 \cdot 10^{-6}$ atm/cc/s.

Figure 13.13 is the schematic representation of the instrument. The quart tube, less than one millimeter in diameter, has a wall thickness of a few thousands of a centimeter and a total active external surface of approximately 2 cm². It is surrounded by a heater, made of a noble metal wire, to raise the temperature of the quartz tube at prescribed intervals of time to increase the rate of permeation during measurement. When helium penetrates into the ion pump (called "pressure detector" in Figure 13.13) the current momentarily increases before the helium is pumped away. The instrument includes an automatic zeroing feature, so that the detection can be made even in a high helium background environment. The time response of this sensing process is about 2 seconds when using a 1.5 meter probe line.

Helium (Mass Spectrometer) Method

This leak detection method uses helium (the lightest inert gas) as a "tracer" and detects it in quantities as small as 1 part in 10 million. It is discussed in Section 13.4 in greater detail because it is the most important method of leak detection used in high-vacuum technology.

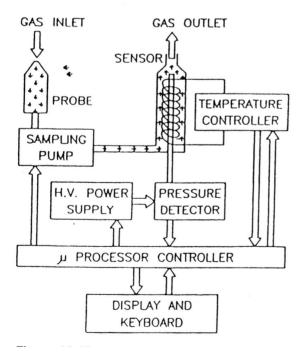

Figure 13.13 Schematic representation of the permeability leak detector (Helitest).

13.3.3 Leakage Specification Formulation and Test Method Selection

The leakage-specification for any product or component should be one of the initial parameters established for the design. The maximum allowable leak rate will indicate the appropriate choice of the test method. In its simplest terms, the selection is derived from a few basic questions:

1. What type of leakage will damage the product or process?
2. How much material flow can be permitted?
3. How long must the product (or warranty) last?
4. How much pumping speed is available?

The answers to these questions will be stated as a volume or mass flow over a time period. For example:

6 oz. of R-12 (or a similar gas) in 4 years, or
10^{15} molecules of water in 2 years, or
2×10^{-7} torr · L/s of air

The answers also may be hard to determine. For example, if your product is a tank in dynamic use, which is allowed to leak its liquid contents but not permitted to show a stain (i.e., leakage cannot exceed evaporation rate), some calculations and assumptions must be made. Similarly, if your product is a refrigeration system, the allowable total loss of gas and lifetime may be straightforward, but the reasonable allocation of permitted leakage among the several components and joints may be challenging.

Having established the rate at which some material will flow through a leak, it is next necessary to convert this flow rate to one that can be practically measured. This step can be taken only when the proposed leak detection method has been selected. For example, if the permitted leakage is to be 3 oz of a refrigerant in 5 years at 250 lbs gauge pressure (internal) and the leak detection method chosen is the helium method, using a technique of evacuating the exterior of the part and surrounding it with helium, there are several conversions to be made.

First, the value of leakage should be expressed in units convenient to the final test method. Referring to the leakage rate conversion nomograph in the Appendix, we find that 3 oz of R-12 in 5 years (0.6 oz/yr) is equivalent to 1.0×10^{-4} std cm^3/s. The next step is to make the pressure conversion. Referring to equation (13.5), it can be seen that in this range of leak rates, the leakage from a high pressure (P_1) into the atmosphere (P_2) relates to the leakage from a lower pressure (P_3) to a vacuum (P_4) as

$$\frac{Q_1}{Q_2} = \frac{(P_1)^2 - (P_2)^2}{(P_3)^2 - (P_4)^2} \tag{13.5}$$

Using this relationship, the 1×10^{-4} std cm³/s leak at 250 psig (1717 kPa) would pass 3×10^{-7} std cm³/s of R-12 from atmospheric pressure to a "complete" vacuum. So

$$Q_2 = 1 \times 10^{-4} \frac{(15)^2 - (0)^2}{(265)^2 - (15)^2} \tag{13.6}$$

Finally, leakage of R-12 at this rate must be converted to a rate of helium leakage. The conversion is dependent on the viscosities (η) of the two gases:

$$\frac{Q_{R-12}}{Q_{He}} = \frac{\eta_{He}}{\eta_{R-12}} \tag{13.7}$$

Since the viscosity of R-12 is 0.0118 cP and that of helium is 0.0178 cP, this yields

$$Q_{He} = 2 \times 10^{-7} \text{ std/cm}^3/\text{s} \tag{13.8}$$

Each of these conversions has its own inaccuracies. First, it is difficult to guarantee that an imperfection which leaks at 250 psi will still leak at 15 psi, especially with the pressure applied from the other side. Second, the formulas for leakage conversion with pressure and with gas composition are only truly applicable in clear and known flow regimes, and real-life leaks are neither. Nonetheless, if it is not practical to measure the leakage with specified substance and pressure, such conversions are necessary. They have, in fact, been used in industry for years with general success. Most industrial users apply a safety factory by specifying a maximum leak rate as much as a factor of 10 smaller than the calculated value.

Leak testing is clearly a time-consuming and frequently expensive task. It must be undertaken, however, when necessary to protect product integrity or personnel safety. It is recommended that the most cost-effective method for making the measurements be selected. Start by establishing a specification of acceptable leak rate, then choose the most effective and reliable leak testing system that is suitable for the required sensitivity and the necessary production rate.

Leaks by Molecular Diffusion

In most of the cases discussed above, the gas flow in the leak is considered to be effected by the presence of a total pressure difference across the leak path.

But flow in leaks can also be caused by molecular diffusion in the absence of a total pressure difference (as noted in Section 3.9.3). We shall consider two situations. If the flow is judged to be molecular, then the presence of the other gas does not influence the rate of flow. Otherwise, for simple first approximation estimates, the flow can be expressed in throughput units by

$$Q_L = A\, D\, \Delta P_g / L \tag{13.9}$$

where Q_L is the leak rate (torr · L/s or std cc/s), A is the cross sectional area of the leakage path (cm), D is the diffusion coefficient (cm^2/s, or rather cm^3/cm · s), ΔP_g is the partial pressure difference for the leaking gas (torr or atm), and L is the length of the leakage path (cm). For helium in air at room temperature, D is approximately 0.5 cm^2/s.

Usually, the flow rate by molecular diffusion is lower than the flow due to a total pressure difference. Therefore, conservative assumptions can be made during leak testing with a method that employs a pressure difference during the test procedure. For illustrative purposes, consider the following nominal estimates. Assume a hole diameter of 1×10^{-4} cm, in a wall 0.01 cm thick, and enclosing an object that contains 1 torr of helium and 759 torr of air. For the case of molecular flow, the conductance of such passage for helium is

$$C = 2.7 \cdot 12.1\; d^3/L = 3.3 \cdot 10^{-9} \text{ L/s}$$

then

$$Q_L = C \cdot \Delta P_g = 3.3 \cdot 10^{-9} \text{ torr L/s} \tag{13.10}$$

Next, assuming viscous flow at 1 atm pressure difference, the conductance for air (which is the main gas) will be, in units of cm and torr

$$C = 179 \cdot P_{ave} \cdot d^4/L = 6.8 \times 10^{-10} \text{ L/s}$$

and the flow

$$(Q_L = C \cdot \Delta P) \tag{13.11}$$

$$Q_{L(\text{helium})} = 6.8 \times 10^{-10} \text{ L/s} \cdot 760 \text{ torr} \cdot (1/760) = 6.8 \cdot 10^{-10} \text{ torr L/s} \tag{13.12}$$

Finally, using the above equation and assuming diffusion flow, with an area of 7.85×10^{-9} cm^2,

$$Q_L = \frac{7.85 \times 10^{-9} \text{cm}^2 \times 0.5 \text{ cm}^2 \times 1 \text{ torr} \times \text{Liter}}{0.01 \text{ cm} \quad \text{sec.} \quad 1000 \text{ cm}^3} = 4 \times 10^{-10} \text{ torr L/s} \tag{13.13}$$

13.4 HELIUM MASS SPECTROMETER LEAK DETECTORS

13.4.1 Basic Considerations

A helium mass spectrometer leak detector is essentially an ionization gauge designed to measure only the partial pressure of helium in a vacuum system. A leak detector consists of a mass spectrometer tube, a vacuum system capable of maintaining a total pressure of about 10^{-5} torr or less in the tube, a very sensitive and stable amplifier, and suitable valving means and auxiliary pumps to permit rapid attachment of the part or device to be tested. The amplified output of the spectrometer tube is read on an output meter. Modern leak detector meters are calibrated in std cm^3/s despite of the fact that the actual parameter being measured is helium partial pressure within the spectrometer tube. This is made possible when the pumping speed is known and is constant. A helium partial pressure increase of 10^{-10} torr, for instance, in a leak detector system having a speed of 10 L/s would indicate an in-leakage of 10^{-9} torr · L/s (1.3×10^{-9} std cm^3/s) of helium into the leak detector. Direct interpretation of partial pressure as leak rate is very convenient for moderately sized test devices, since equilibrium helium pressure occurs in about four time constants (V/S seconds) after applying helium to the leak, where V is the evacuated volume and S is the pumping speed (assuming insignificant delays in the leak passage).

A leak detector having a sensitivity of 1×10^{-10} std cm^3/s at a pumping speed of 10 L/s would have a partial pressure sensitivity of 8×10^{-12} torr, that is,

$$\frac{8 \times 10^{-11} \text{torr} \cdot \text{L/s}}{10 \text{ L/s}} \tag{13.14}$$

Since room air normally contains 1 part helium in 200,000, the background helium pressure at 10^{-5} torr total system pressure would be 5×10^{-11} torr (10^{-5} torr/200,000) under ideal conditions (clean system—no trapped helium, etc.). If the background is considered as noise, this gives a signal-to-noise ratio of 1/6 for a 1×10^{-10} cm^3/s leak. Thanks only to meticulous electronics and spectrometer tube design, this is acceptable, since the background can be zeroed out as long as it is constant. Degradation of minimum detectable leak size begins with varying total pressure caused by contaminants and by added helium entering through polymer seals, through the exhaust of the mechanical pumps, and from trapped areas such as O-rings that have previously been sprayed with helium, or from previously found large leaks.

However, as the test device becomes large or as the gas load of a system under test becomes so great that larger pumps must be used (since the leak detector pumps alone cannot keep the total spectrometer tube pressure below 10^{-5} torr), the direct readout of the meter in cm³/s becomes meaningless unless recalibrated for the given system sensitivity. For instance, on a 1000-L system pumped only by the leak detector, it would take constant application of helium to the leak area for 500 s (5 × 1000/10) to reach direct readout (equilibrium) conditions.

If auxiliary pumps of 1000 L/s speed are necessary to keep the spectrometer tube pressure below 10^{-5} torr, helium need be applied for only 5 s (5 × 1000/1000) to reach equilibrium. However, the output reading achieved would be 1/100 of the scale reading or 10^{-12} cm³/s for what is in fact a 10^{-10} cm³/s leak. This is because the pumping speed of 1000 L/s is 100 times that of the leak detector. In the two examples above, it may be preferable to use an arbitrary scale, together with a standard leak placed in the system (as far from the leak detector as possible, or, ideally, in the vicinity of areas where leaks are suspected to be present).

The entire system sensitivity can now be obtained by applying helium at the standard leak, either for a fixed time and noting the output reading, or by applying helium to the leak until a steady-state reading is obtained. In either case, application of helium to areas to be tested must be for the same length of time. If a reading of 30 is obtained for a standard leak of 1.7×10^{-6} cm³/s, it may be assumed that a reading of 60 from an unknown leak would mean that the unknown leak was twice the size of the standard leak, or 3.4×10^{-6} cm³/s.

Thus far, we have dealt with familiar leak detection applications. As we enter high-production testing of identical parts, we must take into consideration in our calculations that the spectrometer tube is simply a helium partial pressure gauge, not a flowmeter. In rapid cycle industrial test systems it is impractical to evacuate the test chamber to very low pressures. Pressures of 0.1 torr or higher are usually used. The test chamber must be cycled in such a manner that the helium partial pressure produced by a minimum acceptable leak from the test part must be at least as great as the residual helium partial pressure in the test chamber in order to achieve a signal-to-noise ratio of at least 1. For example, the residual chamber helium partial pressure at 0.1 torr total pressure will be 5×10^{-7} torr (0.1/200,000).

If the test part has a minimum acceptable leak of 1×10^{-6} cm³/s (7.6×10^{-7} torr · L/s), the helium equilibrium partial pressure will be 3.8×10^{-8} torr (achieved during pump-down) if the system is pumped at 20 L/s. An increase in helium partial pressure of at least 4.6×10^{-7} torr ($5 \times 10^{-7} - 0.4 \times 10^{-7}$) is required to achieve the necessary $S/N \geq 1$ for minimum reliable detectability. To accomplish this, pumping of the chamber can be stopped, and the part allowed to leak for a predetermined time into the chamber before a sample is taken. If the tare volume (chamber volume less part volume) is 10 L, the helium partial pressure will increase at the rate of $7.6 \times 10^{-7}/V$ · torr/s. About

6 s ($4.6 \times 10^{-7}/7.6 \times 10^{-8}$) of hold time is needed before taking a sample in order to have an unambiguous reading.

The calculation necessary for predicting system performance can be reduced to the following rather simple formula for determining the hold time necessary to obtain a satisfactory reading on this high-speed leak detection system:

$$T_H = \left(\frac{P}{L} - \frac{1}{S}\right) V \qquad (13.15)$$

where T_H is the hold time, P the total chamber pressure when pumping is stopped, L the leak rate, S the pumping speed, and V the volume. It may be added that if $1/S$ is greater than P/L, no hold time is necessary ($T_H = 0$), and P can be raised to obtain a minimum pumping time. Since the total cycle consists of load time plus pump time plus hold time plus unload time, P and T_H should be selected so that the system pumping time to pressure P plus the hold time T_H add to a minimum.

The awareness that the detector measures only the helium partial pressure can also be applied for evaluating leaks found when using "probe" techniques. The probe technique consists of admitting air at a constant rate to the leak detector through a leak in the end of a fixed or hand-held probe. When the normal partial pressure of helium in air (4×10^{-3} torr) is increased due to the proximity of a leak, the partial pressure of helium in the spectrometer tube also increases. A leak may thus be located by moving the probe tip until a maximum signal occurs on the output meter. Quantitative readings are quite impossible unless all the helium escaping from the unknown leaks actually enters the probe.

However, if small devices pressurized with helium can be placed in close-fitting containers for a predetermined time, the container can be probed to determine the increase in helium partial pressure during the accumulation interval, and the unknown leak rate of the device under test can be determined as follows. The arbitrary reading on the leak rate meter when the probe is exposed to the sample gas in the container (R_2), divided by the arbitrary reading when the probe is exposed to fresh air (R_1), and multiplied by 4×10^{-3} torr (normal partial pressure of He in air) is equal to the final partial pressure of helium in the container, P_{He}:

$$P_{He} = 4 \times 10^{-3} \left(\frac{R_2}{R_1}\right) \text{ torr} \qquad (13.16)$$

The unknown leak, then, is equal to

$$\frac{P_{He} V}{t}$$

where V is the volume of the container less the test part volume in liters (tare volume) and t is the storage time in seconds.

For example, a 25-cm³ volume part is filled with helium and placed in a 35-cm³ container. Five minutes later, the container is probed and the leak detector output reading rises from 10 to 70.

$$P_{He} = \frac{70}{10} \times 4 \times 10^{-3} = 2.8 \times 10^{-2} \text{ torr} \qquad (13.17)$$

and the leak must be equal to

$$\frac{2.8 \times 10^{-2}(35 - 25/1000)}{5 \times 60} \qquad (13.18)$$

or 9.3×10^{-7} torr · L/s (1.2×10^{-6} std cm³/s). It may be noted that the sensitivity of the test is determined primarily by the *tare volume* and the *storage time* rather than by the basic sensitivity of the leak detector.

A mass spectrometer leak detector is an excellent tool for locating minute leaks, but if it is necessary to ascertain the rate of leakage from an unknown leak, it is often impossible to make direct interpretation of the output reading in std cm³/s. In those cases, calculations of the unknown leak rate become far easier if the leak detector is understood to be a helium partial pressure gauge.

13.4.2 Helium As a Tracer Gas

Since a mass spectrometer may be tuned to virtually any mass, the choice of a tracer gas is "limitless." However, the particular application of leak testing and desired characteristics for a tracer gas narrow this choice. A tracer gas for leak testing should have the following characteristics:

1. It should be nontoxic.
2. It should diffuse readily through minute leaks.
3. It should be inert.
4. It should be present in not more than trace quantities in the atmosphere.
5. It should be relatively inexpensive.
6. It should have a low absorption rate.

Helium satisfies these requirements and is the most commonly used tracer gas.

The diffusion rate of a tracer gas depends on pressure differential, temperature, and molecular size. Normally, the lighter gases have the highest diffusion rates and will therefore give the best sensitivity. Only hydrogen (mass $H_2 = 2$) is lighter than helium (mass $He = 4$). Of the inert gases, helium is by far the lightest. Because it is inert and monatomic, it also has the lowest absorption rate. Neon, which is the next heaviest of the inert gases (mass 20), is not used in leak detection because of its high cost.

The tracer gas should be inert so that it does not react with materials in the system. It should also be nontoxic and nonflammable. The tracer should be present in the atmosphere in not more than trace quantities so that residual air in the system is not detected as a leakage. Helium concentration in the atmosphere is only 5 parts per million.

After helium and neon, the next heaviest inert gas is argon (mass 40). Argon has a large mass relative to those gases present in the atmosphere (N_2 = 28, O_2 = 32), so that as a tracer it may not flow through leaks as readily as helium. Argon is also present in large quantities in the atmosphere (1%). Even with these undesirable properties, it is used as a tracer gas in cases where helium is undesirable and in cases where specifications call for argon backfilling, as is sometimes required for hermetically sealed relays. Sensitivities obtained with argon leak detectors are approximately 10^{-8} std cm^3/s. If helium is used as a tracer gas but air leaks into or out of the product, what is the relationship between the two?

The mass spectrometer leak detector is capable of indicating leak rates only in the tracer gas to which it is tuned and calibrated (nearly always helium). It cannot indicate leak rates in air. There is, however, an approximate relationship which is derived from the kinetic theory of gases.

For molecular flow conditions, the kinetic theory of gases predicts that the relative flow rate of two gases through (an ideal leak under molecular flow conditions, roughly, below one millionth of atmospheric pressure) will be inversely proportional to the square root of the average molecular weight. Since the molecular weight of air is about 29 and helium is 4, helium will flow 2.7 times as fast as air through this leak. The implication is that the minimum detectable leak in terms of air is 1/2.7 of that for helium. (In fact, some manufacturers actually rate their helium leak detectors in air leakage equivalents because the figure is lower, and it makes the leak detector look more sensitive.)

In the real world, most leaks are tortuous, sometimes multiple, paths of great length compared to cross section. Although air leakage rates can vary from many times smaller up to almost equal to the helium rates, a leak rate measured in helium is conservative; the air rate will not be greater. For viscous flow conditions, see Section 13.3.3.

A helium mass spectrometer leak detector (MSLD) is a complete system for locating and/or measuring the size of leaks into or out of a device or a container. In use, this method of leak detection is initiated when a tracer gas, helium, is introduced to a test part that is connected to the MSLD system. The helium leaking from the test part diffuses through the system, its partial pressure is measured, and results are displayed on a meter. The MSLD operating principle consists of ionization of gases in a vacuum and accelerating of the various ions through a voltage drop and a magnetic or RF field. The helium ions are separated and collected, and the resulting ion current is amplified and indicated on the meter.

An MSLD consists of a spectrometer tube, quantitatively sensitive to the presence of helium; a vacuum system, to maintain adequately low operating pressure in the spectrometer tube; mechanical pump(s), to evacuate the part to be tested; valves, to transfer the connection of the evacuated part from the mechanical roughing system to the spectrometer vacuum system; amplifier and readout instrumentation, to monitor the spectrometer tube output signal; electrical power supplies and controls, for valve sequencing, protective circuits, and so on; and fixturing, for attachment to the part to be leak tested.

13.4.3 Major Components

Spectrometer Tube

The heart of the MSLD is the spectrometer tube, which is essentially a partial pressure ionization gauge that measures only the helium pressure in the system rather than the total pressure. A spectrometer tube is illustrated schematically in Figure 13.14. In spectrometer tube operation, electrons produced by a hot filament enter the ion chamber and collide with gas molecules, creating within the chamber ions quantitatively proportional to the pressure in the ion chamber. These ions are repelled out of the ion chamber through the exit slit by the repeller field. The combined electrostatic effect of the repeller, exit slit, focus plates, and ground slit collimates the ion beam so that it enters the magnetic field as a straight "ribbon" of ions.

At the entrance to the magnetic field, ions of all gas species are present in the ion chamber, and if an ion collector were placed at this point, its current would be proportional to the total pressure in the ion chamber. However, as the ions pass through this magnetic field, they are deflected in proportion to their mass-to-charge ratio. The spectrometer tube is typically designed and adjusted so that hydrogen ions are deflected $135°$, helium ions $90°$, and all heavier species less than $90°$. Consequently, only helium ions pass through the field exit slits and arrive at the collector. The collector current is therefore proportional to the partial pressure of helium in the spectrometer tube and, within the normal operating pressure range of the MSLD, is not affected by the pressure of the other residual gases. The collector current is measured by an electrometer amplifier and displayed on the meter.

The ion source is often a replaceable assembly, usually with two filaments that provide a source of electrons; an ionization chamber into which the electrons are beamed to ionize gas molecules; a repeller plate that guides the posi-

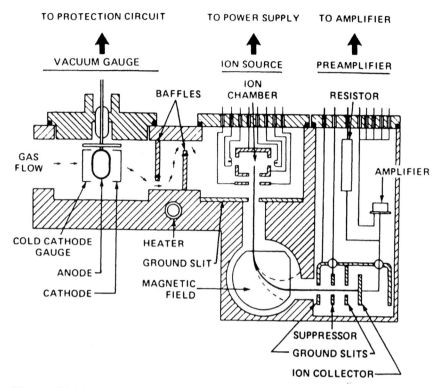

Figure 13.14 Cutaway view of a spectrometer tube (magnetic bending). The vacuum gauge may be placed elsewhere.

tive ions through an exit slit; and two focus plates that direct the ion beam toward a slit in the ground potential plate.

The preamplifier assembly contains an ion collector assembly to translate helium ions into an electrical signal. The preamplifier circuit consists of a high-grain operational amplifier with a feedback resistor. The circuitry is designed to protect the electrical signal from external interference and stabilize the meter reading.

Vacuum System

In all conventional helium leak detector, the spectrometer tube is mounted on the inlet side of the high-vacuum pump, which is connected to a mechanical pump (forepump). The high-vacuum pump keeps the spectrometer tube at the proper vacuum level. In addition, conventional leak detectors utilize a liquid nitrogen trap to provide additional pumping for condensable gases, such as water vapor. The trap also keeps the spectrometer tube clean by collecting condensable contamination from the test piece. A second mechanical pump (roughing pump) may be used to provide faster pumpdown to the roughing vacuum in the test port. A valving manifold is used to effect transfer to the high-vacuum pump and thereby bring the test piece to the same pressure as the spectrometer tube.

13.4.4 Basic Types of Helium Leak Detectors

There are four different types of leak detectors: (1) an automatic conventional leak detector, (2) an ultrasensitive version of the conventional leak detector, (3) an automatic ContraFlow leak detector, and (4) a portable version of the Contra-Flow leak detector.

Conventional Leak Detector

The automatic conventional model is shown schematically in Figure 13.15. Vacuum is obtained and maintained in the spectrometer tube using a 100 L/s air-cooled high vacuum pump backed by a small mechanical pump. A large

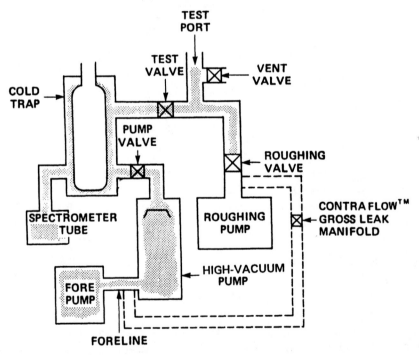

Figure 13.15 Automatic conventional leak detector.

LEAK DETECTION

liquid nitrogen trap pumps condensable vapors at a rate of about 300 L/s. The spectrometer tube pressure is maintained at less than 0.2 mtorr during use. A roughing pump is provided to evacuate the test port to a vacuum level that will not disrupt the high vacuum pump operation (approximately < 100 mtorr). This is typical of most MSLDs used in production. A gross leak manifold allowing testing at higher inlet pressures at reduced sensitivity is available in some leak detectors.

An integral rough vacuum system is included in the automatic MSLD together with semiautomatic valves and programming and protection circuits. With an MSLD such as this, the part to be tested is simply coupled to the test port. A switch is manually actuated whereupon the part is automatically evacuated to a suitable pressure and then valved off from the rough vacuum system and onto the MSLD spectrometer tube vacuum system. A lamp indicates when this process is complete, and leak testing can be started. At the conclusion of leak testing, the switch is returned to its original position, the test port is automatically isolated from the MSLD and vented to air, and the test piece is ready to be uncoupled from the leak detector.

The ultrasensitive model is identical to the conventional model except it has an automatic orifice plate which reduces the pumping speed by a factor of 3. This allows more helium molecules to enter the spectrometer tube and increases sensitivity, for example, from 8×10^{-11} to 2×10^{-11} atm · cm^3/s. The orifice plate open automatically between tests for rapid removal of helium from the spectrometer tube.

ContraFlow Leak Detector

A newer development is the ContraFlow leak detector (Figures 13.16 and 13.17). This technique takes advantage of the differences in compression ra-

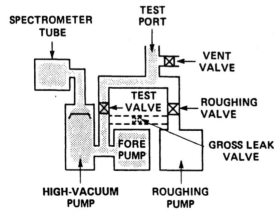

Figure 13.16 Automatic Contra-Flow leak detector.

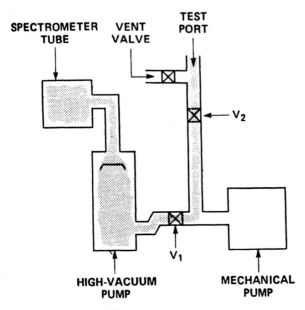

Figure 13.17 Portable ContraFlow leak detector.

tios (outlet pressure divided by inlet pressure) produced by the vapor jet pump or the turbomolecular pump for gases of different molecular weights. For example, the maximum compression ratio for helium may be 10 or 100, while for oxygen, nitrogen, and other gases contained in air, the ratios are normally far in excess of 1 million.

This principle is implemented in the leak detector by introducing helium (and other inlet gases, such as those resulting from a leak in the test piece) into the high vacuum pump outlet (foreline) rather than into the "normal" pump inlet, as in conventional leak detectors. Helium, having a much lower maximum compression ratio than other gases contained in air, diffuses backwards through the high vacuum pump to reach the spectrometer tube where it is detected in the normal manner. Although the mechanical pump is also attached to the high vacuum pump foreline and removes all inlet gases, including helium, there is no appreciable loss of sensitivity in the ContraFlow leak detector. The typical sensitivity of the automatic conventional model is 10^{-11} std cm^3/s while the sensitivity of the ContraFlow is 10^{-10} std cm^3/s. However, at higher test pressures this reverses in favor of the ContraFlow method.

Automatic trapping action. By optimizing the compression ratio for helium and the other gases of heavier molecular weights, the high-vacuum pump acts as a filter that prevents the other gases and contamination, such as water vapor, from reaching the spectrometer tube. These gases and contaminants are introduced by the connection of the test piece to the leak detector. The filtering action of the high-vacuum pump eliminates the need for any cryogenic trapping.

A high-vacuum pump used in this fashion also acts as a buffer that protects the spectrometer tube from pressure bursts that would normally endanger the mass spectrometer tube and trigger protective devices. Interruption of testing due to pressure bursts is less frequent, and the unit can be used at higher pressures, up to approximately 300 mtorr (700 mtorr in the gross leak mode), allowing the measurement of gross leaks (leakage rates generally more than 10^{-3} std cm^3/s) without the need for special test techniques.

Adjustable sensitivity. The compression ratio of helium can be varied by changing the pumping action of the high-vacuum pump. A control is provided to allow the variation of this compression ratio and thus increase or decrease the sensitivity of the leak detector as required by the application.

Advantages of the ContraFlow principle. Advantages of the ContraFlow principle include the elimination of liquid nitrogen, thus saving cost and removing a hazardous material. In addition, pump times are reduced, as tests can be made at pressures ranging from 700 to 10 mtorr without adverse effect on the high-vacuum pump. In a conventional helium leak detector, the high-vacuum pump must be protected from exposure to pressures above 10^{-1} torr. Particular advantages are achieved when testing large systems that cannot be evacuated to low pressures. This type of leak detector is available as an automatic system (Figure 13.16) or as a portable unit (Figure 13.17), which is manually controlled. A disadvantage is that the test port is never at a high vacuum and normally should not be connected to any piece which is at a high vacuum.

Also a disadvantage is the possibility of contaminating the parts to be leak tested with mechanical pump vapors. The best approach for controlling or eliminating this possibility is to use zeolite traps or oil-free pumps. In conventional leak detectors, the sensitivity decreases at higher pressures, because the admittance of test gas to the detector must be restricted by the use of throttling valves (Figure 13.18).

The ContraFlow leak detector, on the other hand, shows an almost constant sensitivity, independent of total pressure. The operation of the ContraFlow leak detector can be, of course, extended to still higher pressures by using a throttling valve after the tolerable forepressure of the high-vacuum pump is approached. In this case, the advantage over the conventional detector is preserved, the two response curves continuing parallel to each other toward higher pressures.

Thus, under carefully controlled conditions, the sensitivity of the ContraFlow leak detector can be as high as the conventional unit (for the same spectrom-

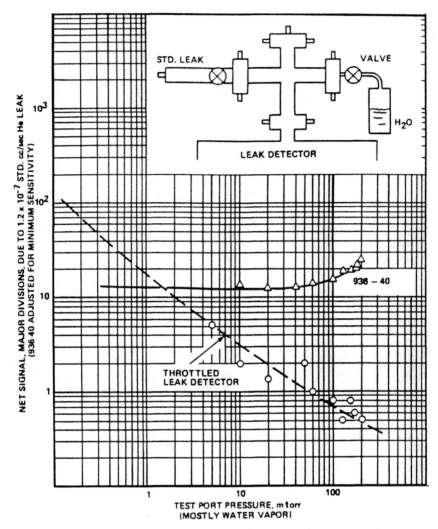

Figure 13.18 Comparison of conventional and ContraFlow leak detectors at higher test pressures.

eter and electronics). However, the main advantage is the simplicity of operation and the higher sensitivity at the higher test pressures. The advantage is particularly prominent when the gas producing the high pressure is noncondensable.

Beyond the conveniences of equipment design, simplicity, and perhaps cost of operation, the question remains: What are the basic advantages and disadvantages of the two methods; or, which method is suitable for which applications?

For precise measurements (repeatability and accuracy) of very small helium flow rates, the conventional leak detector has an overall advantage because of the promise of better linearity under molecular flow conditions. An example of such an applciation would be permeability measurements of helium through "porous" solids. For applications in which the primary object is to find leaks, particularly with systems and objects that are difficult to pump into the high-vacuum range, the ContraFlow method provides a very useful and more sensitive technique.

Test procedures are the same for the automatic system, whether it is conventional or ContraFlow. The manual system (Figure 13.17) differs in that it has only a single mechanical pump, and actuation of the valve must take into consideration the fact that the one pump must share the tasks of evacuating the test port and also be the forepump to the high vacuum pump. The instrumentation and the valve configuration of the portable system indicate the proper vacuum level for opening the test port to the foreline.

The portable unit is particulary useful for testing large systems that already have vacuum pumping capability and where its small size and light weight permit it to be brought to places of difficult access. It is also particularly well suited to probe-mode applications.

13.4.5 Hybrid Leak Detectors for Testing Small Parts

Hermetically sealed small objects are often leak tested by placing the items into a vessel filled with tracer gas under high pressure. The presence of a tracer gas (helium), which may penetrate into the object, is later revealed in a subsequent leak testing procedure. The actual leak rates can be correlated with the "bombing" pressure, time of exposure, and elapsed time between the two procedures. Unfortunately, a very large leak may be confused with a nonleaking part because the tracer gas can escape before the final test. To prevent this possibility, a pressure decay test is often incorporated into helium leak detectors to monitor the time to reach a certain pressure during the initial evacuation. If the chosen pressure is low, it will be affected by unpredictable parameters such as the variation of water vapor content and the condition of the rough vacuum pump. If the chosen pressure is high, the very small ratio of the additional added volume of a leaking small part to the comparatively large volume of the pumping system can make the discrimination rather uncertain. This problem can be solved by making an initial pump-down comparator, which at first is isolated from the rest of the vacuum system as shown in Figure 13.19. At the beginning of the test, the $V2$ valve is closed and the expansion of the gas into

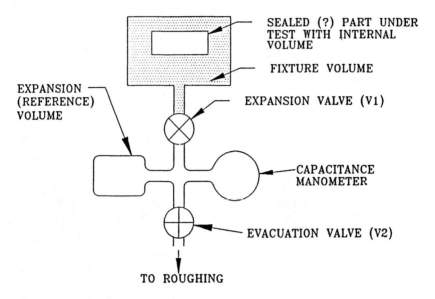

Figure 13.19 A device for detecting sealed objects with very large leaks.

the small expansion volume is monitored by the capacitance pressure gauge. Suitable volumes and timing can be chosen to produce unambiguous results.

13.4.6 Sensor Location and Component Leak Detector

Leak detectors are usually made as table-height cabinet models (containing elements as in Figure 13.15) or as smaller cart-top or table-top portable cabinets (for example, as in Figure 13.16), although recent models often incorporate both types. Smaller objects are tested by directly attaching the item to the test port. Larger vacuum systems are tested by connecting the test port (with a suitable hose) to a suitable location on the vacuum chamber. A clean metallic connector with a minimum of couplings is preferable. The best location is the one that gives the highest signal in the sensor. This is not always obvious. For ContraFlow leak detectors, the most logical choice is to connect the leak detector test port to a location between the exhaust of the high-vacuum pump and the inlet of the backing pump. Often, conventional leak detectors are also connected in the same location but then the signal is attenuated by the necessity of throttling (see Figure 13.18). In that case, the signal level and response time will depend on the volume of the chamber and the relevant pumping speed. It is possible to obtain a higher leak detection sensitivity by placing the sensor (a small mass spectrometer) directly onto the vacuum chamber similar to an ordinary high-vacuum gauge. Leak detectors that can be used in this way are available in the form of a set of components as shown in Figure 13.20, where each separate section is outlined with dashed lines. Mechanical pumps can also be any suitable pump available on the system. A compact arrangement of a small mass spectrometer, which is directly attached to the turbomolecular pump, is shown in Figure 13.21.

13.4.7 Oil-free Pumping Systems

Whenever oil-sealed mechanical and vapor jet pumps are used in leak detectors, the possibility of contamination to the test object (due to backstreaming) as well as the clean room in which the leak detector may be placed (due to the exhaust from the mechanical pump) exists. Therefore, the current trend is to use turbomolecular high-vacuum pumps and oil-free coarse vacuum pumps. The basic requirement is clear. The oil-free pumping system must have a sufficient compression ratio for helium to produce a low helium background signal before each test. Generally, the common oil-free pumps tend to have a compression ratios 100 to 1000 times lower than oil-sealed pumps. For conventional leak detectors the lower compression ratio can be compensated by using hybrid turbomolecular pumps without further complications. In ContraFlow leak detector systems, the clean-up of helium after each test must be accomplished by the rough vacuum pump. Normally, to obtain a high sensitivity, the compression ratio of the high-vacuum pump must be reduced but to obtain low background it must be increased. Therefore, for high sensitivity detection, it may be necessary to use two turbopumps or two oil-free pumps in series, or special oil-free coarse vacuum pumps that have higher compression ratios for helium.

13.4.8 Review of Pumping System Requirements

When using leak detectors in component form, it is useful to consider general arrangements of associated vacuum systems.

There are several basic reasons for having high vacuum in MSLD leak detectors.

1. To permit electron emission without destroying the filament in the ion source.
2. To permit long travel distances for electrons and helium ions without collisions.
3. To activate the leak flow, which results from one atmosphere pressure difference.
4. To permit fast helium travel from the leak site to the sensor.

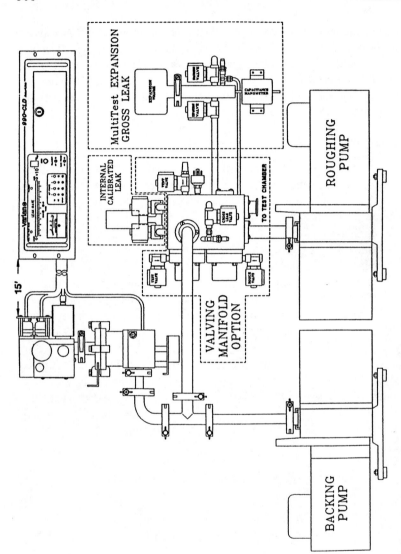

Figure 13.20 Functional sections of a component ContraFlow leak detector.

LEAK DETECTION

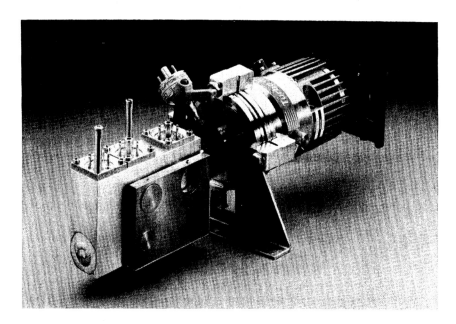

Figure 13.21 A compact helium detector (mass spectrometer) directly attached to a turbomolecular pump.

There are a few basic pumping system requirements. During the test, the pumping speed for helium should be low in order not to dilute the helium from the leak. During the subsequent clean-up period, helium should be removed quickly to be ready for the next test.

The general requirements are different for the conventional and counterflow leak detectors. The basic design issues can be appreciated from the expression for pressure at the inlet of a pump given in Eq. 4.17. As noted elsewhere, the leak detector sensor does not measure a leak rate as such. It measures only a partial pressure of helium in the sensor itself, not in the tested object. The leak rate is obtained by calibration and comparison to a known (standard) leak or by calculation by knowing the pumping speed at the sensor and how it relates to the test object.

In leak detection practice, the most relevant gases are helium, water vapor, and to some extent, hydrogen. Near high-vacuum water vapor is the dominant gas in the system and it affects the total pressure. Hydrogen is relevant because its heavy isotope (mass 3) may interfere with helium (mass 4). Hydrogen is particularly important for ContraFlow leak detection because its compression ratio in a high-vacuum pump is lower than helium. It should be noted that in counterflow leak detection, the test object is pumped not by the high vacuum pump but by the backing pump. Therefore, to lay out a pumping system for a leak detector, we should trace the helium presence through both pumps during the test and during the clean-up period. Some simple requirements then become apparent.

Total pressure in the sensor (when activated) must not exceed about 2×10^{-5} torr. If the helium compression ratio for the high-vacuum pump is set at 100, the hydrogen compression ratio is likely to be 70. This means that the foreline cannot have more than about 1.5×10^{-3} torr of hydrogen. Hydrogen is usually present due to water vapor breakdown on hot filaments and it is a product of pump oil breakdown. In oil-free pumps, the pressure ratio for hydrogen is likely to be the same as for helium, which is low compared to oil-sealed pumps. Oil-sealed pumps, however, have the disadvantage of the presence of absorbed helium (see Figure 5.8).

Pumping speed for other gases generally should be high. In conventional leak detectors, liquid nitrogen traps are used mainly as efficient cryogenic pumps for water vapor. In counterflow detectors, the compression ratio for helium should be low during test and high during clean-up. Some of these requirements are contradictory and usually are not achievable without the use of many valves. Because many leak detection systems are required to detect leaks in a few seconds, it is better to reduce the number of valves, which will also affect the cost and reliability issues.

There are additional complications. The compression ratio may be nearly constant in the molecular flow range but is rapidly decreased at viscous flow. When discharge pressure of a high-vacuum pump is relatively high, in viscous flow, gas interactions can occur such that the presence of another gas can affect the compression ratio for helium. This is particularly important for ContraFlow detectors because the ducts between the test object and the outlet of the high-vacuum pump are often in the viscous flow range. This is often the case when oil-free backing pumps are used. These considerations indicate that the pumping speeds and compression ratios of various pumps, conductances of various ducts, and the place where helium is introduced into the pumping train, are very important for an optimized system.

To get a low helium background signal before the next test, it is desirable to reach 10^{-10} torr partial pressure of helium. There is about 5 mtorr of helium in atmosphere (possibly 10 mtorr due to helium accumulation in the ambient air near leak detectors). If the high-vacuum pump is set to have a compression ratio for helium of 100, then the backing pump must have a compression ratio of 10^6 (with the total being 10^8). Usually, oil-free backing pumps have only 10^4 to 10^5. For conventional leak detectors, hybrid turbopumps provide the possibility to bridge this gap. The only remaining issue is, where should helium be introduced for best results?

During counterflow detection, the pumping speed for helium is zero (independent of the size of the pump!). For fast evacuation after test, both high pumping speed and high compression ratios are required. For example, the time needed to reduce the helium pressure (a nonadsorbing gas) by 7 orders of magnitude (in 1 liter of volume, with a pumping speed of 10 L/s), is only 2 seconds. High pumping speed, however, is necessary only in the first few stages of a turbopump. To prevent helium dilution, it should be introduced into the pumping train at a point where helium pumping speed is low. For optimized design, it is desirable to have a step change of pumping speed at the point of helium introduction. This, for example, is done in ContraFlow leak detectors where helium is introduced between the high-vacuum pump and the backing pump.

If helium is introduced at some intermediate point in the body of a conventional turbomolecular pump, the change of speed between stages is insufficient. Even if the helium entrance is arranged between a set of bladed rotors with different blade angles, the change of pumping speed is rather small. Therefore, it would make sense to use a molecular drag pump (with low pumping speed) between the relatively opaque (for helium) turbomolecular pump and the backing pump. Helium would then be introduced ahead of the drag pump. The low pumping speed and the high compresion ratio of the drag pump prevents helium dilution, bridges the compression ratio gap toward the backing pump, and reduces the effect of dissolved helium in oil-sealed backing pumps. There is the additional advantage of using a drag pump between the turbomolecular pump and the backing pump. The small channels of drag pump increase the bulk velocity of flow and enhance the gas interaction effect. Thus, helium is mixed with other gases and has less opportunity to diffuse back against the flow. The hybrid turbopump (for example, Varian Macrotorr) is more suitable for this application because it combines in one body a relatively low-compression turbomolecular pump and a relatively low pumping speed drag pump. Helium can be easily introduced between the two sections.

13.4.9 Supplementary Aids to Leak Detector Operation

Many MSLD manufacturers offer a number of supplementary aids designed to increase the usefulness of the leak detector and provide operating convenience. Most of the devices can be used with either the conventional or Contra-Flow detectors. Custom-tailored devices can also be provided to meet the needs of special test requirements or to further extend the usefulness of the equipment for general testing.

Helium Spray Probe

In order to locate leaks when the leak detector is used to evacuate the test piece, it is necessary to have a controllable source of helium so that the helium can be directed at the point of leakage in small quantities. This procedure uses a spray probe and is an example of the outside-in test technique. The spray probe assembly consists of a flexible hose connected to a regulated helium supply, a spring close valve, and a fine capillary nozzle to direct the helium to a small area.

Calibration Leak

This device is an external reference standard that permits setting up the leak detector to read the leak rate directly on the leak rate meter (Figure 13.22). It is a necessity if leak rates are to be measured or if the leak detector is to reject leaks in excess of a predetermined value. The calibration leaks are normally supplied in the 10^{-7} or 10^{-8} std cm^3/s ranges. The leaks can be conveniently built into the machine, so that calibration can be verified by actuating a switch.

Tuning Leak

This device, which is an accessory for the ContraFlow leak detector, can be adjusted to allow sufficient ambient air into the leak detector inlet port so that the trace amount of helium in the air may be used to tune the leak detector. With this leak open, the mechanical pump maintains vacuum levels of 10 to 300 mtorr. The tuning leak takes advantages of the helium in the air to tune the leak detector. It is also useful for checking that the various functions of the leak detector are operating correctly. In a well ventilated room, a response equal to 5×10^{-6} std cm^3/s helium is typically indicated on the leak rate meter.

Sniffer Probe

The sniffer probe is used in the leak location technique known as the detector probe mode. This device attaches to the inlet of the leak detector with a long flexible tube and is used to pinpoint leaks from test parts pressurized with helium. The probe at the end of the tube is a small orifice which allows vacuum

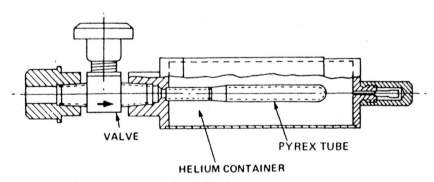

Figure 13.22 Reservoir or permeation leak.

to be maintained in the tube at 100 mtorr in a ContraFlow system or much lower in a conventional leak detector. This probe may be a fixed orifice matched to the pump size or a variable orifice (needle valve) that can be adjusted over the operating range of the pumping system. Any leakage in excess of 10^{-5} std cm^3/s can be readily located.

Audible Alarm

This alarm gives an audible signal that varies in pitch as the indicated leak rate increases. It is useful when the test method makes it difficult to perform the test and watch the leak rate meter at the same time.

REFERENCES

1. AVS 2.1 REV 1973, "Calibration of Leak Detectors of the Mass Spectrometer Type," American Vacuum Society, New York, 1973.
2. ANSI/ASTM E 425, "Standard Definitions of Terms Relating to Leak Testing," *Amer. Soc. Test. & Materials,* Philadelphia, (1973).
3. ASTM E 493-73, "Standard Methods of Test for Leaks Using the Mass Spectrometer Leak Detector in the Inside-Out Testing Mode," American Society for Testing and Materials, Philadelphia, 1973.
4. ASTM E 498-73, "Standard Methods of Testing for Leaks Using the Mass Spectrometer Leak Detector or Residual Gas Analyzer in the Tracer Probe Mode," American Society for Testing and Materials, Philadelphia, 1973.
5. ASTM E 499-73, "Standard Methods of Testing for Leaks Using the Mass Spectrometer Leak Detector in the Detector Probe Mode," American Society for Testing and Materials, Philadelphia, 1973.
6. L. C. Beavis, *Vacuum,* 20, 6, 233 (1970).
7. R. B. Bird, W. E. Stewart, and E. N. Lightfoot, *Transport Phenomena,* John Wiley & Sons, Inc., New York, 1960.
8. M. H. Hablanian, *Vacuum,* 35, 3, 119 (1985).
9. H. F. Kinney, "Practical application of leak detection methods," *J. Vac. Sci. Technol.,* 6(6), 958 (1969).
10. R. W. Kraus, *Introduction to Mass Spectrometry and Its Applications,* Prentice-Hall, Inc., Englewood Cliffs, N.J., 1965.
11. F. J. Norton, *Trans. 8th Nat. Symp. 2nd Int. Congr. Vac. Sci. Technol.,* 1, 8, Pergamon, N.Y. (1962).
12. A. Roth, *Vacuum Sealing Techniques,* Pergamon Press Ltd., Oxford, 1966.
13. D. J. Santeler and T. W. Moller, Fluid flow conversion in leaks and capillaries in *2nd Natl. Symp., Vac. Technol. Trans.,* E. S. Perry and J. H. Durant, eds., Pergamon Press, Elmsford, N.Y., 1956, p. 29.
14. N. G. Wilson and L. C. Beavis, *Handbook of Vacuum Leak Detection,* American Vacuum Society, New York, 1976.
15. C. R. Winkelman, *J. Vac. Sci. Technol.,* 12(1), 435 (1975).
16. R. C. McMaster, Ed., *Nondestructive Testing Handbook, Vol. 1,* Amer. Soc. f. Nondestructive Testing, Publ. by ASNT/ASM, 1982.

17. N. E. Wilson and L. C. Beavis, *Handbook on Vacuum Leak Detection*, American Vacuum Society, New York, 1979.
18. M. H. Hablanian and W. E. Briggs, *Proc. 7th Intern. Vac. Congr.*, Vienna, (1977).
19. M. H. Hablanian, *Vacuum*, *35*,3,p 119 (1985).
20. M. H. Hablanian, *J. Vac. Sci. Technol.*, *A9*(3) 2039 (1991).
21. S. Ruthberg, *IEEE Trans. Comp. & Mfg. Technol.*, CHMT-3,4 (1980).
22. D. G. Mahoney, *J. Vac. Sci. Technol.*, *A12*(4) 1740 (1994).
23. P. J. Abbot and S. A. Tison, *J. Vac. Sci. Technol.*, *A14*(30) 1242 (1996).
24. M. Audi, *Vacuum*, *41*, 7–9, 1856 (1990).

Appendix

UNITS AND FORMULAS

Units of Pump Speed (Volume/Time)
    1 L/s = 2.12 ft^3/min = 3.6 m^3/h
    1 ft^3/min = 0.47 L/s = 1.69 m^3/h
    1 L/m = 0.035 ft^3/min = 0.06 m^3/h

Units of Gas Quantities PV
    1 molar volume = 22.41 L (at standard conditions—STP)
    1 mol = 6.023 × 10^{23} molecules
    1 L·atm = 2.69 × 10^{22} molecules
    1 std cm^3 = 2.69 × 10^{19} molecules
    1 torr·L = 3.53 × 10^{19} molecules
    1 std cm^3 = 0.76 torr·L
    1 std.cm^3 = 1 atm · cm^3
    1 std ft^3 = 7.6 × 10^{23} molecules

Standard conditions are 1 atm at 273°K.

Units of Throughput ($Q = PV$/Time)
    (PV = work; work/time = power)
    1 std cm^3/s = 760 μm (Hg)·L/s
               = 1.6 torr·ft^3/min
    1 torr·ft^3/min = 0.62 std cm^3/s
                  = 472 μm (Hg)·L/s

$1 \, \mu\text{m (Hg)} \cdot \text{L/s} = 1.32 \times 10^{-3} \text{ std cm}^3/\text{s}$
$= 2.12 \times 10^{-3} \text{ torr} \cdot \text{ft}^3/\text{min}$

Vacuum Formulas

A. Conductance = PV rate of gas flowing through orifices or pipes per unit time and per unit pressure difference.

1. Molecular flow for air through orifices

$$C_0 = 11.6 \times \text{Area} \quad \text{L/s, cm}^2$$

$$C_0 = 11.6 \frac{\pi D^2}{4} \quad \text{L/s} \quad \text{for a circular orifice}$$

where D = diameter (cm).

2. Molecular flow for air through long circular pipes

$$C_p = 11.6 \frac{D^3}{L} \quad \text{L/s}$$

where D = diameter (cm), and L = length (cm).

3. Viscous flow ($\sim 10^{-1}$ to 100 torr) through long circular pipes

$$C_v = 180 \frac{D^4}{L} \bar{P} \quad \text{L/s}$$

where D = diameter (cm), L = length (cm), and $\bar{P}$ = average pressure (torr).

4. Conductances in parallel, C_{total}

$$C_t = C_1 + C_2 + \cdots + C_n$$

5. Conductances in series, C_{total}, for long tubes

$$\frac{1}{C_t} = \frac{1}{C_1} + \frac{1}{C_2} + \cdots + \frac{1}{C_n}$$

6. Conductances in series for short tubes of the same diameter

$$\frac{1}{C_t} = \frac{1}{C_1} + \frac{1}{C_2} + \cdots + \frac{1}{C_n} - \frac{n-1}{C_0}$$

7. Resultant pumping speed at a vacuum system

$$\frac{1}{S_n} = \frac{1}{S_p} + \frac{1}{C}$$

where C = conductance (for compatible geometric elements) (L/s), and S_p = pump speed (L/s).

B. Fundamental equation for a system in equilibrium

$$Q = PS$$

where Q = system gas load (torr·L/s), P = equilibrium pressure (torr), and S = volume flow (pumping speed) (L/s). Q is also called throughput and is work per unit time (PV/t).

C. Ideal gas equation

$$PV = nRT$$

where P = pressure (torr), V = volume (L), n = number of moles, R = molar gas constant, and T = temperature (°K).

D. Mean free path for air

$$L = \frac{5 \times 10^{-3}}{P_{torr}} \quad \text{cm}$$

E. Molecular velocity

1. $\overline{V} = 1.5 \times 10^4 \sqrt{\dfrac{T}{M}}$ cm/s

   where T = temperature (°K), and M = molecular weight.

2. $\overline{V} = \sqrt{\dfrac{8kT}{\pi m}}$ cm/s

   where k = Boltzmann's constant, T = temperature (°K), and m = molecular mass (g).

F. Pump-down (roughing) formulas (excluding outgassing)

1. $P = P_0 e^{-St/V}$

   where P_0 = original pressure (torr), P = pressure (torr), S = constant net pumping speed (L/s), V = volume (L), and t = time (s).

2. $t = \dfrac{V}{S} \ln \dfrac{P_1}{P_2}$ (for a constant S)

   or

   $t = 2.3 \dfrac{V}{S} \log \dfrac{P_1}{P_2}$

TENTATIVE STANDARDS OF THE AMERICAN VACUUM SOCIETY*

2.1 Calibration of Leak Detectors of the Mass Spectrometer Type
2.2 Method for Vacuum Leak Calibration
2.3 Procedure for Calibrating Gas Analyzers of the Mass Spectrometer Type
3.1 Unbaked, Ungrooved, Bolted Vacuum Connection Flanges, Nominal Sized 4 in. to 24 in.
3.2 Flanges Bakeable to 500°C
3.3 Method for Testing Flange Seals to 500°C
3.4 Dimensions for Unbaked Flanges, Light Series
3.6 Procedure for Rating All-Metal Valves Bakeable to Above 250°C
4.1 Procedure for Measuring Speed of High-Vacuum Pumps; (see also M. Hablanian, *J. Vac. Sci. & Technol.*, *A5*(4) (Jul/Aug 1987).
4.2 Procedure for Measuring Throughput of High-Vacuum Pumps; (see also M. Hablanian, *J. Vac. Sci. & Technol.*, *A5*(4) (Jul/Aug 1987).
4.3 Procedure for Measuring the Forepressure Characteristics of Oil Diffusion Pumps (Vapor Jet Pump)
4.4 Procedure for Measuring the Ultimate Pressure of Oil Diffusion Pumps (Vapor Jet Pumps)
4.5 Procedure for Measuring Backstreaming of Oil Diffusion Pumps (Vapor Jet Pumps)
4.6 Procedure for Measuring the Warmup and Cooldown Characteristics of Oil Diffusion Pumps (Vapor Jet Pumps)
4.7 Procedure for Measuring the Ultimate Pressure of Pumps Without Working Fluids
4.8 Procedure for Measuring Speed of Pumps Without Working Fluids; see also M. Hablanian, *J. Vac. & Sci. Technol.*, *A5*(4) (Jul/Aug 1987).
4.10 Determining the Refrigerant Consumption and Temperature Characteristics of Baffles and Traps
5.1 Measurement of Blank-off Pressure (Permanent Gases) of Positive Displacement Mechanical Vacuum Pumps
5.2 Presentation of Pumping Speed Curves of Mechanical Pumps; see also B. R. F. Kendall, *J. Vac. Sci. & Technol.*, *A7*(3) (May/Jun 1989).
5.3 Method for Measuring Pumping Speed of Mechanical Vacuum Pumps for Permanent Gases, see also B. R. F. Kendall, *J. Vac. Sci. Technol.*, *A7*(3) (May/Jun 1989).
6.2 Procedure for Calibrating Vacuum Gauges of the Thermal Conductivity Type

---

*120 Wall Street, New York, NY 10005   (212) 248-0200.

APPENDIX 519

6.4 Procedure for Calibrating Hot Filament Ionization Gauges Against a Reference Manometer in the Range 10^{-2}–10^{-5} Torr
6.5 Procedures for the Calibration of Hot Filament Ionization Gauge Controls
7.1 Graphic Symbols in Vacuum Technology
9.1 Reporting of Outgassing Data
9.2 Reporting of Thermal Degassing Data
American Vacuum Society Recommended Practices S-2, Vol. 1.

ISO STANDARDS RELATED TO HIGH-VACUUM TECHNOLOGY

1607/I and II.2 (1977)	Positive Displacement Vacuum Pumps—Measurement of Performance Characteristics—Part II: Measurement of Ultimate Pressure
1608/II and II.2 (1977)	Vapor Vacuum Pumps—Measurement of Performance Characteristics—Part II: Measurement of Critical Backing Pressure
1609.3 (1978)	Vacuum Technology—Flange Dimensions
2861/II.2 (1978)	Vacuum Technology—Quick Release Couplings—Dimensions—Part II: Screwed Type
3529/I.2 (1978)	Vacuum Technology—Vocabulary—Part I: General Terms
3529/II.2 (1978)	Vacuum Technology—Vocabulary—Part II: Vacuum Pumps and Related Terms
3529/III (1978)	Vacuum Technology—Vocabulary—Part III: Vacuum Gauges
3530.2 (1978)	Vacuum Technology—Mass Spectrometer-Type Leak-Detector Calibration
3556/I (1974)	Sputter-Ion Pumps—Measurement of Performance Characteristics—Part I
3567 (1974)	Vacuum Gauges—Calibration by Direct Comparison with a Reference Gauge
3568 (1974)	Ionization Vacuum Gauges—Calibration by Direct Comparison with a Reference Gauge
3570/I (1975)	Vacuum Gauges—Standard Methods for Calibration—Part I: Pressure Reduction by Continuous Flow in the Pressure Range of 10^{-1} to 10^{-5} Pa (10^{-3} to 10^{-7} torr)
3669 (1975)	Vacuum Technology—Bakeable Flanges—Dimensions
5300 (1976)	Vacuum Gauges of the Thermal Conductivity Type—Calibration by Direct Comparison with a Reference Gauge

ASTM STANDARDS RELATED TO HIGH-VACUUM TECHNOLOGY

ASTM E296 (1970)	Ionization Gauge Application to Space Simulators
ASTM E297 (1970)	Calibrating Ionization Vacuum Gauge Tubes
ASTM E425 (1971)	Standard Definition of Terms Relating to Leak Testing
ASTM E427 (1971)	Testing for Leaks Using the Halogen Leak Detector (Alkali-Ion Diode)
ASTM E432 (1971)	Selection of a Leak Testing Method
ASTM E479 (1973)	Preparation of a Leak Testing Specification
ASTM E493 (1973)	Leaks Using the Mass Spectrometer Leak Detector in the Inside-Out Testing Mode
ASTM E498 (1973)	Leaks Using the Mass Spectrometer Leak Detector or Residual Gas Analyzer in the Tracer Probe Mode
ASTM E499 (1973)	Leaks Using the Mass Spectrometer Leak Detector in the Detector Probe Mode
ASTM F78 (1971)	Calibration of Helium Leak Detectors by Use of Secondary Standards
ASTM F97 (1972)	Determining Hermeticity of Electron Devices by Dye Penetration
ASTM F98 (1972)	Determining Hermeticity of Electron Devices by a Bubble Test
ASTM F134 (1972T)	Determining Hermeticity of Electron Devices with a Helium Mass Spectrometer Leak Detector
ASTM D2715 (1971)	Measurements of Volatilization Rates of Lubricants in Vacuum
ASTM E295 (1967T)	Measured Speed of Oil Diffusion Pumps (vapor jet)
ASTM F14 (1972)	Making and Testing Reference Glass-Metal Bead-Seals
ASTM F19 (1971)	Tension and Vacuum Testing Metalized Ceramic Seals
ASTM E294 (1972)	Effective Pumping Speed of Vacuum Chamber Systems

ATMOSPHERIC PRESSURE (0°C)

14.7 psia = 760 torr = 1 std Atm = 101,000 Pascal

Atmospheric pressure at sea level (45° N. Lat.) is 14.7 psia, 760 torr, or 101,000 Pa. Vacuum processes are usually done at pressures much lower than atmospheric pressure. Atmospheric pressure changes with distance above sea level.

APPENDIX

Table A.1 Pressure at Various Altitudes

Altitude (feet)	Pressure torr	Altitude (feet)	Pressure (torr)	Altitude (feet)	Pressure (torr)
−1,000	787.87	7,000	586.49	25,000	282.40
−500	773.83	7,500	575.45	27,500	253.00
0	760.00	8,000	564.58	30,000	226.13
500	746.37	8,500	553.88	35,000	179.33
1,000	732.93	9,000	543.34	40,000	141.18
1,500	719.70	9,500	532.97	45,000	111.13
2,000	706.66	10,000	522.75	50,000	87.497
2,500	693.81	11,000	502.80	55,000	68.889
3,000	681.15	12,000	483.48	60,000	54.236
3,500	668.69	13,000	464.76	70,000	33.662
4,000	656.40	14,000	446.63	80,000	21.010
4,500	644.30	15,000	429.08	90,000	13.208
5,000	632.38	17,500	387.65	100,000	8.356
5,500	620.65	20,000	349.53	120,000	3.446
6,000	609.09	22,500	314.51	140,000	1.508
6,500	597.70				

Source: U.S. Standard Atmosphere, 1962 (NASA).

Table A.2 Partial Pressures of Gases in Air

Gas	Symbol	Percent by Volume	Partial Pressure torr	Partial Pressure Pascal
Nitrogen	N_2	78	593	79,000
Oxygen	O_2	21	159	21,000
Argon	Ar	0.93	7.1	940
Carbon dioxide	CO_2	0.03	0.25	33
Neon	Ne	0.0018	1.4×10^{-2}	1.8
Helium	He	0.0005	4.0×10^{-3}	5.3×10^{-1}
Krypton	Kr	0.0001	8.7×10^{-4}	1.1×10^{-1}
Hydrogen	H_2	0.00005	4.0×10^{-4}	5.1×10^{-2}
Xenon	Xe	0.0000087	6.6×10^{-5}	8.8×10^{-3}
Water vapor	H_2O	Variable,	(comparable to Argon).	

At standard conditions (760 torr, 0°C), each gas exerts a pressure relative to its percent of the total volume: for example, $N_2 = 78\% = 0.78 \times 760 = 593$ torr. (Values are for dry air.)

GAS SENSITIVITY TABLES FOR IONIZATION GAUGES

The following table lists relative gauge sensitivities for various gases. The values are derived by empirical methods substantiated by measurements reported in literature. The multiple values found in the table represent several different references. This table has been compiled and published by Robert L. Summers of Lewis Research Center, NASA Technical Note TN D-5285, National Aeronautics and Space Administration, Washington, D.C., June, 1969.

To convert readings from the ionization gauge control (normally calibrated for nitrogen), divide the indicated pressure by the number listed in the third column for the particular gas (S/S_{n_2}).

Table A.3 Gas Sensitivity for Ionization Gauges

Substance	Formula	Relative Ionization Gage Sensitivity, S/S_{N_2}
Acetaldehyde	C_2H_4O	2.6
Acetone	$(CH_3)_2CO$	3.6
		4.0
		3.6
Acetylene	C_2H_2	1.9
		2.0
Air		1.0
		0.98
Ammonia	NH_3	1.3
		1.2
		1.3
Amylene:		
iso-	iso-C_5H_{10}	5.9
cyclo-	cy-C_5H_{10}	5.8
Argon	Ar	1.3
		1.1
		1.2
		0.9
Benzene	C_6H_6	5.9
		5.8
		5.7
		5.9
		6.0
Benzoic acid	C_6H_5COOH	5.5
Bromine	Br	3.8
Bromomethane	CH_3Br	3.7

Table A.3 Continued

Substance	Formula	Relative Ionization Gage Sensitivity, S/S_{N2}
Butane:		
n-	n-C_4H_{10}	4.9
		4.7
iso-	iso-C_4H_{10}	4.6
		4.9
Cadmium	Cd	2.3
		3.4
Carbon dioxide	CO_2	1.4
		1.4
		1.5
		1.5
		1.4
Carbon disulfide	CS_2	5.0
		4.7
		4.8
Carbon monoxide	CO	1.05
		1.05
		1.1
Carbon tetrachloride	CCl_4	6.0
		6.3
Cesium	Cs	4.3
		2.0
		4.8
Chlorine	Cl_2	0.68
		2.6
		1.6
Chlorobenzene	C_6H_5Cl	7.0
Chloroethane	C_2H_5Cl	4.0
Chloroform	$CHCl_3$	4.7
		4.8
		4.8
Chloromethane	CH_3Cl	2.6
		3.2
		3.1
Cyanogen	$(CN)_2$	2.8
		3.6
		2.7

(*continued*)

Table A.3 Continued

Substance	Formula	Relative Ionization Gage Sensitivity, S/S_{N_2}
Cyclohexylene	C_6H_{12}	7.9
		6.4
Deuterium	D_2	0.35
		0.38
Dichlorodifloromethane	CCl_2F_2	2.7
		4.1
Dichloromethane	CH_2CL_2	3.7
Dinitrobenzene:	$C_6H_4(NO_2)_2$	
o-		7.8
m-		7.8
p-		7.6
Ethane	C_2H_6	2.6
		2.8
		2.5
Ethanol	C_2H_5OH	3.6
		2.9
Ethyl acetate	$CH_3COOC_2H_5$	5.0
Ethyl ether	$(C_2H_5)_2O$	5.1
		5.1
Ethylene	C_2H_4	2.3
		2.4
		2.2
		2.2 to 2.5
Ethylene oxide	$(CH_2)_2O$	2.5
Helium	He	0.18
		0.15
		0.13
		0.12
Heptane	C_7H_{16}	8.6
Hexadiene:		
1,5-	$1,5\text{-}C_6H_{10}$	6.4
cyclo-	$cy\text{-}C_6H_{10}$	6.0
Hexane	C_6H_{14}	6.6
Hexene:		
1-	$1\text{-}C_6H_{12}$	5.9
cyclo-	$cy\text{-}C_6H_{10}$	6.4
Hydrogen	H_2	0.46
		0.38
		0.41
		0.45

Table A.3 Continued

Substance	Formula	Relative Ionization Gage Sensitivity, S/S_{N2}
Hydrogen bromide	HBr	2.0
Hydrogen chloride	HCl	1.5
		1.6
		2.0
		1.5
Hydrogen cyanide	HCN	1.5
		1.6
Hydrogen floride	HF	1.4
Hydrogen iodide	HI	3.1
Hydrogen sulfide	H_2S	2.2
		2.2
		2.3
		2.1
Iodine	I_2	5.4
Iodomethane	CH_3I	4.2
Isoamyl alcohol	$C_5H_{11}OH$	2.9
Isobutylene	C_4H_8	3.6
Krypton	Kr	1.9
		1.7
		1.7
Lithium	Li	1.9
Mercury	Hg	3.6
Methane	CH_4	1.4
		1.5
		1.6
		1.4 to 1.8
		1.5
		1.5
Methanol	CH_3OH	1.8
		1.9
Methyl acetate	CH_3COOCH_3	4.0
Methyl ether	$(CH_3)_2O$	3.0
		3.0
Naphthalene	$C_{10}H_8$	9.7
Neon	Ne	0.30
		0.31
Nitrobenzene	$C_6H_5NO_2$	7.2
Nitrogen	N_2	1.0
Nitrotoluene (o-, m-, p-)	$C_6H_4CH_3NO_2$	8.5

(*continued*)

Table A.3 Continued

Substance	Formula	Relative Ionization Gage Sensitivity, S/S_{N2}
Nitric oxide	NO	1.3
		1.2
		1.0
Nitrous oxide	N_2O	1.5
		1.7
		1.7
		1.3 to 2.1
Oxygen	O_2	1.0
		1.1
		.9
		.9
Pentane:		
n-	n-C_5H_{17}	6.2
		6.0
		5.7
iso-	iso-C_5H_{17}	6.0
neo-	$(CH_3)_4C$	5.7
Phenol	C_6H_5OH	6.2
Phosphine	PH_3	2.6
Potassium	K	3.6
Propane	C_3H_8	4.2
		3.7
		3.7 to 3.9
		3.6
Propene oxide	C_3H_6O	3.9
Propene:		
n-	n-C_3H_6	3.3
		3.2 to 3.7
cyclo-	cy-C_3H_6	3.6
Rubidium	Rb	4.3
Silver perchlorate	$AgClO_4$	3.6
Sodium	Na	3.0
Stannic iodide	SnI_4	6.7
Sulfur dioxide	SO_2	2.1
		2.3
Sulfur hexafloride	SF_6	2.3
		2.8
Toluene	$C_6H_5CH_3$	6.8
Trinitrobenzene	$C_6H_3(NO_2)_3$	9.0

Table A.3 Continued

Substance	Formula	Relative Ionization Gage Sensitivity, S/S_{N2}
Water	H_2O	1.1
		1.0
		0.8
Xenon	Xe	2.9
		2.2
		2.4
Xylene:		
o-	o-$C_6H_4(CH_3)_2$	7.8
p-	p-$C_6H_4(CH_3)_2$	7.9

RECOMMENDED PROCEDURE FOR MEASURING PUMPING SPEEDS[a]

The American Vacuum Society's tentative standard for measuring pumping speed was first issued in 1963 and published in a revised form in 1971. The newly established Recommended Practices Committee of AVS felt that it was appropriate to review the standard to reconcile it with modern developments in technology and technique. These include among others: new units for pressure measurement, establishment of new pumping methods, recognition of emergence of international standards, and general improvements in high-vacuum practices. The main substantial change involves the positioning of the pressure gauge in the measurement chamber at a distance from the pump entrance equivalent to 0.5 of pump inlet diameter. This will reduce the measured speed values by 10% to 15%, depending on pump size and inlet geometry, but will correspond more closely to the definition of pumping speed in relation to the time of evacuation of a given volume.

Introductory Comments

The purpose of this article is the introduction of the new Recommended Practice for measuring pumping speed of high-vacuum pumps.

The text has been developed by a group which is part of the AVS Committee for Recommended Practices. It is presented here in order to solicit comments from a wider audience before the text is published in a final version with the

(*text continues on p. 535*)

---

[a]Reprinted with permission from *J. Vac. Sci. Technol., A5 (4)*, (Jul/Aug 1987). Copyright 1987 American Vacuum Society.

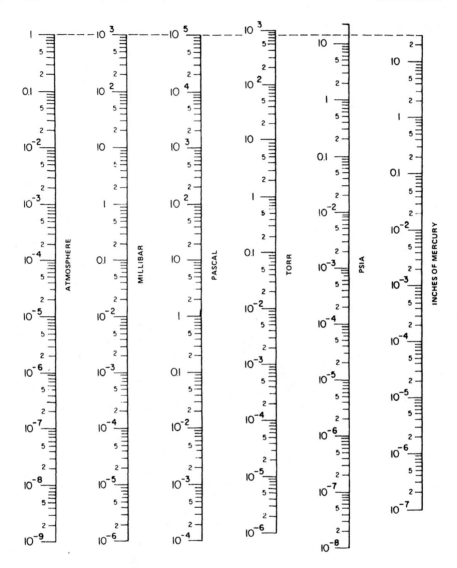

HORIZONTAL LINES REPRESENT EQUAL PRESSURE VALUES
1atm = 760 Torr = 1013mb = 101325Pa = 14.7psia = 29.9in Hg

Figure A.1 Units of pressure conversion nomograph.

APPENDIX

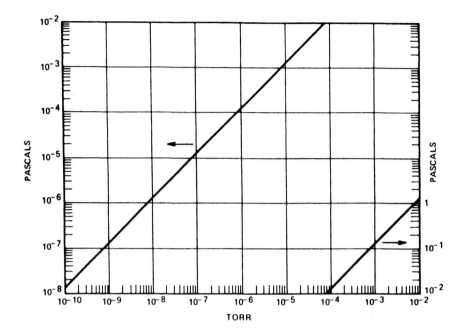

	Pascal (N/m²) (Pa)	Torr	Standard Atmosphere (atm)	Millibar (mbar)	Dyne per Square Centimeter (dyn/cm²)
1 Newton per square meter (N/m²) = Pascal =	1	7.5 × 10⁻³	9.87 × 10⁻⁶	10⁻²	10
1 Torr = 1mm Hg =	133	1	1.32 × 10⁻³	1.33	1,330
1 standard atmosphere (atm) =	101,000	760	1	1,010	1,010,000
1 millibar (mbar) =	100	0.75	9.87 × 10⁻⁴	1	1,000
1 dyne/square centimeter (dyn/cm²) =	10⁻¹	7.5 × 10⁻⁴	9.87 × 10⁻⁷	10⁻³	1

Figure A.2 Units of pressure.

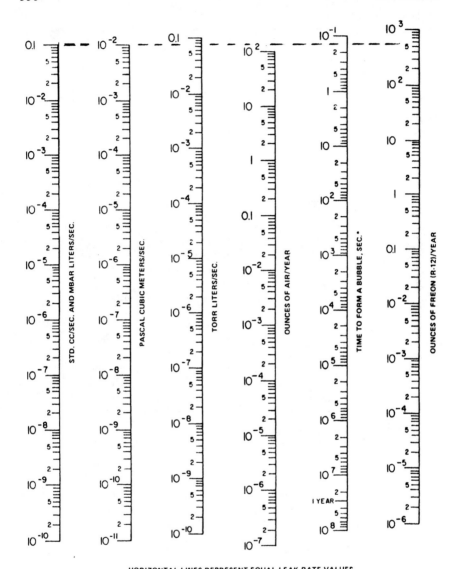

Figure A.3 Units of leakage rate conversion nomograph.

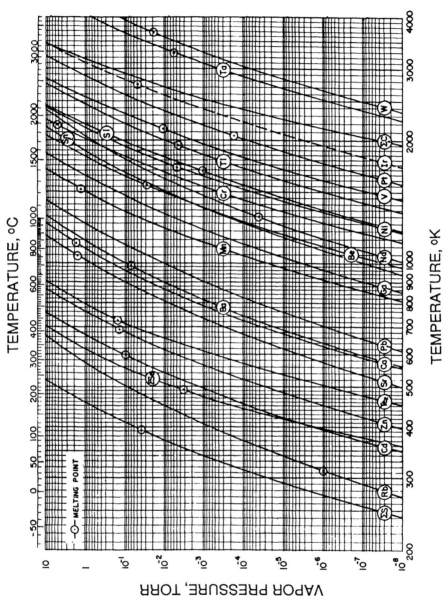

Figure A.4 Vapor pressures of common elements at various temperatures (RCA Review, R. E. Honig, 1957).

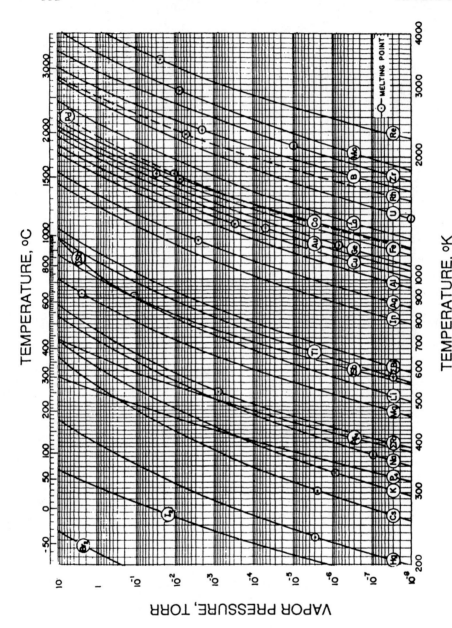

Figure A.4 Continued.

APPENDIX 533

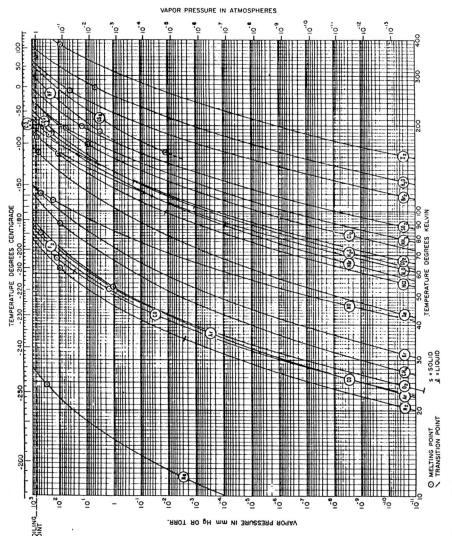

Figure A.4 Continued.

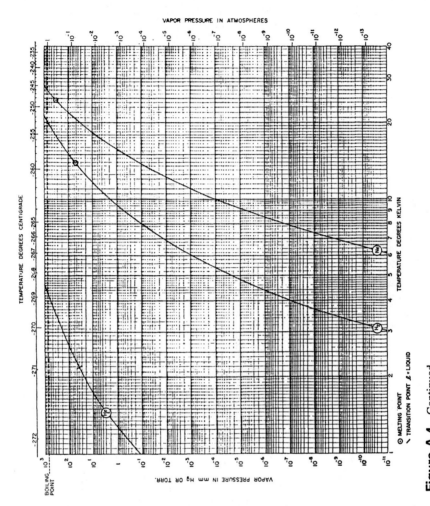

Figure A.4 Continued.

APPENDIX 535

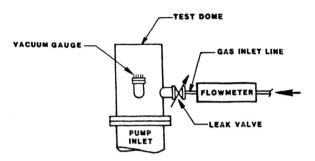

Figure A.5 Measurement apparatus for the flowmeter method.

AVS Committee approval (John Sullivan, MKS, Chairman). The composition of the group which developed the present procedure (the Subcommittee) is as follows: M. D. Boeckman (Martin Marietta Energy Systems); B. B. Dayton (Consultant, East Flat Rock, NC); M. H. Hablanian (Chairman, Varian/VPD); T. A. Howard (Perkin Elmer Vacuum Products); R. W. Hyland (NBS); B. R. Kendall (Pennsylvania State University); J. F. O'Hanlon (IBM, Yorktown Heights, NY); D. J. Santeler (Process Applications); G. J. Still (Secretary, EG & G, San Ramon, CA); and I. Warshawsky (NASA, Cleveland).

Basic Features

The following recommended procedure is intended to replace the former AVS Standards 4.1, 4.2, and 4.8: The major substantive change involves the location of the pressure gauge in the test chamber (dome) used to make the measurements. The gauge location is moved away from the pump inlet from one-

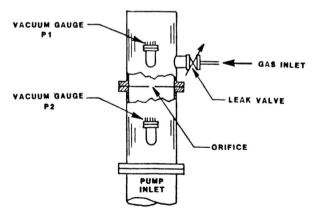

Figure A.6 Measurement apparatus for the orifice method.

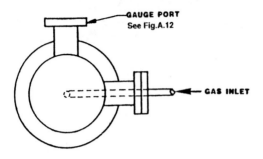

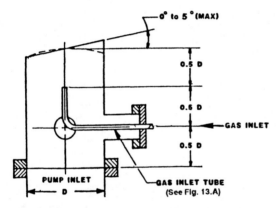

Figure A.7 Test dome for flowmeter method.

quarter to one-half of the pump inlet diameter. This change is in agreement with the ISO (International) Standard 1608/1. It will result in somewhat lower pumping speed values compared to the old AVS Standard. For pumps of conventional inlet geometry the reduction will be approximately 10% to 15%.

The change in gauge location produces measurements which correspond better with the basic definition of pumping speed as it relates to the rate of pressure reduction in a large vacuum chamber [1]. The issue of gauge location is associated with the lack of isotropic conditions in the vicinity of inlets to high-vacuum pumps. Other changes are mainly of editorial nature. Pressure units have been changed to correspond to the ISO Standard. Torr units are used in parentheses, although for the purposes of discussion of pressure ranges, etc., the mbar units can be used instead of torr.

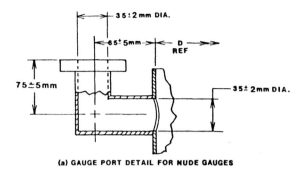

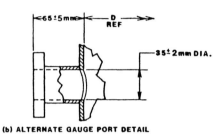

Figure A.8 Gauge port details.

References to ultimate pressure have been omitted because it is a somewhat elusive concept. It is often very difficult to separate the limitations of the pump and the effects of the outgassing in the test apparatus.

Other changes involve clarifications in figures and text to remove ambiguities in interpretation. The ambient temperature at which the tests should be

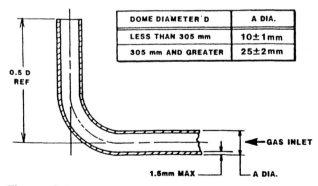

Figure A.9 Inlet tube detail.

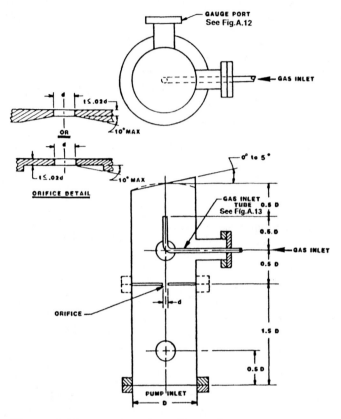

Figure A.10 Test dome for orifice method.

conducted was changed slightly to 22 ± 3°C because the committee felt this represented a likely average in laboratories and industrial environment in the United States. (Corrections should be made if the temperature is significantly different.)

The question raised may be why is there a need for an AVS procedure while an ISO Standard exists? The answer is that ISO 1608/1 is somewhat more general in its recommendations than we would like to see in order to assure that truly comparative results are achieved by users. In principle, there are no conflicts between the two procedures. The AVS Recommended Practice may be viewed as a further elaboration attempting to standardize the test apparatus and measurement procedures in more detail.

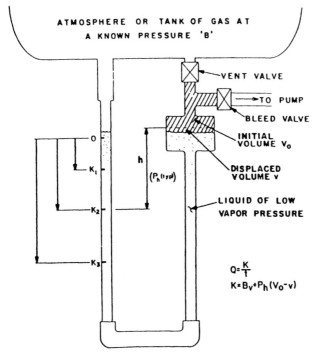

Figure A.11 Device for measuring flow rate.

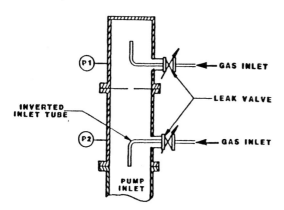

Figure A.12 Arrangement for in situ comparison.

New AVS 4.1. Procedure for Measuring Pumping Speed and Throughput of High Vacuum Pumps (Formerly AVS 4.1, 4.2, and 4.8)

Scope

This procedure describes methods for measuring the speed and throughput of vacuum pumps capable of producing ultimate pressures less than 1×10^{-2} Pa ($\approx 1 \times 10^{-4}$ torr) in the molecular flow region. The apparatus and methods described here may also be used to measure net speed above baffles, traps, or valves.

Definitions

Pumping speed. The volume of specified gas at 22°C which flows in unit time through the pump inlet under ideal conditions. The units adopted for pumping speed are cubic meters per second at 22°C (m^3/s at 22°C) or liters per second at 22°C (1/s at 22°C).

Throughput. The quantity of gas in pressure–volume units at 22°C flowing per unit time through (or captured by) a pump under steady-state conservative conditions. The units adopted for throughput are the Pascal-cubic meter per second at 22°C (Pa m^3/s at 22°C) or the torr 1 per second at 22°C (torr 1/s at 22°C).

Measured pumping speed. The quotient of the throughput (Q) of an introduced test gas and the equilibrium pressure (p) of the introduced test gas at a specified position in a given test dome under specified conditions of operation.

Test dome. A chamber of specified form and dimensions attached to the inlet of the pump through which a measured flow of gas is admitted to the pump, and which is equipped with means of pressure measurement.

Methods

Flowmeter method. The rate at which the test gas enters the test dome is measured via a suitable flowmeter (Fig. A.5). The resultant throughput in conjunction with the resultant equilibrium pressure within the dome is used to calculate the measured pumping speed.

Orifice method. When the pumping speeds and flow rates are sufficient to provide molecular flow conditions across an orifice, the throughput of the test gas is determined by measuring the pressure drop within the dome across an orifice (Fig. A.6). This throughput in conjunction with the resultant equilibrium pressure on the downstream side of the orifice is used to calculate the measured pumping speed.

Apparatus

Test dome–flowmeter method. The test dome is to be of circular cross section and of inside diameter D as shown in Figure A.7. For pumps with inlet diameters of 50 mm or greater, the inside diameter D is to be equal to that of the inlet of the pump to within $\pm 2\%$. For pumps with inlet diameters less than 50 mm, the inside diameter of the test dome is to be 50 ± 1 mm. The total mean height of the test dome is to be $1.5D$. The top of the dome is to be flat and perpendicular to the dome axis, slightly dished outward, or inclined to an angle of 5° maximum.

The dome is to be equipped with two cylindrical ports projecting perpendicularly outwards from the dome wall, with axes at a height of $0.5D$ above the bottom flange. One port, per Figure A.8, is fitted with a vacuum gauge. The gauge connection should be a metal-sealed flange to reduce outgassing in the vicinity of the gauge. The other port is fitted with a gas inlet tube containing a right-angle bend upward. The inlet tube terminates along the central axis of the dome at a distance of $0.5D$ below the top of the dome. A variable leak valve is mounted at the entrance of the gas inlet tube. Details of the inlet tube are shown in Figure A.9. The basic geometry of the dome must not be distorted substantially by the addition of auxiliary ports, such as for roughing. The entire dome setup can be horizontal or at an angle for pumps that can be used in such positions.

Note: The only substantial difference between this dome and the one specified in the tentative standard AVS 4.1, Procedure for Measuring Speed of Oil-Diffusion Pumps (1963), is the vacuum gauge location.

Test dome-orifice method. The test dome is to be of circular cross section and of inside diameter D as shown in Figure A.10. For large pumps with inlet diameters of 50 mm or greater, the inside diameter D is to be equal to that of the inlet of the pump to within $\pm 2\%$. For pumps with inlet diameters less than 50 mm, the inside diameter of the test dome is to be 50 ± 1 mm. The total height of the test dome is to be $3D$. The top of the dome is to be flat, and perpendicular to the dome axis, slightly dished, or inclined to an angle of 5°maximum. All features of the upper dome to be the same as in the previous section.

A thin flat plate furnished with a central circular orifice (Fig. A.10) of diameter d is mounted at a height of $1.5D$ above the bottom flange of the dome. The diameter of the orifice d is selected in relation to the expected nominal pump speed such that the indicated pressure ratio in the two sections of the test dome ($P1/P2$) is between 50 and 100. The thickness t should always be less than 0.02 times the diameter d.

The section of the dome above the orifice is to be equipped with two cylindrical ports projecting perpendicularly outwards from the dome wall, with axes at a height $D/2$ above the plane of the plate containing the orifice. One port, per Figure A.8 is fitted with a vacuum gauge, designated $P1$. The other port is fitted with a gas inlet tube containing a right-angle bend upward. The inlet terminates along the central axis of the dome at a distance $0.5D$ below the top of the dome. A variable leak valve is mounted at the entrance of the gas inlet tube. Details of the inlet tube are shown in Fig. A.9.

The lower section of the test dome is equipped with one cylindrical port projecting perpendicularly outwards from the dome wall, with its axis at a height of $0.5D$ above the bottom flange of the dome. This port, per Figure A.8, is fitted with a vacuum gauge, designated $P2$. The gauge connections for both $P1$ and $P2$ should be metal-sealed flanges to reduce outgassing in the vicinity of the gauges. Gauges $P1$ and $P2$ should be of the same type and model.

The basic geometry of the dome must not be distorted substantially by the addition of auxiliary ports, such as for roughing. The entire dome setup may be horizontal or at an angle for pumps that can be used in such positions.

Pressure measurements. Adequate vacuum gauges are to be used for the measurement of total pressure over the range of 1×10^{-2} Pa ($\approx 1 \times 10^{-4}$ torr) to ten times less than the lowest pressure for which the speed is to be measured. The gauges are to be calibrated using the test gas of interest under conditions that approximate the measurement realm as closely as possible. Existing standards for gauge calibration (AVS, ISO, etc.) are to be used.

Leak valve. The variable leak valve should provide stable control of flow and flow changes as reflected in equilibrium pressures within the pressure range of interest. It should be bakable for measurements below 10^{-6} Pa.

Flow measurement devices. Application of the flowmeter method requires an accurate measurement of the test gas flow into the dome. Any precision (calibrated) device may be used. One example is presented below in "Flow Measurement Devices."

Determination of Measured Pumping Speed

Flowmeter method. A continuous flow of the test gas is admitted to the dome via the variable leak valve to produce a pressure p as measured with the gauge. The throughput of the test gas Q that produces the pressure p is measured via the flowmeter. The measured pumping speed S_p at the pressure p is then derived from the expression

$$S_p = Q/(p - p_0)$$

where p_0 is the indicated pressure when the leak valve is closed.

Orifice method. A continuous flow of the test gas is admitted to the dome via the variable leak valve to produce an equilibrium pressure p_1 above the orifice and an equilibrium pressure p_2 below the orifice as measured with the $P1$ and $P2$ gauges, respectively.

If C is the calculated molecular conductance for the test gas of the orifice (in which calculation has included the appropriate Clausing factor correction for the thickness to diameter ratio of the orifice used) and Q is the throughput of the test gas entering the dome, then

$$Q = C[(p_1 - p_{01}) - (p_2 - p_{02})] \tag{A.1}$$

where p_{01} and p_{02} are the indicated pressures on gauges $P1$ and $P2$ when the leak valve is closed. The measured pumping speed at the pressure p_2 is

$$S_{p2} = Q/(p_2 - p_{02}) \tag{A.2}$$

Substitution of Eq. (1) in Eq. (2) gives

$$S_{p2} = C\left(\frac{p_1 - p_{01}}{p_2 - p_{02}} - 1\right) \tag{A.3}$$

Note: The pressure p_1 must always be such that C represents true molecular conductance. (p_1 must be low enough so that the mean free path is always greater than the diameter d).

Note: See also "In Situ Gauge Comparison" on page 545.

Procedure

Base pressure measurement (prior to test). The pump under test is mounted to the test dome per Figures A.5 or A.6, and the pump is started. With the variable leak valve closed, the pump and dome are allowed to reach an equilibrium pressure that is at least ten times less than the lowest pressure for which speed is to be measured. This pressure (p_{01}, p_{02}) is then recorded. If ionization gauges are used and required degassing, this should be accomplished at least 30 min prior to recording the ultimate pressure.

Bakeout of the pump and/or dome may be necessary to achieve the required base pressure. If bakeout is performed, the apparatus must have cooled to 22 ± 3°C prior to recording the system base pressure and starting speed measurements.

Stabilization of capture pumps Capture pumps, such as sputter-ion pumps and cryopumps, are to be stabilized prior to taking measurements. To stabilize the pump, a quantity of test gas (in pressure–volume units) is pumped which is in excess of $4S_n$ Pa 1 (0.03 S_n torr 1), where S_n is the nominal pump speed in liters per second. After pumping this quantity of test gas, the pump is allowed to reach its ultimate equilibrium pressure without bakeout. This pressure must not exceed 0.1 of the lowest pressure at which the pumping speed is to be measured.

Speed measurement. The inlet pressure is increased via the variable leak valve in steps to establish pressures over the range starting from the lowest value up to the highest value. A minimum of three steps per decade, roughly doubling the pressure between each step, is recommended. The pressure should then be decreased in steps to check hysteresis and any variation in speed measurements. A period of 10 min should be allowed between when the pressure is changed and when readings are taken to ensure that equilibrium conditions prevail. For measurements in the ultrahigh vacuum range, it may be necessary to wait a longer period of time between pressure settings to achieve equilibrium. For this purpose, equilibrium conditions are said to have been established when any increasing or decreasing trend of the measured speed or of the indicated pressure does not exceed $\pm 5\%$ in a period of time which may be 5 min at 10^{-5} Pa or 1 h in the ultrahigh vacuum range. If ionization gauges are used, they shall not be degassed during these measurements. Residual gas analyzers, etc., are not to be used during these measurements to avoid interaction.

Presentation of Results

For each test gas under consideration, the measured speed of the pump will be displayed on a graph on which the speed is plotted as a function of inlet pressure. The throughput Q may be displayed on a graph on which Q is plotted as a function of inlet pressure. The pressure p_0 prior to the time the measurements were made shall be indicated. The precise identity of the test gas is to be reported. The pressure gauge type and methods used for calibration should be included in the test report. The prevailing temperature during the measurements must be reported.

The following information is also to be reported wherever applicable: (a) type and speed of the forepump system (if applicable); (b) type and quantity of diffusion-pump fluid (for diffusion pumps); (c) power input to the pump; (d) cooling water flow rates (if any), input temperature, and discharge temperature; (e) ion pump currents and voltages for each speed point (for ion pumps); (f) type and identity of power supply (for ion pumps) or compressor (for cryopumps) used to drive the pump during the test; and (g) type of flowmeter(s) used.

Note: It is recommended also to specify the maximum permissible steady-state throughput capability of the high-vacuum pump, operating without an overload condition. Such overload condition may be manifested by sudden reduction of pumping speed, excessive motive fluid loss, reduction of rotational speed, overheating, excessively short regeneration periods, etc. A plot of pressure as a function of throughput can be used to indicate the overload region.

Flow Measurement Devices

Constant pressure displacement meters of many types have been used for measuring small flow rates. An example is shown in Figure A.11.

The tubes used for measuring the displaced volume should be precision burettes of glass tubing selected for uniformity of bore and having accurately measured inside diameters (accuracy 0.25%, commercially available). The instruments should be designed, calibrated, and used in such a way as to measure the actual quantity of gas transferred to the test dome in some conveniently measurable time. Ambient temperature during the measurement should be 22 ± 3°C. For very small flows, the entire apparatus must be insulated from draft and kept at constant temperature.

Referring to Figure A.11 a leak rate is determined by observing the time i required for the displaced fluid to rise (or fall) through some arbitrary distance h in the displacement tube. V_0 is the volume and B the pressure of gas filling the displacement meter at time zero, $V_0 - \upsilon$ is the volume and $B - P_h$ the pressure of the gas remaining at time t (P_h is the pressure due to fluid head h). Then the leak rate Q in PV units per second is

$$Q = BV_0/t - (B - P_h)(V_0 - \upsilon)/t$$
$$= B\upsilon/t + P_h(V_0 - \upsilon)/t \qquad (A.4)$$

Displacement devices may be designed so that the quantity $P_h(V_0 - \upsilon)$ is negligibly small compared with the quantity $B\upsilon$. In such cases, Eq. (1) reduces to $Q = B\upsilon/t$. If it is not convenient to make $P_h(V_0 - \upsilon)$ negligibly small, Eq. (1) may be used to construct a displacement scale which reads quantity change directly (in Pa 1 or torr 1).

1. Measurement procedure

(1) Open vent valve and adjust bleed valve to obtain desired pressure at pump inlet.
(2) Close vent valve and measure time period for liquid level to drop from mark zero to K_1, K_2 or K_3.
(3) Open vent valve to return liquid back to zero level. Allow time between measurements for the liquid to drain fully.

In Situ Gauge Comparison

In the orifice method, gauge $P1$ can be compared *in situ* to gauge $P2$. Hence it may only be necessary to calibrate gauge $P2$ in advance of the test. This comparison is accomplished by running the entire system at the same pressure and determining a correction factor to be applied to subsequent $P1$ readings. This is readily accomplished by installing an inverted gas inlet tube in the lower dome section as shown in Fig. A.12. The inverted tube is identical in size and form as the upper inlet tube.

With the test gas being bled through the lower inverted tube only, the relative calibration of the $P1$ gauge with respect to the $P2$ gauge can be determined for each pressure point i. The pressure produced in this manner must be at least ten times the lowest (ultimate) pressure. A correction factor K_i can then be derived for each i as follows:

$$K_i = \frac{P_{2i}}{P_{1i}}$$

where P_{2i} and P_{1i} are the *indicated* pressures at point i.

Note: K_i should be determined for several pressures over the range of interest since it may not be constant.

The measured pumping speed can then be found from the expression

$$S_p = C\left(\frac{K_i(p_1 - p_{01})}{p_2 - p_{02}} - 1\right)$$

where P_1 and P_{01} are nominal (uncorrected) gauge readings.

A parallel "Recommended Practice for Testing Mechanical Vacuum Pumps" by B. R. F. Kendall was published by *J. Vac. Sci. Technol., A7*(3), 2404–2407, (May/Jun 1989) and is available from American Vacuum Society, 120 Wall Street, New York, NY 10005; (212) 248-0200.

REFERENCES

1. M. D. Boeckman, *J. Vac. Sci. Technol., A4*, 353 (1986).
2. *Journal of Vacuum Science and Technology*, published by the American Institute of Physics for the American Vacuum Society (120 Wall Street, 32nd floor, New York, NY 10005-3993).
3. *Vacuum*, published by Pergamon Press Ltd. for the British Vacuum Council, Oxford, England.
4. The American Vacuum Society publishes a number of monographs on subjects relevant to high-vacuum technology. It also conducts an annual symposium, numerous short courses, and stages equipment exhibitions.

Index

Adsorption (and desorption), 25-30, 59, 63, 64
   heat of, 27
   isotherm, 26
Alumina, activated, 4
Anodized aluminum, 26
Atmosphere, composition, 34, 38-39
Atoms, number in cm^3, 16-18, 31, 455

Backstreaming, *see* Pumps
Baffles and traps impedance, 223, 254-255
Baking (vacuum chambers), 406-407
Bernoulli's principle, 166

Capacitance manometers, 435
Compression ratio, 62-65
Compressors, 4, 62
Condensation, 25, 319
Conductance
   molecular, 55-58, 96-112
   molecular series elements, 105, 108, 110-113

[Conductance]
   short tubes, 101-104
   viscous flow, 113-115
Conductivity, thermal of gases, 20, 21
ConFlat flanges, 407-409
Convection, thermal, 20, 21
Crossover, between pumps, *see* Transition
Cryopumps, *see* Pumps, cryogenic

Diaphragm gauges, capacitance manometers, 149, 434-436
Diaphragm pumps, 196, 199-203
Density gradient, 52
Density, particle, 1, 16, 430
Desorption (*see also* Adsorption), 27, 28, 32, 70-71, 87-88
Diffusion, of one gas in another, 47, 71-72
Diffusion pumps, *see* Pumps, vapor jet
Displacement, mechanical pumps, 144

547

Distance between collisions, 1, 17, 18, 20, 46, 62, 72

Ejector pumps (venturi), 166–168
Energy efficienty, of vacuum pumps, 65
Evacuation time, 59, 83–92, 128
Evaporation, rate of, 24, 25
Exposure to air, of vacuum chamber, 116

Finger print outgassing, 96
Flow
 delay due to adsorption, 68–70
 mass flow, 45
 molecular, 46, 50–52, 55, 61–62
 regimes (types), 45–46, 67
 through an orifice, 54–55, 58, 97
 through nozzles, 47–50
 velocity, 43–44, 55
 volumetric, 43, 55, 65
Flow of gasses
 electrical analog, 41, 64
 heat transfer analog, 42
 in small passages, capillary, 66–67, 69

Gas ballast, mechanical pumps, 151–154
Gas content, of materials, 31, 78–79
Gas laws, 17
Gas load measurement, 126–127, 392
Gauges
 Bourdon type, 433–434
 calibration, 452–454
 choice of, 451
 high-vacuum accuracy and precision, 454
 high-vacuum degassing, 457–459
 high-vacuum effect of gauge location, 456
 high-vacuum interaction between gauges, 459
 ionization, 4, 444–451
 ionization, Bayard-Alpert type, 445–449

[Gauges]
 ionization, cold cathode, 444–445
 ionization, extractor type, 449
 ionization, long-term sensitivity changes, 457
 ionization, "millitorr" type, 447–448
 ionization, "nude", 446, 448
 ionization, pumping speed of, 450–451
 ionization, sensitivity to various gases, 449–450, 453, 522
 McLeod, 149, 436–438
 mechanical manometer, 432–435
 partial pressure, 462–469
 Pirani, 440–441
 spinning rotor, 442–443
 thermal, constant temperature operation, 441
 thermocouple, 327, 328, 369, 439, 440, 484
 thermocouple, sensitivity to various gases, 440
 vacuum, 4
Getter pumps, *see* Pumps, gettering

Heat transfer, 21
Helium, tracer gas, 471, 496–497

Interactions between gases (in pumps), 301
Ion-getter pump, *see* Pumps, ionization

Leak detection
 acoustical, 481–482
 calibrated leaks, 512
 dye penetration, 486
 halogen type, 485
 helium mass spectrometer, 493
 gas bubble testing, 483
 ionization gauges and pumps, 485, 488–489
 mass spectrometer leak detector comparison between conventional and ContraFlow types, 503
 ContraFlow type, 501–505
 conventional type mass spectrometer, 500

INDEX

[Leak detection]
  probe technique, 495-496
  spectrometer design, 498-499
  measurements, units, 476
  methods, 479
  pressure decay or rise, 384
  radioisotope type, 387-388
  requirements, 473-475
  system sensitivity, 494
  test method selection, 490-491
  thermocouple gauges, 484
Leak rate units, see Appendix
Leaks, sources of, 474
Liquid ring pumps, 167-168

Magdeburg hemispheres, 5
Mass flow, 45
Mass spectrometers, 6, 493
Mean free path, see Distance between collisions
Mechanical pumps, see Pumps, mechanical
Microelectronics, 3
Molecular beams, 59
Molecular collisions, on a surface, 18, 25
Molecular weight, 19
Monolayer formation time, 29, 405
Monolayers, of gas absorbed, effect of, 26, 27

Oil-free coarse vacuum pumps, 170-203, 342
Orifice, see Flow, through an orifice
O-ring compression seal (gas source), 96, 191, 235
Outgassing, 31-37, 79, 87-88, 95, 115, 128, 217, 405, 406-407
Overloading of pumps, see Pump, overload and Chapter 10

Permeability, permeation, 37-38, 47, 73-75
Pressure
  absolute, 15
  crossover, see transition
  distribution (in tubes), 99-100

[Pressure]
  inlet, 87, 131
  gradient, 52
  units, 15, 16 (see also Appendix)
  ultimate, 95-96
Pump effectiveness, in pressure regions, 138
Pump overload, 118-119, 125-127, 221, 225, 239, 266, 293, 383
Pumping speed, 58-61, 85, 146, 162
Pumping speed, early pumps, 9
Pumping trapped volumes, 92-94
Pumps
  basic limitations, 65-66, 80
  cryogenic pumps
    argon capacity, 335
    compressor construction, 327-330
    cool-down time, 324-325
    expander construction and operation, 326-329
    exposure to air, 335-336
    cryogenerators, helium, 322-332
    hydrogen capacity, 332-334, 341
    large sizes, 345-346
    maintenance, 331
    nitrogen capacity, 338
    Polycold pumps, 346, 349-350
    power consumption, 337
    pumping periods, 337, 341
    pumping speed, 332-334
    pumps, cryogenic, 4, 12
    regeneration, 441-444
    throttling, 339-340
    ultimate pressure, 95, 325
    vibration, 325-326
  cryosorption, 4, 319-322, 380
  definition, 62
  diaphragm pumps, 196, 199-203
  diffusion pumps, see Pumps, vapor jet
  gettering pumps, 4, 353-360
    capture probability, 355
    non-evaporated getters, 372-375
    pumping speed for various gases, 355, 358
    titanium sources, 354, 359
  ion-gettering, 4, 11, 353-360

[Pumps]
  mechanical pumps, 7–9, 79, 80, 137, 145
    anti-"suck back" devices, 143–144
    backstreaming of oil, 143–144, 155–160, 240, 300, 342
    maintenance, 159–162
    oil-free, 170–203
    operation at 50 or 60 Hz., 161
    ultimate pressure, 147–151
  molecular drag pumps, 271
  piston pumps, 187–190
  screw pumps, 183–185
  scroll pumps, 190–198
  sputter-ion pumps, 11, 360–370
    argon instability, 366
    cathode erosion, 362
    exposure to air, 369
    maximum throughput, 363, 378, 397
    pre-evacuation, 376, 378
    pressure indication, 372
    pumping speed for various gases, 365–367
    Starcell pumps, 368
    starting pressure, 369–370
    types of pumps, 365–368
    voltage and magnetic field, 365
  turbodrag pumps, compression ratio, 280–285
  turbomolecular pumps, 4, 10, 11, 286–288
    advantages, 314–315
    backstreaming, 311–312
    compression ratio, 299–302
    compound pumps, 304–305
    cooling, water or air, 313–314
    disadvantages, 315
    hybrid pumps, 303, 308
    hydrogen pumping, 299–302
    lubrication, 293, 295, 311–312
    maintenance, 308
    number of stages, 290, 292
    role of stator, 290–292
    venting to air, 312
    vibration, 294–296

[Pumps]
  vapor jet pumps, 4, 10, 80, 207, 386, 394
    backing pump, 217, 227, 229
    backstreaming, 227, 233, 235
    boiler design, 205, 210, 213, 217, 227, 245, 248–249
    cold caps, 251–253
    compression ratio, 212, 228, 230–232
    cooling water temperature, 265–266
    design features, 244
    ejector stage, 215, 234
    liquid/vapor separation, 215
    loss of fluid, 247
    operation at lower power, 224, 263–265
    maintenance, 262
    maximum gas load, 223–226, 266
    mercury fluid, 220
    power variation effects, 230, 265
    pressure distribution (vapor and gas), 211–212
    pressure stability, 249–250
    pressure ratio, 212, 231
    pumping fluids, 218, 219
    pumping speed, 222–225
    throughput, 216, 224
    ultimate pressure, 231–235
    vapor cycle, 209, 248
Pumps, history, 6–12
Pumps, rotary vane, 137, 142–144
Pumps, sorption, 169

Reciprocating piston pump, oil-free, 187–190
Residence time, of gas molecules on surfaces, 27, 69, 322
Residual gases, 115, 148, 158, 245, 303, 468, 469
Reynolds number, 46
Roots pumps (blowers), 164–167
Rotary piston pumps, 162–163

Safety, vacuum equipment, 77, 135 (Ref. 19), 155, 160, 267, 284